Queen Bess

Queen Bess

Daredevil Aviator

Doris L. Rich

Smithsonian Institution Press

Washington and London

This book was edited by Initial Cap Editorial Services and designed by
Janice Wheeler.

The paper used in this publication meets the minimum requirements of the
American National Standard for Permanence of Paper for Printed Library Materials
Z39.48-1984.

Library of Congress Cataloging-in-Publication Data

Rich, Doris L.
    Queen Bess : Daredevil aviator / Doris L. Rich.
      p. cm.
Includes bibliographical references and index.
ISBN 1-56098-265-9 (alk. paper)
1. Coleman, Bessie, 1896–1926. 2. Afro-American air pilots-
-Biography. 3. Women air pilots—United States—Biography.
I. Title.
TL540.CB46R52   1993
629.13'092–dc20
[B]   93-14785
                                          CIP

Printed in the United States of America.

10 9 8 7 6 5 4 3 2 1 99 98 97 96 95 94 93

FOR CELIA AND RUBÉN

Because of Bessie Coleman, we have overcome that which was much worse than racial barriers. We have overcome the barriers within ourselves and dared to dream.

Lt. William J. Powell
Founder, Bessie Coleman Aero Clubs

Note numbers are not used in this book. Instead, as a pleasurable convenience to the reader, notes are printed at the back of the book and are identified by page number and an identifying phrase or quotation from the text.

CONTENTS

ACKNOWLEDGMENTS

Bessie Coleman in life was a challenge to the status quo—a constant struggle against the myriad of limitations conventional society erected against anyone who dared to be different. In death she presents an equal challenge to the biographer, for she left little of the material necessary for a reconstruction of her life.

African American intellectuals of her day left behind published works, minutes of meetings, correspondence, reminder notes on scratch pads and calendars for would-be literary detectives to examine. Similarly, black society of her time—many educated at Howard, Spelman, or Hampton—handed down a wealth of miscellaneous source material such as wills, financial records, photograph albums, and invitations to weddings, debuts, benefit balls, dinner parties, and graduation ceremonies. A Chicago manicurist raised in the cotton fields of Texas, Bessie read the works of the former and did the nails of the latter but belonged to neither group. She left behind virtually no personal memorabilia and all but a few of the letters she wrote have been lost. Thus the biographer attempting to reconstruct her life is primarily dependent on two sources—the recollections of her contemporaries and reports in the media.

Each source holds both rewards and limitations.. The brief unpublished memoirs of her oldest sister Elois provided an invaluable starting point. But Bessie was born a century ago and the few people who knew her and are still alive today are now in their seventies and older. While many of their remembrances are vivid, precise dates and places are often vague and conflicting.

"The media" in Bessie's case essentially means black weekly newspapers. The white press generally ignored Bessie, as they did most blacks unless they were actors or athletes, or involved in sex, crime, or violence. Fortunately, however, Bessie was considered an entertainer (aviation was yet to be elevated to a profession) and a celebrity. So the black media interviewed her and covered her activities. Microfilm and a few originals of these newspapers were a principal source of information on Bessie's career.

Other material in this book derives from special collections and the archives of the following libraries and special collections: Library of Congress, Martin Luther King Memorial Library, and the Moorland-Spingarn Research Center at Howard University, all in Washington, D.C.; Black Archives Research Center and Museum, Florida A&M University; Chicago Historical Society; City of Dallas Public Library; Du Sable Museum of African-American History, Chicago; Eartha White Collection, Thomas G. Carpenter Library, University of North Florida; Ellis County Museum, Inc.; Eugene C. Barker Texas History Center, University of Texas at Austin; Florida Collection, Jacksonville Public Libraries; Fullen Library, Georgia State University; Houston Metropolitan Research Center, Houston Public Library; Jacksonville Historical Society; Lilly Library, Indiana University; Memphis Shelby County Public Library; Microfilms Collection, Pennsylvania State University; Nicholas Sims Library in Waxahachie; Rosenberg Library, Galveston Public Library; Schomburg Center for Research in Black Culture, New York Public Library; Soper Library, Morgan State University; State Historical Society of Missouri; State Historical Society of Wisconsin; William Henry Smith Memorial Library, Indiana Historical Society. In all of them the staffs were patient and helpful. I wish to thank in particular Eileen D. Brady, Brenda T. Brown, Theresa Christopher, Barbara W. Clark, Jan Curry, Brian Dirck, Joan Dobson, James N. Eaton, Ralph L. Elder, Wilma Gibbs, Maureen Hady, C. H. Harris, Ara Kaye, Reba King, Margaret Koyne, Gary F. Kurutz, Patricia M. LaPointe, Genette McLaurin, Shirlene Newman, Shannon Simpson, Janet Sims-Wood, Martha Smith, Stephen C. Stappenbeck, Saundra Taylor, and Kenneth J. Whisenton.

During the two years that I worked on this book I spent many days in the library of the National Air and Space Museum of the Smithsonian Institution where I was given advice and encouragement by Tim Cronen, Tom Crouch, Robert Dreeson, Phil Edwards, Dan Hagedorn, Melissa Keiser, Mary Pavlovich, David Spender, Patricia Williams, and Larry Wilson. I thank them all.

I am especially grateful for the interviews given me by Bessie Coleman's nieces, Marion Coleman and Vera Buntin; her nephews, Dean Stallworth and Arthur Freeman; and other members of the family, Vera Jean Ramey, Jilda Motley, Fredia Delacoeur, and Alberta Lipscombe.

Others who gave me interviews were Inez Bentley, Mattie Borders, Eugenia Brown, A'Lelia Bundles, Dr. Margaret T. Burroughs, Cornelius R. Coffey, Patricia Fletcher, Beulah Florence, Jean Albright Gilley, George E. Haddaway, Bernice Hamilton, B. K. Hargrove, Ben J. Henderson, John P. Ingle, Marion Jeffers, Frances Johnson, William T. Johnson, Doris Jones, Dr. Marjorie Stewart Joyner, Georgia Lewis, Neal Loving, John C. McDonald, Hettie Mills, Mrs. W. C. Pittman, Annie Pruitt, G. Edward Rice, Theodore W. Robinson, Jacqueline Smith, Arthur Spaulding, Audrey Tillinghast, Camilla Thompson, Rutha Waters, Seth Williams, and Jean Yothers.

Information through correspondence came from Dr. Johnnetta

Betsch Cole, Georgia Conlon, Dr. Gabriele Dankert, Frances E. Davenport, Marie-Josèphe de Beauregard, Diana Estudillo, Stuart L. Faber, Doris H. Farr, H. Oakley Haynie, Joan Hrubec, Lenore Kieiling, Barbara Kozuh, Edward L. Leiser, Paul McCully, Wolfram Müller, Dr. Lorna Marie Polk, Deborah Palacios, Henry Snyder, Bernice T. Steadman, Victoria K. Steele, Dempsey J. Travis, and John Underwood.

Paulette Floyd, C.S.W., read early versions of the manuscript and provided valuable suggestions.

Offering shelter and good advice during research trips were Maria and Malcolm Bellairs, Carol Covington, Barbara Gault Hayes, Mary and William North, Claudia Oakes, Theresa and N. G. "Pat" Payne, Chris Rich, and D. Vicky Spencer-Burrows.

I am indebted to Dr. Mari Evans for the conclusion of my story. Only a poet as gifted as she could evoke the essence of the irrepressible Bessie.

For their editorial guidance I thank Felix C. Lowe, Ruth Spiegel, and Therese D. Boyd. And finally my thanks to Stanley Rich, my husband, who ferreted out flawed writing with gimlet-eyed efficiency but spoke of it gently to this writer. He should have been named co-author but modestly declined.

Queen Bess

CHAPTER 1

The Reluctant
Cotton Picker

On October 15, 1922, eight-year-old Arthur Freeman stood at the edge of a runway at Chicago's Checkerboard Airdrome, his head thrown back, looking wide-eyed into the sky at the Curtiss Jenny performing a figure eight. At the top of the eight the plane seemed suddenly to heel over and plunge downward, gaining speed as it hurtled toward earth. Just 200 feet above the runway, the aircraft slowed, shuddered, then slowly nosed up, soaring back into the sky before circling the field and coming in for a perfect landing.

"That's my aunt!" Arthur shouted.

It's doubtful anyone heard him. The 2,000 people in the bleachers right behind him were making enough noise of their own, a din of yelling, clapping, and whistling as the plane rolled to a stop and the pilot climbed down from the cockpit, pushed oil-smeared goggles up over a leather helmet, and smiled at them.

Arthur's aunt was Bessie Coleman. the first African American to earn an international pilot's license and the first black woman in the world to fly an airplane.

For a moment she stood by the plane, her long coat open, her uniform tailored like those worn by Canadian aviators of World War I. Puttees, cloth strips wound spirally around her legs from ankle to knee, topped her shiny leather boots, and a Sam Browne officer's belt, with its over-the-shoulder strap, circled her trim waist. Not only did Arthur's aunt know how to fly. She was also beautiful and small with a perfect figure and a flair for dramatic dress.

Bessie beamed at the adoring crowd, savoring the fact she had finally made it home in triumph as she had always said she would.

It had not been easy. Born in a one-room cabin, raised in a single-parent family, and educated in a school for black children that closed whenever the cotton needed picking, she went from doing laundry in Waxahachie, Texas, to manicuring nails in Chicago. Her decision to learn to fly presented more obstacles than those she had already faced. Black aviators couldn't teach her because there weren't any and white aviators wouldn't. But, someone told her, if her own countrymen couldn't help her others would. After learning French at a Michigan Avenue language school, Bessie made her way to Europe and earned a flying license from the Fédération Aéronautique Internationale, honored and recognized by every nation in the world.

Now, finally, she was home. "I've shown them all," she thought, supporters and critics, believers and doubters alike. She had proved herself to all Chicago, the South Side of her own people and the white world outside. As she walked proudly toward the stands she winked at Arthur and the rest of her family who had come to see her perform. The black weekly Chicago Defender *had already proclaimed her "Queen Bess." She had proved herself to all Chicago. Now she would prove herself to the rest of the world.*

The baby was a girl, tiny but perfectly formed, a beautiful copper color. Her little legs and arms waved vigorously, and her cries filled the air of the one-room, dirt-floored cabin as her mother wrapped her in a worn cotton quilt and laid her on the straw-filled mattress. With a sigh of relief Susan Coleman looked down at the infant. A healthy one, she thought. I can get back to work by morning.

The tall, slim black woman whose handsome face was marked more by care than laughter was no stranger to childbirth. She married in 1875 at age 20, later than most black women in the South. In the following twenty-three years she had thirteen children, four of whom did not survive childhood.

The baby Susan had just laid on the bed, born on January 26, 1892, was named Bessie. Her arrival was not recorded on a birth certificate or in a family Bible. Neither Susan nor her husband, George Coleman, could read or write. It is not known if Bessie's parents, both born in Texas, were slaves before the Civil War. It is likely that Susan was because both of her parents were originally from Georgia, migrating to Texas before her birth. George Coleman, from whom Bessie inherited her copper-colored skin, may have been born free. Three of his grandparents were Indian—probably Choctaw or Cherokee. George's father was from Indian territory, his mother from Missouri.

When Bessie was born, George and Susan were living in Atlanta, Texas, less than ten miles west of where the borders of Texas, Arkansas, and Louisiana meet. The town had been named by migrating Georgians for the city they had left when they traveled west to build the Texas and Pacific Railroad. Although it was founded only three years before the Colemans were married in 1885 and its residents numbered fewer than 1,000, Atlanta was a place where fortunes could be made in railroads, oil, and lumber.

On the main street, shaded by pin oaks arching overhead, citizens gathered at the general store run by R. S. Allday, who had opened his "mercantile establishment" before the railroad had reached the city, back when goods were hauled by ox cart from the river boats that docked at Jefferson, twenty miles southeast of Atlanta. The town's shady main street and general store were not part of the Colemans' world, however. Theirs was one of dirt roads at the edge of town, of tenant farms, and tiny, crowded cabins and incessant labor.

This was the world Bessie entered, one not only of poverty but of repressed rage and fear. Two years before her birth the state of Mississippi had begun the process of disenfranchising African Americans by legal means, a process soon followed by all of the Southern states. Three months after Bessie's birth a black postal employee in Memphis, Tennessee, and his two partners who had financed a small grocery store were taken by a mob from a Memphis jail and killed a mile outside the city. Their offense was to defend their property from an armed attack by white night raiders.

A year later, at a lynching in Paris, Texas, a black man accused of raping a five-year-old girl was first tortured with red-hot irons, then burned alive. Schoolchildren were excused from their classrooms to witness the burning and the railroads ran excursions for spectators from nearby communities. After the body was reduced to ashes, a mob fought over the bones, teeth, and buttons. Before Bessie was a year old, this and similar incidents had launched crusading black journalist Ida B. Wells into an investigation of lynchings throughout the South—728 of them in a single decade.

African Americans were not the only victims of violence and injustice. In Texas, for example, violence was common in both the black and white communities. In the last week of January 1892, Dallas—then a city of 40,000—was the scene of eight murders, three fires, two rapes, and four robberies. At the same time, the entire nation was plagued by a severe economic depression. High tariffs enabled millionaire industrialists to create monopolies in trade and industry while laborers worked twelve-hour days, seven days a week. When the workers unionized and called strikes, the industrialists hired armed men as strike breakers. In March of 1893, when Bessie was fourteen months old, panic in the financial centers of the country sent stocks plunging. Banks failed. Farmers lost their land. Prices fell, factories closed, and unemployment increased.

Somehow, in spite of the nationwide depression, hardworking, frugal George Coleman managed to save enough money from his wages as a day laborer to buy a small plot of land in Waxahachie, thirty miles south of Dallas. George Coleman may have chosen Waxahachie, a town of fewer than 4,000 inhabitants, because he saw an opportunity there to make more money than he had been making in Atlanta. The seat of Ellis County, self-proclaimed "largest cotton-producing county in the United States," Waxahachie somehow prospered and remained relatively untouched by the national depression. Oil had not yet come to dominate the Texas economy; cotton was still king. Straddled by two railroads—the Fort Worth and New Orleans, and the Missouri-Kansas-Texas (which would become the Chicago–Rock Island)—Waxahachie was a teeming hub of cotton yards, cotton warehouses, and cotton mills.

As in most cities and towns of the South, housing was segregated, the railroad tracks dividing the west side, where the whites lived, from the east, where the property George Coleman brought was located. Ellis County records show that on October 15, 1894, Coleman bought "one quarter of an acre, more or less," on Mustang Creek. A justice of the peace and notary public, Y. D. Kemble, notarized the bill of sale on January 1, 1895, "at 2½ o'clock P.M." The price was twenty-five dollars. The lot was four miles from the center of town on Palmer Road, near the railroad tracks on the east side where black workers had settled along East Main and Wyatt streets. Here small homes, single-cell and two- and three-room houses, lined the unpaved streets. It was more or less a separate community within Waxahachie where blacks established their own religious, commercial, and social institutions.

On his quarter acre, a wedge of land fronting sunbaked Palmer Road before plunging in the rear to tree-shaded Mustang Creek, Bessie's father built a three-room "shotgun" dwelling, characterized by a series of doors from one room to the next through which one could "shoot a shotgun the length of the house." It was T-shaped, with the rectangular front room placed crosswise, the back two lengthwise, and a porch added to both front and back.

Bessie was two when the family moved into the new dwelling. Her early childhood was a happy one, spent playing on the front lawn edged by red and yellow roses her mother had planted. When it rained she took cover on the front porch and watched small puddles form near the steps below. These would be her "lakes" for sailing leaf boats and later there would be mud for making "marbles" and mud pies.

Sundays were spent at church, morning and afternoon. Awakened at dawn, bathed and dressed in her best frock, Bessie was handed over to one of her two older brothers, Isaiah or John, to be kept clean and out of mischief until Susan was ready. The church was African American Baptist. If anything like the one Susan chose years later in Chicago, it was conservative. Members might sing enthusiastically but there would be no dancing in the aisles. To Susan Sundays were the Lord's and her children were to worship the Lord as she did, frequently and with great respect.

In 1894, soon after the Colemans moved to Waxahachie, Susan had another daughter, Elois. Two more girls followed, Nilus, born in 1896, and Georgia, in 1898. As John and Isaiah began to find work in the fields, Bessie was expected to keep an eye on her sisters, a responsibility assigned the eldest girl in many rural families. She also watered her mother's plants on the front porch and weeded the back garden where Susan raised corn, peanuts, and vegetables. The pear, peach, and plum trees were too tall for her to climb but she harvested the windfall fruit as it ripened.

There were also household chores for her in a family that numbered eight by 1900. Three of the older children no longer lived at home. Lillah, age 25, Alberta, 21, and Walter (also known as "Bud" or "Samy"), 17, had all left the family home. Isaiah (called "Osa" or "Ozzie"), 15, and John, 11, remained.

Bessie was 6 when she started school. Two years before, in 1896, the U.S. Supreme Court in the case of *Pessy v. Ferguson* had established the legality of "separate but equal" schools. Her separate-but-equal school was a one-room wooden building on Aiken Street in the already segregated black residential district of Waxahachie. Hot in the summer and cold in the winter, the room held students in grades one through eight. There was one teacher for all, one who probably lacked minimum qualifications for teaching. (As late as 1922 a survey by a professor of rural education at Columbia University claimed there were 15,000 black teachers in rural schools with no more than a sixth-grade education.)

Bessie walked four miles from her home to school, where she was taught reading, writing, and arithmetic, often without textbook or enough paper and pencils. Intelligent, uninhibited, and eager to learn, she quickly established herself as the star pupil in math.

In 1901 Bessie's life—until then a relatively happy one with ample time for play between school and chores—underwent a dramatic upheaval. George Coleman left his family. As a day laborer, he was denied the hoped-for prosperity promised by Waxahachie's booming cotton mills and warehouses. Being black, he was barred from voting by discriminatory poll taxes and "literary" tests. He was denied representation in state or local government and participation in land-grant programs. Recently passed Jim Crow laws forbade his riding in railroad cars with whites. He was denied the use of public facilities used by whites—restrooms, restaurants, hotel rooms, and even water fountains. If he protested he risked being whipped, tarred and feathered, or even lynched.

He had had, he told Susan, enough of Texas. His decision could have been prompted by news from Brenham, seventy miles northwest of Houston, that a mob of whites had rioted for two days to protest the employment of a black brakeman by a railroad company. Or by the fact that there had been 115 lynchings in the South the previous year.

George could not escape the bondage of race in Texas. Not only was he black but he came from a lineage of three Indian grandparents. Ironically, in Texas his Indian blood put him in even more jeopardy than did his being black. How most white Texans of the time viewed Indians is reflected in a history published in 1894 describing Ellis County Indians as "savage and treacherous" and predicting it was "doubtful whether one of their race will be living one hundred years hence."

In Oklahoma, however, that same Indian lineage could offer George an escape from the double bondage in Texas. If the family moved to Indian territory in Oklahoma, he told Susan, they would enjoy the full rights of citizens. He could go if he wanted to, Susan replied, but she was neither pioneer nor squaw. She and her children would remain in Waxahachie.

Soon after George left, John and Isaiah also departed. John, by then 15, joined his older brother Walter in Chicago where the latter worked as a Pullman porter on the Chicago–St. Paul run. Isaiah went to Canada where he became a successful farmer in the Amber Valley, 100 miles from Edmonton.

At 45, Susan Coleman was left with four small girls, the oldest age 9, at a time when 85 percent of black households were headed by males and single mothers were a decided minority. She had no kinfolk to help her as did most African American families in the South. Proud, inflexible, a loner sustained by her faith in the Lord and herself, Susan was on her own.

Within days of George's leaving she found work as a cook-housekeeper for a white couple, Mr. and Mrs. Elwin Jones. Considering how most black domestics were treated in Texas at the turn of the century, the Joneses were generous employers, willing to hire someone who could not "live in" and thus could not be on call twenty-four hours a day.

Mrs. Jones provided most of the food consumed by Susan and her children, sacks of flour and meat, and on bread-baking days told Susan to "bake more so you'll have enough to take home." The plants on the Coleman porch were given her by Mrs. Jones, who also clothed the Coleman girls with hand-me-downs from her own daughters' wardrobes. Some of these dresses had hardly been worn, given to Bessie and her siblings because they had admired them.

While her mother worked at the Jones house, Bessie took over as surrogate mother and housekeeper at the Coleman home on Mustang Creek. She washed, cooked, ironed, and cleaned, all without running water, electricity, or plumbing. Water was drawn from an outdoor well. Laundry was done in an iron tub, and meals prepared on a wood-burning stove that also heated the house.

Seven-year-old Elois noticed and recalled years later that "Bessie had little time to be a carefree child. She seldom took an interest in dolls though she would watch us play at times." There were days, even weeks, when Bessie missed school because she was needed at home to look after Nilus and Georgia. Only when these two reached school age were all four girls able to make the daily four-mile trek to school.

Apart from the rudimentary instruction in the "three Rs" she received from her one-room-school teacher at Aiken Street, Bessie's principal training and instruction came from her proud, hardworking, deeply religious mother. Susan made attendance at the Missionary Baptist Church mandatory for all the children. As soon as Bessie learned to read she was assigned a reading from the Bible every night after dinner.

"Even though Mother had to pinch pennies," Elois recalled, "she managed to get books from a wagon library that passed the house once or twice a year, telling of the accomplishments made by members of our race . . . Of course we learned Harriet Tubman at Mother's knees."

Susan also had hopes of her children achieving worldly success. For this she advocated their emulating educated white people like the Joneses. She observed the manners and speech of her employers, then brought her observations home to her children. Bessie and her sisters ate at a table covered with a cloth, and they ate with knife, fork, and spoon, just as the Joneses did, with many a "please," "thank you," and "excuse me," under the watchful eye of their mother.

Every year Bessie's carefully set routine of school, household chores, and church attendance was shattered by the cotton harvest. While cotton was king, Waxahachie's black schoolchildren clearly were his subjects and school shut down as soon as cotton picking began. Depending on the weather, this was usually late July or August. And no matter how long the harvest season might last, school stayed closed until all the cotton was picked, meaning that sometimes the fall school semester was delayed until November or even December.

It was a mutually beneficial arrangement. The cotton growers of Waxahachie needed every hand—man, woman, and child—and the Colemans and their neighbors needed the money. From the time the plants dropped their pink and white flowers and the cotton bolls showed their white fruit, Susan and her girls were in the fields. Excused from her work at the Joneses, Susan dragged a long bag, fastened by a strap around her neck, along each row, picking and putting the cotton into the bag until she reached the end of the row where she emptied it into a basket.

Until they left home Isaiah and John joined her and as each of the girls became strong enough to drag a sack they worked with their mother as pickers. The younger ones tagged along, fetching water or running errands for the harvesters. Backs bent and fingers bleeding, adults averaged between 150 and 200 pounds of cotton a day. One picker recalled, "Every colored person worked from the time he or she was old enough to drag a sack through the cotton fields. The work was back-breaking, exhausting and sometimes degrading."

At 10 Bessie was a reluctant, evasive picker. The bright, eager scholar who secretly dreamed of a better life she was not yet able to define showed her aversion to the debasing labor by deliberately lagging behind. On one occasion she was even caught riding on the sack of the picker in front of her.

Susan was aware of Bessie's errant behavior but was forced to overlook it because Bessie was the family accountant, the only Coleman who could record the weight of each sack accurately and make sure that the foreman totaled them correctly. Better yet, she dared to increase that total and add to the day's earnings by pressing a discreet foot on the platform of the weighing scale whenever the foreman happened to be looking the other way.

There was one break during the cotton-picking season—the day the circus came to town. Elois remembered how "we laid our cotton sacks aside and took a day off to go to the Ringling Brothers Circus with a shiny half dollar each to spend as we wished, for pink lemonade, and popcorn, balloons . . . . "

The circus came to Waxahachie only once a year. For the rest of the time Bessie was the family's chief source of entertainment. She would read aloud every night by the light of an oil lamp from the books that Susan rented from the "wagon library." Imitating the mellifluous voices of the black Baptist preachers she heard on Sundays, clever and talented Bessie lent added drama to the stories about "the Race." She read first from the Bible, at Susan's insistence, then the books she herself loved best, stories of African American heroes—Booker T. Washington, Paul Lawrence Dunbar, and Harriet Tubman among them. Her favorite book was fiction, *Uncle Tom's Cabin*, although she obviously thought that Tom and Topsy were feckless cowards. The night she finished reading the book for the first time she announced to her spellbound audience, "*I'll* never be a Topsy *or* an Uncle Tom!"

Young Bessie displayed an impressive self-confidence despite living in a world that seemed to demand her abasement as both a black and a female. To some degree her sense of worth was the result of being born healthy, intelligent, and beautiful. Some stemmed from her successfully performing the work of an adult even though still herself a child, looking after a home and three siblings. And some came from her certainty that since she had mastered reading and math there was nothing in the world she could not learn.

Underlying all of this was the example set by her mother whose own sense of worth was rooted in faith. To Susan's mind God created all people equal. So while rendering necessary respect to the white Caesars she never felt less than equal in the sight of God.

At 12, Bessie was formally accepted into the Missionary Baptist Church. (Later, as an adult and free from her mother's authority, Bessie attended church service less frequently, not because she had lost her basic belief but because God's heaven in the hereafter was not as immediately intriguing as the world around her.) In Waxahachie, however, she became an enthusiastic church fund-raiser, and when a hand organ was offered as a prize to the person selling the most raffle tickets, it was Bessie who won.

Bessie completed all eight grades of Waxahachie's one-room black school. Limited and sporadic though her education was, it fed her growing hunger for more. And whatever that inchoate more was, Bessie suspected it was not to be found in Waxahachie. She had, in fact, been certain of this for some time and, with Susan's strong encouragement, began working as a laundress to earn money for whatever was to come.

In 1910 she took her savings and left for Langston, Oklahoma, to enroll in the Colored Agricultural and Normal University as "Elizabeth Coleman," a more elegant name than the one first given the census taker by her parents. The town of Langston, forty miles northeast of Oklahoma City, was founded in 1891. Named after John Mercer Langston, uncle to poet Langston Hughes's mother, it was the first of several all-black municipalities in Oklahoma. Seven years later the university opened. Belying its title, Langston, a state land-grant institution, was more a vocational school, offering four-year courses in education, agriculture, health services (home economics), and mechanical arts. Under the leadership of Dr. Inman Paige, still president when Bessie enrolled, the university also included a preparatory school for new students who fell short of full entrance requirements. There, at the age of 18, Bessie was placed in the sixth grade.

Bessie completed only one term before her money ran out and she was forced to return to Waxahachie. Whatever personal disappointment she must have suffered she kept to herself. Informed that the Missionary Baptist Church was planning to give her a homecoming party, she brought the entire Langston band home with her to provide the music. The little girl champion raffle ticket seller had matured into a young woman with a celebrity's talent for self-aggrandizement. She came home the conquering hero.

This brief moment of glory over, Bessie was back where she had started in east Waxahachie. If she had chosen to marry, as did most black women of her age and circumstances, her husband most likely would have been a tenant farmer or a cotton factory laborer. But she did not marry and resumed her work as a laundress.

As did most others like her Bessie collected and delivered the dirty laundry of her clients once a week, walking roughly five miles past the Romanesque-revival Ellis County Court House, whose pink and red limestone is still a landmark attraction, into west Waxahachie. She worked at home where she boiled the clothing in a tub in her backyard, scrubbed it on a washboard, rinsed, starched, and wrung it out, and hung it on clotheslines to dry. She ironed with a heavy iron heated on the top of the stove. On Saturdays she delivered her work, "keeping her place" by bringing it to the back doors of the west-side mansions. Elois said later that this and other humiliations failed to embitter Bessie who, like her mother Susan, never felt less than anyone else's equal.

While she worked Bessie dreamed of escaping Waxahachie. At night she read. Elois was often awakened at 2 or 3 o'clock in the morning for conversation that Bessie called "thinking." As Elois recalled, "Bessie would introduce a subject and I was supposed to respond."

"Thinking" was essential to the young woman so anomalous to the community in which she lived. Much of her "thinking" was stimulated by the brash exhortations of the African American weekly, the *Chicago Defender*, distributed by porters working on the southbound trains. The paper's advice was to "leave that benighted land [the South] . . . You are free men," it adjured. "To die from the bite of the frost is far more glorious than that of the mob." There were, it was claimed, high wages in the great terminus of the north-south railroads—as much as eight dollars a day.

In 1912 Bessie made less than that in a week. In Dallas a rented room was $2.00 a week, a hotel room with a telephone $1.00 a night, a silk dress $5.45, a leather purse $1.35, a Ford automobile $590, and an eight-room house with two baths $7,500.

There are no letters to prove it but Bessie, ever the opportunist, must have written to her brother Walter for help. In 1915, when she was twenty-three, Walter told her she could come to Chicago and stay at his apartment while she looked for work. Here, at last, was the chance for her to "amount to something," a goal Susan aspired to for all her children. "You can't make a race horse out of a mule," her mother told Bessie. "If you stay a mule, you'll never win the race." Bessie was certain that Chicago was the right track. She was going to be the race horse who came in first, the one that would "amount to something."

CHAPTER 2

That Wonderful Town

In 1915 Bessie Coleman left Waxahachie for Chicago. She took the Rock Island train after buying her ticket on the "colored side" of the depot off South Roger Street. Like most of her fellow pilgrims seeking the promised land of Chicago she was dressed in her "Sunday best," carrying all of her possessions in suitcases and bundles along with a sack of food for the twenty-hour ride ahead.

In those days blacks could travel in only two sections of a train, the all-black car or the Jim Crow section of a men's smoker. The former had a single toilet that would become unbearably filthy by the end of the trip. The Jim Crow section on the back of the smoker was separated from the white male passengers by a shoulder-high partition. A blue fog of smoke clung to the ceiling, the odor of ashes permeating the stale air.

The car on which Bessie sat on a hard wooden bench exemplified the oppression she longed to escape, a world of rear seats on buses and balcony seats in theaters, of forbidden public restaurants, water fountains, and lavatories, a world in which the old taboos and fears were now being augmented by a new one, the resurrection of the Ku Klux Klan, first in the South and soon after throughout the nation.

Even as Bessie rode north, movie director D. W. Griffith's epic film *Birth of a Nation* was playing to packed theaters throughout the nation. Glorifying the anti-black, anti-Catholic, anti-Jewish Klan, the film was applauded by the white Protestants of Chicago. Its message, somewhat disguised in a coating of compassion for a South humiliated by its defeat in the Civil War, was that *only* white Protestants were genuine Americans, fit to run the country. Except for this elite in the big, ever-expanding metropolis of Chicago, ruled by millionaires and the political bosses they controlled, the Klan was a threat not only to blacks but to many of the city's immigrants. They were Germans, Poles, Irish, Hungarians, Russians, Greeks, Italians, and Slavs, mostly either Catholic or Jewish. Each national group formed a cultural enclave with its own schools, churches, synagogues, associations, and clubs. In time there was assimilation, but not for African Americans, whose skin color proved an effective barrier. Shunned by all the other groups, blacks erected their own enclave. Pride, habit, and the need for mutual protection led to the establishment of a self-segregated ghetto with its own churches, clubs, and fraternal organizations.

Fifteen years after Bessie arrived in Chicago the attitude of most whites throughout the nation toward African Americans was still so demeaning that a reputable New York firm published a book by a Chicago historian describing the typical black as "a virtual savage from the cotton fields." It went on to say that although he might "don clothes and something of the manner of an urban dweller, . . . the Negro, half-child, half-man, was still the under dog. He was still a 'separate' being . . . He threw out his chest and jingled his silver, to keep up his spirits. He was delighted when white people treated him nicely."

Bessie was leaving the segregation of the South for the ghetto of the North. But within its borders, black Chicago offered the excitement and opportunities of big city life.

No one knows how Bessie made her way from the South Side station to her brother Walter's apartment at 3315 Forest (now Giles) Avenue, but she did and was welcomed. The first of George and Bessie Coleman's children to join the northern exodus, Walter had lived in Chicago for more than a decade. A gentle, quiet man of 35 who was called "Bud" by his siblings and friends, he was married to an acid-tongued woman named Willie from Tyler, Texas. Living with them was brother John, six years Walter's junior, and John's wife, Elizabeth. Both couples were childless.

Walter was a Pullman porter, a respected position within the black community but on the road a difficult and demanding job in which his race was a constant burden. On the train he washed in a separate washroom, drank from a separate fountain and slept, when he slept at all, on inferior blue sheets because the white sheets were "for white bodies only." The rules for porters were issued by the president of the Pullman Company, Robert Todd Lincoln, son of the late president.

Bessie quickly recognized that Walter was the main provider for the crowded Coleman household. John, a charming extrovert given to drinking bouts, was frequently unemployed. And it was Walter who had found the apartment and paid the rent. The Coleman housing situation was typical of the Chicago of their time, confined to the "Black Belt" by racial and social pressures and limited in their choices by the incredible influx of Southern migrants, who doubled the city's black population between 1910 and 1920. By the end of that decade, in fact, 90 percent of the African American population of the city would be crammed into housing in an area bordered by Twelfth and Thirty-ninth streets on the north and south and Lake Michigan and Wentworth Avenue on the east and west. Unlike the urban ghetto it has since become with its disproportionate number of impoverished, undereducated, and underemployed, in Bessie's time the area consisted of a number of mixed and balanced communities in which the wealthy and the well-educated, the large middle-class and the poor and the hard-working coexisted in generally law-abiding harmony.

Bessie clashed frequently with her sister-in-law Willie, whom she considered "too bossy" and, she said, dominated everyone in the household—everyone, that is, except Bessie. But she continued to live there because she loved Chicago and was determined to stay.

From almost the moment she arrived Bessie started looking for a job. In 1915 most black women who worked outside the home were domestics, although a few escaped that fate by becoming school-teachers. Not until the First World War would factory jobs open up for black women. But Bessie had not left Texas to be a maid, cook, or laundress. After a lengthy search she decided to be a beautician. Though perhaps somewhat comparable to domestic service in that she would be providing personal services, at least it would keep her out of the kitchen. Best of all, there were plenty of openings available in the numerous beauty shops of Chicago's South Side.

The thriving profession Bessie chose was the target of considerable criticism from African Americans who considered much of it simply a hypocritical attempt by blacks to imitate and emulate a ruling white society. One such critic, noted educator Nannie Helen Burroughs, wrote, "What every woman who bleaches and straightens out needs is not her appearance changed, but her mind changed . . . If Negro women would use half the time they spend on trying to get white, to get better, the race would move forward a pace."

Although the black press overflowed with similar exhortations to take pride in "the Race," many advertisements were for skin lighteners and hair straighteners. Two-column ads urged readers to use Nadinola Bleaching Cream ("See how it lightens and whitens your skin overnight"). Others advocated using Pluko White hair dressing, Arroway hair cream for men, and caps for making hair straight and smooth. The firm of Poro Products for hair and skin frequently offered to train sales personnel. Most prominent and plentiful of all were the ubiquitous solicitations of saleswoman supreme Madame C. J. Walker whose "wonderful hair and skin preparations" had already made her one of the wealthiest woman entrepreneurs in the country—black or white.

The debate continued into almost the next decade with sociologist Frances Marion Dunford warning in the *Journal of Social Forces* that the penchant of the black male for "using materials which tend to make him more like his white brother" would be the downfall of the race. An outraged *Chicago Defender* writer's response was that black people follow fashion just as most people in any society do. The progress of the race, he wrote, would be determined by its ability to assimilate within the environment of which it is a part. Segregation would remove the race from the environment, making assimilation, and progress for the race, impossible.

Bessie was untroubled by the debate that swirled around her. She was going to be a beautician because it was the most interesting and appealing employment she could find. Her own appearance changed as she experimented with makeup, hairstyles, and clothing. Photographs and the recollections of her two nieces, Marion and Vera, indicate that her choices were eclectic. She used whatever was most becoming to her and seemed most to enhance her own figure, face, and personality.

Bessie enrolled in the Burnham School of Beauty Culture for a course in manicuring—or so she wrote to Elois, who was still in Texas. Chicago's city directory of 1915 identifies Edward Burnham as a distributor of hair-care goods with offices on North State Street and West Washington Street. But there is no Burnham School listed. Regardless of where Bessie took her apprenticeship, she did learn the beautician's trade. A year later she sent Elois a clipping from the *Chicago Defender* describing her as the winner of a contest to discover the best and fastest manicurist in black Chicago.

She certainly was one of the shrewdest. By confining her work to manicures she avoided the time-consuming lessons that would have been required of her before she could work as a hairdresser. As a manicurist she could do men's nails in a barbershop, where the customers appreciated her looks and charm and expressed their admiration in generous tips. She worked first in a shop at 3447 State Street and later in a barbershop owned by John M. Duncan at 206 East Thirty-sixth Street. At Duncan's she sat at a table in the window where her customers could enjoy being seen having their nails done by a very pretty woman.

The shop on State Street where Bessie first worked was in the area known as "The Stroll," eight blocks of State between Thirty-first and Thirty-ninth streets. The eight blocks were best described by Chicago historian Dempsey Travis as a "black Wall Street and Broadway." It was a place in which to promenade, to make bets, to sell goods, legal or illegal, to start a business, to talk politics, to bank, to go dancing, to watch a show, and to see and be seen. It was downtown, but with an ethnic identity. On the sidewalks people in colorful, often expensive clothing greeted one another warmly or gathered on corners for exuberant, sometimes noisy conversation. Along with obvious communal friendliness and uninhibited laughter there was, for many, a firm belief in luck, made evident by the sale of "dream books." These offered guidance to picking a number in an illegal lottery, the "numbers game." Sales of some—"The Three Witches," "Gypsy Witch," "Japanese Fate" and "Aunt Della"—were topped only by sales of the Holy Bible.

Bessie knew every inch of this territory and explored it all. She went to the dozens of nightclubs squeezed among banks, bars, shops, and restaurants along the Stroll. She worked next door to the Elite Club #2 owned by Teenan Jones. Jones had opened the original Elite Club in Hyde Park at Fifty-sixth and Lake streets but had been forced to close it by anti-black neighborhood groups in

1910. He moved it to State Street and added the second club to compete with other popular spots such as the Royal Garden, Pekin Café, and Dreamland. Bessie went to all of them and saw, among the great black performers of the day, "King" Joe Oliver, Louis Armstrong, Bessie Smith, Ethel Waters, and Alberta Hunter.

These nightspots were open to both races. Dancing was uninhibited and, even after the Eighteenth Amendment banned its sale in 1919, liquor flowed freely. Typical was the Pekin Café owned by undertaker Dan Jackson, where a large amount of illegal whiskey was discovered after two white detectives from Detroit were found murdered on the premises. The African American *Birmingham Reporter* called the Pekin "a Chicago rowdy house where whites and blacks meet." The same newspaper noted that Dreamland was temporarily closed soon after during one of Chicago's sporadic "rigid law and order movements" to clean up the "Great Light Way" where "amusements and other things" were offered.

Bessie's brother John was well acquainted with the patrons, both black and white, of places such as Dreamland. He worked as a cook for several bootleggers and gangsters. Bessie also knew them through her work. Men who wanted their appearance to reflect their status spent lavishly on their attire and grooming and came to Bessie at the shop on State Street for their manicures.

But not all of Bessie's continuing education was gained in the shop or on the Stroll. She remained an avid reader, her favorite source of information now being the *Chicago Defender*; its editor and publisher, Robert Abbott, became her idol. Spokesman for the race and owner of a newspaper whose readers would soon number a half million, Robert Sengstacke Abbott was a handsome, elegantly dressed, still-youthful man in his mid forties. He would stand on the corner of Thirty-fifth and State, cigar clamped between his teeth, and hold court as he chatted with community leaders. Abbott told them what they ought to do and they often did it. Bessie knew all about him and had seen and heard him often before the two actually met.

At that same corner of the Stroll Bessie mingled with others of the rich and powerful in black Chicago, among them Jesse Binga, whom she would later know intimately and who, along with Abbott, would influence her career. The notorious Binga, an ex-barber and Pullman porter, began his fortune in 1907 as a "block-buster" real estate dealer. He would buy one house on a white block and sell it to a black family, then buy up the remaining houses at lower and lower prices as the white owners rushed to sell them cheaply and move out.

Unloved, but respected and even feared by his own people, Binga also had white enemies who bombed his $30,000 house in the Englewood district seven times in less than two years. "As usual," one report said, "the pillars of the front porch were blown out of place and scores of window panes in the neighborhood shattered. Binga and his family were out of the city and the only one in the home was a maid who locked herself in and refused to open the door."

Binga took his money to the bank in a Model-T truck with one man driving and another armed with a sawed-off shotgun. "He was a mean son-of-a-bitch," one of his employees said later, adding nonetheless that invitations to his "high-class" parties were coveted by many. By 1925 Binga had a million-dollar bank and office building at the corner of Thirty-fifth and State. A reporter for a Baltimore black weekly wrote, "Walk down State Street and you will find two highly illuminated colored banks, with electric signs sufficient to attract you a mile away. On the signs you will read these words in color: Douglas National Bank. Binga State Bank."

Binga frequently met another of Bessie's acquaintances, his friend Oscar DePriest, Chicago's only black alderman and, later, U.S. congressman. DePriest, who also dealt in real estate, was tried but acquitted in 1917 on charges of corruption. Chief witness against him was nightclub owner Teenan Jones, who claimed he was DePriest's bagman and head of a gambling combine.

In addition to Binga and DePriest, Bessie knew Dan Jackson,

lawyer Edward Wright, and Anthony Overton, a cosmetics manufacturer and, in time, founder of a bank that would fail when all of its capital was found to have been invested in one of his own businesses.

For more than three years while she manicured nails and added to her circle of friends Bessie showed no particular interest in any one man. Then, on January 30, 1917, four days after her twenty-fifth birthday, the manicurist who loved the bright lights of the Stroll, the woman who had spoke so often of her determination to "amount to something," married Claude Glenn, a friend of her brother Walter's and fourteen years her senior.

The marriage must have astonished the members of her family— if they knew about it at all. Sister Elois never mentioned it in her memoirs of Bessie's life. When queried some seventy years later, none of Bessie's four nieces and nephews knew anything about it. The two nieces who had lived with Bessie in Chicago did recall that "Uncle Claude," a tall, dark-skinned man, had been a caller at both Susan's and Bessie's apartments. But they never suspected he was anything more than a family friend. Nor did their Aunt Bessie ever tell them she was Glenn's wife.

The only real evidence of the marriage is a license issued on December 30, 1916, by Cook County Clerk Robert M. Switzer. Added to the bottom of the license is a record of the marriage ceremony performed by Baptist minister John F. Thomas at 3629 Vernon Avenue—probably the minister's home—a few blocks from Walter's apartment. So far as is known, Glenn never lived at the same residence as Bessie. Soon after the marriage Bessie moved from Walter's apartment to one of her own at 3935 Forest Avenue. But the city directory never listed Claude Glenn at that address—not that year or later.

The circumstances surrounding Bessie's marriage are unclear. Whatever her reasons, Bessie certainly didn't marry Glenn for his money; he was no wealthier than Walter. Glenn was neither handsome nor dynamic, just a pleasant, quiet, older man. Nor had

Bessie ever shown any evidence of sexual attraction for him during his frequent calls at her apartment or Susan's. It is possible that Bessie found in him a dear friend when she needed one and that she married believing that friendship was all she needed.

During the next year it became increasingly clear that for Bessie the vow that took precedence over all others was the one she had made as a child to "amount to something," to be noticed and respected by as wide a circle of admirers as possible. Marriage, especially in that era, required far more subservience than she could bring herself to allow.

Bessie's mother and three sisters were equally strong-willed, all abandoning relationships they found distasteful or unbearable. Soon after Bessie's marriage Susan arrived in Chicago with her youngest daughter, Georgia, and Georgia's one-year-old daughter, Marion. Georgia may not have been married. When asked by Marion years later her mother refused to say whether or not she had wed Marion's father. Her only comment was, "I couldn't stand him."

In 1918 Elois was the next to come to Chicago from Texas. She arrived with four children, Eulah B., Vera, Julius, and Dean, fleeing the beatings of her cruel, hot-tempered spouse, Lyle Burnett Stallworth. Not long after Elois, the last sister, Nilus, came to the city with her four-year-old son, Arthur, leaving her husband, Willy Freeman, in Oklahoma. All three sisters, their children, and their mother, Susan, stayed briefly with either Walter or Bessie until they could find places of their own.

Three months after Bessie's marriage the United States declared war on Germany. Both Walter and John were members of the all-black Eighth Army National Guard. Although both men ultimately were sent to France, Bessie was more worried about attacks on them by white racists than she was about their becoming casualties at the hands of the Germans. Segregated and often mistreated by an Army command that used them as stevedores, black men in uniform were perceived by many whites as threatening. Those who regarded African Americans as inferior refused, often using violence, to acknowledge the equality the uniform seemed to demand. In Houston that summer twelve civilians died in a riot that started when black soldiers attempted to board a streetcar reserved for whites. In the aftermath, thirteen soldiers were sentenced to death and another fourteen to life imprisonment. In Spartanburg, South Carolina, black musician Noble Sissle, in Army uniform, was struck repeatedly by whites after he failed to remove his cap in a hotel lobby. The Army forestalled retaliation by hastily issuing a directive ordering the black unit to break camp and leave for France immediately.

John and Walter both served in France and survived the war unharmed. Their outfit, the Eighth Army National Guard, was part of the 370th Infantry, whose men had won twenty-one American Distinguished Service Crosses and sixty-eight French War Crosses. When the 370th arrived in Chicago, the entire regiment paraded down Michigan Avenue in full uniform, but received little or no applause from Chicago's white onlookers. Returning white veterans, many from ethnic groups whose economic and social status was just one step above the African Americans, discovered their jobs had been taken by blacks. Twenty percent of the meatpacking force was now black. In eighteen months the city's black population had increased from 50,000 to 150,000. Some of the migrants displaced men who had enlisted or been drafted into the armed forces. Others were needed for new jobs created by the increased production demands of the war. Still others had been hired as "scabs" to replace striking workers.

In March of 1919 the re-election of Chicago's mayor William Hale Thompson was made possible by the votes of the city's two black wards on the South Side. "Big Bill," six feet, two inches tall and 260 pounds wide, had wooed and won black voters with rousing speeches in which he called them "brothers" and "sisters" and kissed their babies. Yet he did nothing to abate the rampant racial hatred aroused in jobless whites caught in a postwar depression.

Four months later Chicago was battered by the worst race riot in its history, touched off on July 27, 1919, when a black youth on a homemade raft drifted into an area of Lake Michigan customarily used only by whites. He was stoned by a group of white bathers, fell into the water, and drowned. That night groups of blacks and whites fought in the street with guns and knives. The violence carried over into the next day when gangs of whites, many of them unemployed, dragged blacks off streetcar platforms and beat them.

Just a block from Bessie's apartment 10-year-old Joe Crawford found shelter with relatives after fleeing from the violence in his racially mixed neighborhood. A white woman on the second floor of his apartment building had warned Joe's mother that a group of teenagers planned to give her and her son a public flogging. As Joe and his mother fled out the back they heard voices at the front door shouting, "Niggers, come out and get your asses whipped or stay in there and be barbecued!"

Four days later the National Guard, aided by a torrential rainstorm, restored order to the city. The riots had left 38 dead, 537 injured, and more than 1,000 people homeless. Bessie and her family all escaped injury or property loss.

At the close of that summer Bessie had been in Chicago almost five years. She had moved north, settled in Chicago, learned a trade, married, found her own place to live, seen her brothers go to war, and survived a race riot. But at 27 she was still looking for a way to "amount to something."

The answer came unexpectedly one day that fall when she was working at her manicurist's table in Duncan's barbershop. Her brother John, who had failed to "settle down" after his wartime service, walked in. Far from sober, he began a teasing discourse— one he had given before—aimed at Bessie. His theme was the superiority of French women over those of Chicago's South Side. French women, John said, had careers. They even flew airplanes. "You nigger women ain't never goin' to fly," he told Bessie. "Not like those women I saw in France."

After the laughter from his captive audience of barbershop customers had subsided, John looked over at Bessie. She was smiling
at him. "That's it!" she said. "You just called it for me."

John's comment no doubt rankled, but Bessie's response was
clearly more than just a spur-of-the-moment reaction. During the
war Bessie had read the stories and seen the photographs of aviation heroes. And John had talked about French women fliers before. Clearly Bessie had decided that flying would be her new
vocation. The air would be the arena for her ambitions, a way for
her to be noticed.

CHAPTER 3

Mlle. Bessie Coleman—
Pilote Aviateur

From the moment Bessie decided to become a pilot nothing deterred her. The respect and attention she longed for, her need to "amount to something," were directed at last toward a definite goal. Ignoring all the difficulties of her sex and race, her limited schooling and present occupation, she set off to find a teacher. She approached a number of fliers—all of them white, since there were no black aviators in the area at the time—but they all refused.

Perhaps they did so because her good looks and self-assurance in the presence of whites struck them as brash and suggested to them more a publicity seeker than a dedicated student. Perhaps they turned her down because she was black. Or because she was a woman. Most likely it was a mix of all three with race and gender ranked first and second.

At any rate, Bessie was forced back to the South Side, to her own community and to the man she so admired, Robert Abbott, editor-publisher of the *Chicago Defender*. She wanted to become a pilot, she told him, but no one would teach her. What should she do?

Abbott saw nothing outrageous in Bessie's objective. The paper he had established in 1905 aimed to gain recognition for black Americans as a people worthy of respect by all Americans. Unlike many of his gender, black or white, Abbott didn't think of women as less capable than men. His background wouldn't let him. His aunt, Priscilla Hammond, was a founding member of St. Augustine's Episcopal Church in Atlanta, Georgia. Another aunt, Cecilia Abbott, founded St. Stephen's Episcopal, also in Atlanta. His cousin Roberta Gwendolyn Thomas was a student at American University in Washington, D.C. And when his mother, Flora Abbott Sengstacke, came visiting from Atlanta, she was met by no less than Julius Rosenwald, president of Sears Roebuck and noted benefactor of black schools.

Abbott said Bessie must go to France. The French, he claimed (as did his frequent newspaper editorials), were no racists. They were also leaders of the world in aviation. If she worked hard, saved her money, and learned French, Abbott told her, he would inquire about an accredited aviation school in France and provide her with a reference.

Taking Abbott's advice Bessie enrolled in a language school on Michigan Avenue. She found a better-paying job as manager of a chili parlor at Thirty-fifth Street and Indiana Avenue and began to save her money, putting it in the Franklin Trust and Savings Bank at 100 East Thirty-fifth. Sister Georgia took over her manicurist's table at Duncan's after Bessie trained her. With Georgia now earning money and contributing to the rent of the apartment she shared with their mother at 3757 Indiana, Bessie moved a few blocks up to a new apartment of her own at 4533 Indiana.

It is difficult to believe that Bessie's wages and tips as a chili-parlor manager were enough to pay for passage to Europe and back, as well as room, board, and tuition for the flying lessons. Presumably Jessie Binga, who was already a rich man when Bessie was still a schoolgirl, gave much of the money. Gossip reputed her to be Binga's mistress although there is no real evidence of this.

Robert Abbott's interest was platonic, but he did give her money as well as guidance in his desire to see her become a pilot. That interest arose, one of his reporters said, from his tireless pursuit of increased readership for the *Defender* rather than any real concern for Bessie. As the first African American woman pilot, she would provide a pride-in-race increase in circulation for his newspaper.

There are other possible sources of the funds she needed. In an interview several years later Bessie mentioned an unnamed Spaniard who "made it possible for me to continue my studies in aviation." "She had a lot of men callers," her niece Marion recalled. "Some were black and others were white—and of several nationalities. I remember hearing different languages." While there is no evidence that any money changed hands or that any of these men received sexual favors for their assistance in financing Bessie Coleman's career in aviation, it is clear that Bessie lived as she pleased. The understandings of the times are perhaps the best guide.

On November 4, 1920, Bessie applied for an American passport at the Chicago office. By then, with Abbott's help, she had located an accredited aviation school in France and had learned enough French to read most of the school's first reply. However, her writing was no better than a child's and she left the second "t" out of Atlanta.

She put January 20, 1896, as her birthdate, four years less than her age. On this sworn statement she gave her occupation as manicurist and stated she had never been married. The purpose of her visit to "England, France and Italy" was "to study."

On the application, John Coleman also gave 4533 Indiana Avenue as his address. As a character witness he swore that his sister was an American citizen, born in Atlanta, Texas, on the date she had given. Further identification was provided by a second witness, Mrs. Anna M. Tyson of the same address who gave her occupation as "housekeeper."

The deputy clerk of the U.S. District Court wrote the description of the applicant as "twenty-four; five feet, three and half inches in height; a high forehead; brown skin; brown eyes; a sharp nose and medium mouth; a round chin and brown hair." The brown skin was copper-colored and "sharp" defined a nose more Caucasian than African. The attached photograph is of a very pretty woman.

A week after the passport was issued on November 9, Bessie went to the British and French consulates in Chicago for two visas. The first, from the British, was a month's transit permit; the second, from the French, was a tourist's visa for a year, valid until November 16, 1921. On November 20 Bessie sailed for France from New York City on the S.S. *Imparator*.

In her account of the months she spent in France, Bessie said, "I first went to Paris and decided on a school. But the first to which I applied would not take women because two women had lost their lives at the game." Forced to find another school, Bessie looked for the best and found it "in the Some, Crotoy where Joan of Arc was held prisoner by the English." She had selected France's most famous flight school—École d'Aviation des Frères Caudron at Le Crotoy in the Somme—managed by French aviators and plane designers Gaston and René Caudron. There she completed what she said was a ten-month course (it was actually seven months), "including tail spins, banking and looping the loop."

During her training at Caudron Bessie witnessed a terrible accident in which another student pilot was killed. "It was a terrible shock to my nerves," she said, "but I never lost them; I kept going," although, she added, she had to sign away her life by agreeing to take all responsibility for her injury or death.

Bessie must have longed to see people of her own race, for she mentioned that only two non-Caucasian students were in the class and she did not identify them as black. Her room, she said, was nine miles from the airfield, nine miles that she walked "every day for ten months." Without family or friends, obliged to speak in a language she had not yet mastered, Bessie finished the course in June, her name appearing in the registry of 1921 graduates.

Bessie learned to fly in a French Nieuport Type 82 (School). Just as the Curtiss JN-4, or "Jenny," was a favorite learner's plane in the United States, the Nieuport was frequently used in France for teaching pilots, both before and after World War I. The twenty-seven-foot biplane with a forty-foot wing span was designed by Gustave Delage and manufactured by the Société Anonyme des Establishments Nieuport. It was a fragile vehicle of wood, wire, steel, aluminum, cloth, and pressed cardboard. Structural failure, often in the air, was all too common. Bessie probably heard the pilots joke about the cloth-covered wings that they claimed were made by an insane seamstress whose work tended to peel off the almost forty-foot span of the upper wing.

Each time she took a lesson in the Nieuport with its dual controls, she had to inspect the entire plane first for possible faults—wings, struts, wires, cloth covering, engine, propeller, cowling, and the four-wheel landing gear with pneumatic tires. The plane had no steering wheel or brakes. The steering system consisted of a vertical stick the thickness of a baseball bat in front of the pilot and a rudder bar under the pilot's feet. These were duplicated for the student in the rear cockpit. The stick controlled the pitch and roll of the aircraft. The rudder bar controlled its yaw, or vertical movement. To stop after landing, the pilot would lower the tail of the plane until a rigid metal tail skid dug into the earth.

Once in the cockpit, Bessie had to wait for a mechanic to start the engine, priming it first with castor oil. She could feel the heat from the eighty-horsepower engine and smell the overpowering odor of hot castor oil that covered her face, goggles, and leather coat with a fine yellow mist.

Bessie could not always see what the instructor was doing or hear his comments with the engine's roar drowning out all other sounds. She learned by watching her stick and rudder bar move as the pilot moved his. Soon she was placing her hands on the stick and her feet on the rudder bar, enabling her to get the feel of what the pilot did. This primitive dual-control system posed a risk to both Bessie and her instructor. If she were to "freeze" on the stick, to grasp it so tightly in a panic that the pilot could not regain control of it, both could die, a tragedy that occurred all too frequently for students and instructors at that time.

After seven months of lessons and practice flights Bessie took the test qualifying her for a license from the renowned Fédération Aéronautique Internationale (FAI), the only organization at the time whose recognition granted one the right to fly anywhere in the world. She flew a five-kilometer closed-circuit course twice, climbing to an altitude of fifty meters, negotiating a figure eight, landing within fifty meters of a predesignated point, and turning off the engine before touching down.

Bessie's coveted FAI license is dated June 15, 1921. The document accurately recorded her name (Bessie Coleman), her birthplace (Atlanta, Texas) and, somewhat less accurately, her age (25, the figure Bessie gave the passport authorities in Chicago, instead of the 29 that she really was). It did not record that Bessie was the first black woman ever to win a license from the prestigious FAI and of the sixty-two candidates to earn FAI licenses during that six-month period, Bessie was the *only* woman.

From Le Crotoy Bessie went to Paris in June to take more lessons from, she said, an unnamed French ace who had shot down thirty-one planes in the war. There is no record of Bessie's stay during the next two months but she arrived at a time when summer tourists filled the city's hotels and rooming houses. An acquaintance, Dr. Wilberforce Williams, who was also a columnist for the *Chicago Defender*, didn't reach the city until after Bessie had left but his impressions of France indicate an atmosphere in which Bessie must have reveled. "In France," he wrote, "more than any other country, one finds the privileges of individual freedom and political unity. There is a total absence of racial antagonism."

Aside from her flying lessons Bessie must have done what most tourists did—explored the city, sat at sidewalk cafés, walked a good deal, and perhaps saw some museums and landmarks. She certainly shopped, for she brought home a stunning wardrobe including dresses, a tailored flying suit, and a leather coat.

Bessie left France on September 16, 1921, on the S.S. *Manchuria* from Cherbourg to New York. This time reporters—black and white—met her in New York to interview her. The *Air Service News* reporter stated that this "twenty-four-year-old Negro woman" had returned "as a full-fledged aviatrix, the first of her race," and that she intended to give exhibition flights in the United States. His competitor from the *Aerial Age Weekly* noted that she arrived with credentials from the French "certifying that she has qualified as an aviatrix." Adding to the aviation magazines' stories the *New York Tribune* reported Bessie's claim that she had ordered a Nieuport scout plane to be built for her in France.

For most of the country's black weeklies Bessie was a front-page story. The *Dallas Express* quoted her as saying that few colored people and no women had taken any interest in aviation. Using India as an example, she said, "Out of four hundred million Hindus only one has piloted a plane, and that was a man."

Before leaving New York City for Chicago Bessie was the guest of honor at a performance of the musical *Shuffle Along*. Written and produced by two vaudeville comedians, Flourney Miller and Aubrey Lyles, the all-black musical was first presented in one-night stands with no more scenery than could be moved in one taxi. After every Broadway theater asked had turned it down, the show opened a mile from Broadway at the Sixty-third Street Music Hall, where a special stage had to be built for it. Yet, in a summer plagued by a heat wave, *Shuffle Along* was an instant hit, playing to standing-room-only audiences. The *Shuffle Along* cast included Eubie Blake, Ethel Waters, Florence Mills, Roger Matthews, Lottie Gee, and Noble Sissle, who recorded the music on the Emerson label. The women in the chorus were beautiful and talented. (One aspiring performer who didn't make it, however, was 15-year-old Josephine Baker, turned down because she was "too young and too dark.")

When Bessie appeared on stage to receive a silver cup engraved with the names of the cast the entire audience—whites in the 550 orchestra seats reserved for them and blacks in the balcony and boxes—rose to applaud the only black woman aviator in the world.

Returning to Chicago in October Bessie was interviewed in her apartment at 4533 Indiana Avenue by a reporter from the *Chicago Defender*. Because the *Defender* had used Bessie's picture the previous week, on this occasion they photographed a proud Susan holding the silver cup presented to her daughter by the *Shuffle Along* cast.

For the *Defender* interview Bessie gave a complete account of her flying experiences in France, beginning with her failure to find a teacher in Paris, then her move to Le Crotoy, the lessons and the risks. After passing her tests, she said, she returned to Paris for more lessons with the French ace. She liked to fly high, she said, because "the higher you fly, the better the chance you have in case of an accident," meaning she would have more time to correct a problem.

Asked why she wanted to fly, Bessie said, "We must have aviators if we are to keep up with the times. I shall never be satisfied until we have men of the Race who can fly. Do you know you have never lived until you have flown? Of course, it takes one with courage, nerve and ambition. But I am thrilled to know we have men who are physically fit; now what is needed is men who are not afraid of death." She offered to meet with and provide information on flying schools to anyone fearless enough to meet that challenge.

Bessie told the *Defender* reporter that she had ordered a plane built for her own use, a "Nieuport de Chasse" with a 130-horsepower engine, in which she intended to give exhibition flights "in America and other countries." During her stay in France she had seen, she said, "fine Goliath airplanes, the largest built by the house of Farman, equipped with two Sampson motors, planes shown only to flyers because, although they could carry up to fourteen passengers, they were fitted out as war planes."

The Goliaths, built by the Société des Avions H. & M. Farman, actually carried twelve passengers, a pilot, and a mechanic, and some indeed were fitted out as bombers. So Bessie was not too far off the mark. But the Nieuport she claimed to have ordered was never delivered.

Clearly Bessie was prone to exaggeration and not all of her stories were true. Still, she was no more guilty of enhancing her interviews with falsehoods than most of her pioneer aviator colleagues. She knew instinctively that in this postwar period of avid hero worship for cinema stars, athletes, and aviators, the more sensational the story, the better-known the storyteller. And, with radio in its infancy and television not yet on the horizon, there remained only the newspapers to report and relate their activities to a naive public of eager believers. That same public could not imagine aviation as anything other than wartime weaponry or high-risk amusement—the risk for the flier, the amusement for the observer. With regular commercial flights a decade or more in the future, the only way that most civilian aviators could earn a living was to give air shows for paying audiences. But it took publicity to get those audiences and for that Bessie and her like were almost totally dependent on the press. Only the print media—the daily and weekly newspapers—could create their public image and ensure their livelihood. Bessie, even more than most of her colleagues and competitors, quickly sensed that to succeed in this strange mixed world of business and entertainment and showmanship, a certain amount of embroidering the truth, of flamboyance and flair, was a virtual necessity. And for a black flier, a story *had* to be sensational to grab the attention of the white press.

But for all her embellishment of her own adventures Bessie's pride in her race was deep and genuine. When a reporter from the *Chicago Herald* offered to do a story on her if she agreed to pass as

white, she took her mother and niece along with her for the inter-
view. She was laughing as they walked into the reporter's office.
Pointing to Susan and Marion, who were dark-skinned, she said,
"This is my mother and this is my niece. And you want me to pass?"

Bessie soon realized that it was difficult for any pilot, let alone a
black one, to earn a living flying. Most of them became "barnstormers" in shows called "flying circuses." These aviation pioneers
moved from town to town, renting unused patches of farmland (and
often sleeping in the farmers' barns) to gain paying audiences for
their demonstrations of aerial acrobatics to the curious. In war-
surplus planes of fabric and wood, they "looped the loop," deliberately stalled in midair, dived, and barrel-rolled at the risk of their
lives.

While Bessie was still in France a woman flier, Laura Brownell,
had set a "loop-the-loop" record for women of 199 loops at
Mineola, New York. But most barnstormers were men, with women
consigned to the role of wing-walker or parachutist. A month before Bessie returned from Europe, Phoebe Fairgrave, later a
trophy-winning speed pilot, set a parachute-jump record for
women of 15,200 feet. Fairgrave did the jumps to pay for flying
instruction and later opened an air-show company of her own.
Bessie was back in Chicago only ten days when a Chicago wait-
ress, Lillian Boyer, made her first plane-to-plane transfer and
within the next year developed a stunt in which she transferred to a
plane from a speeding automobile by means of a rope ladder.

Feature stories in the *Defender* and admiration for her Paris
wardrobe were not going to get Bessie a job flying. And her FAI
certificate was of little use since licenses weren't required to fly in
the United States. Nor did it take a genius to see that mere figure
eights and pinpoint landings would never draw a paying audience
looking for thrills and excitement. To join an air circus she would
need more lessons to enhance her skills and her flying repertoire.
There was still no one willing to teach her in Chicago. She would
have to return to France.

Where the money for further lessons came from, and whether it came from one source or more, no one really knows. But in circumstances different from her first departure Bessie now had not only chic Paris gowns and attractive leather flying apparel but much favorable newspaper exposure, especially in hometown supporter Robert Abbott's *Chicago Defender*. In early February she booked passage on the S.S. *Paris* and left New York City on February 22 for more training.

CHAPTER 4

Second Time Around

Bessie arrived in New York a week before she was scheduled to sail on the S.S. *Paris* for Le Havre. It was a week in which she was much sought after and lionized by black New Yorkers. The *Shuffle Along* company had moved on to Boston but others in the vibrant community of Harlem had read her press notices and were eager to meet her. She was the guest of a woman she claimed was her aunt, a Mrs. Robinson, at 36 West 136th Street, a house on a site now occupied by the Harlem Medical Center, across the street from the New York Public Library's branch on Malcolm X Boulevard. This was the heart of Harlem, an area described by African American scholar James Weldon Johnson as "twenty-five blocks northward from 125th Street and Fifth Avenue, bordered on the east by Eighth Avenue and the west by St. Nicholas Avenue." No member of the Coleman family knew of an aunt named Robinson but she may have been an older woman whom Bessie referred to as "aunt" rather than "sister," a common reference in the black community.

Bessie had arrived in New York during the beginning of a creative explosion of African American art. Poets, novelists, artists, and musicians, later dubbed the "Niggerati" by anthropologist, novelist, and playwright Zora Neale Hurston, were all demanding justice and equality for their race. Among the most popular of these social revolutionaries were poets Claude McKay, Countee Cullen, and Langston Hughes, all of whom gave frequent poetry readings. Joining in their call for change was the commanding voice of Marcus Garvey, who advocated the establishment of a new African American nation in Africa.

Yet the new, free voices of Harlem were often ignored by the majority of whites. The white Drama League had voted black actor Charles Gilpin, star of Eugene O'Neill's *Emperor Jones*, one of the ten persons most advancing the American theater. But when Gilpin, then appearing in O'Neill's work at the Maryland Theater in Baltimore, tried to buy an orchestra ticket at the box office, the great star was refused. "Colored" were allowed balcony seats only.

While blacks were refused orchestra seats in white theaters, whites—and their money—were welcomed in Harlem's nightclubs. Anxious to test the new freedoms from prewar social restrictions, some whites used Harlem as their testing ground, believing that there they could behave as they imagined black people did, expressing themselves uninhibitedly in a manner that would not have been sanctioned in their own circles. In this misperception they were no different than the white majority of Chicago. After a three-year investigation of the 1919 riots, for example, the Chicago Race Commission concluded that the "primary beliefs held by whites" were, in order, that blacks are mentally inferior, immoral, predisposed to crime, physically unattractive, and highly emotional, given to "noisy religious expression." And white "secondary" beliefs, the report added, were that "Negroes" are "lazy . . . happy-go-lucky . . . boisterous . . . bumptious . . . overassertive . . . lacking in civic consciousness . . . addicted to carrying razors . . . fond of shooting craps . . . and flashy in dress."

In Harlem, the most popular nightspot for affluent whites was the Cotton Club, for dancing, drinking illegal liquor, and watching talented black performers in floor shows featuring beautiful black women. The customers included white gangsters, some of whom had supplied the bootleg liquor. But only the most influential African Americans were admitted. For the rest, there were plenty of other clubs—as well as illegal bars, gambling dens, and houses of prostitution.

In spite of Harlem's flourishing nightlife, the majority of the residents lived quiet lives, their leisure time spent at home or in church. These were the people to whom Bessie looked for support and if she visited the clubs of Harlem she was discreet about it. She used the New York office of the *Defender* as her headquarters because Abbott had ordered its manager, William White, to act as Bessie's agent. While there she was introduced to George W. Harris, editor of a black weekly, the *New York News*. Harris, a Harvard graduate, was a political power among New York's African Americans and alderman of the city's Twenty-first District.

One Sunday, with the alderman as her escort, Bessie went to the Metropolitan Baptist Church where she had been invited to speak by the pastor, the Rev. W. W. Brown, whose congregation numbered 2,500 members. Harris introduced Bessie to the assembly. In her first speaking engagement before a large crowd, Bessie displayed the eloquence she had learned as a child listening to Baptist preachers in Waxahachie. She described her first trip to France for flying lessons and said she was going back to take delivery of a new Nieuport plane she had ordered. During her three months in Europe, she would also purchase other airplanes for her school of aviation in the United States. And when she returned to the United States, she promised, she would give exhibition flights at Mineola, Long Island, and instructions at her flying school in New York. When she finished speaking, the congregation gave her a standing ovation.

Alderman Harris immediately offered to look after Bessie's interests when she returned from France. Some questioned Harris's honesty. Tammany Hall supporter Fred Moore, editor of another black weekly, the *New Age*, wrote that Harris was so dishonest he was "unfit to hold public office." Meanwhile, the *Washington Bee* carried a Negro Press Service report censuring both Harris and Moore for "washing their dirty clothes" in public. The story noted that while Harris's record as alderman was called into question his defenders accused Moore of shoddy behavior both socially and in business. The following year Harris would be unseated by the Board of Aldermen's Committee of Privilege and Elections. But, honest or not, Harris became Bessie's new manager, taking over from Abbott's New York manager Wright.

On February 28 Bessie's ship docked at Le Havre. There she told a reporter for the Baltimore weekly *Afro-American* that she was en route to the French capital where she would be trying out a Nieuport biplane especially built for her. There is no evidence that she ever owned the Nieuport, but she did take advanced flight instruction in it during the two months she was there.

During this second stay in France Bessie may have encountered the racist attitudes of white American tourists visiting Paris. The Americans' attitude became a much-discussed public issue. Bob Davis, a black theater manager, wrote from Paris, "White Americans must check the 'color line' at the three-mile limit. American capitalists and the petit bourgeois, touring France, have caused a furor." Davis detailed how American tourists resented black French colonials dancing with white women and having equal access to cafés, restaurants, hotels, and buses. After concerts, he wrote, when white Americans stood to sing the American national anthem, black members of the audience, some of them Americans, remained silent until they finished, then sang the "Marseillaise."

The controversy became so heated that white American expatriates soon sprang to the defense of the tourists in a concerted counterattack. One expatriate spokesman even warned that the American colony in Paris would take steps to bring deportation of "Negroes who infest the Montmartre section . . . continually insulting, assaulting and robbing tourists." Bessie must have bridled at such racist attacks but she said nothing about them in interviews, directing all her comments to her accomplishments and hopes for a future in aviation.

Despite the rise in racism, Paris's African American colony continued to thrive, embracing in 1925 Josephine Baker, who became the most famous black American expatriate and an idol of the French. With an aviation school to establish in the United States, Bessie herself would not consider living in France. She stayed for two months—probably the time needed to finish her advanced course—and on April 24 applied for and received a visa for Holland at the Dutch consulate in Paris.

She had decided to call on one of the world's most noted aircraft designers, Anthony H. G. Fokker. Fokker, who was Dutch, had been married to a German woman and was living in Germany at the beginning of World War I. He was either persuaded or coerced into designing planes for the German air force, aircraft that matched or excelled that of the Allies. At the war's end, the wily Fokker moved his entire factory to Holland so suddenly that the Allied victors had no time to confiscate it as enemy property.

Fokker not only received Bessie, she said, but he invited her to visit his plant and to fly some of his planes. He also took her to dinner and, she said, "after considerable persuasion" on her part, promised her he would come to America to build planes and set up an aviation school that would be open to men and women "regardless of race, color or creed."

Bessie left Holland only a few days after Fokker, who had gone to the United States to sell airplanes. Obviously determined to make the most of her limited time, she left Amsterdam and arrived in Germany on the same day, May 24, 1922, which was also the day her Dutch visa expired and her German visa, acquired earlier from the German consulate in Amsterdam, became valid.

Fokker may have given Bessie a letter of introduction but, with or without one, she soon made friends in the aviation circles of

Germany, where she stayed ten weeks. In Berlin Bessie was filmed by Pathé News, first standing by her plane and then flying over the defeated Kaiser's palace. She secured a copy of this newsreel (since lost) and later showed it to audiences during her lecture tours. She also brought home a picture of herself, wearing a leather coat and aviator's helmet, standing in front of an airplane beside a handsome pilot at Aldershof airfield near Berlin. Kurt Schnittke, technical assistant to World War I ace Gen. Ernst Udet, identified the pilot as Robert Thelen, the ninth pilot in Germany to receive a pilot's license, and the plane as an LFG-Roland with a 160-horsepower engine. Schnittke added that Thelen had been teaching Bessie how to fly the plane as part of an agreement by which LFG would deliver thirteen of the aircraft to her in the United States, presumably for use in her aviation school. If, indeed, there had been such a purchase agreement, Bessie never received the planes nor was she held to any contract.

Bessie also displayed to reporters a letter signed by a Captain Keller who she said was a German war ace and a member of the Deutsche Luftreederei, a postwar organization that controlled air traffic. The letter praised Bessie for showing "unusual skill" during fifty flights she had made over Berlin in a German plane equipped with two 220-horsepower motors.

Bessie left Berlin the first week of August and sailed for the United States from Amsterdam on the S.S. *Noordam*. When the ship docked at New York on August 13, a number of reporters were waiting to interview her, including two from white newspapers. Her answers to their questions were designed to create the image of a beautiful, brave black woman who loved flying and wanted to teach her people. Bessie hadn't the slightest doubt that she was that woman. All she had to do now was to convince the public. And that would not be easy because as an aviator she was a threat to whites who cherished their racial superiority, and as a *woman* pilot she threatened the ego of black males.

Bessie realized that to make a living at flying she would first have to dramatize herself, like Roscoe Turner, the great speed pilot who wore a lion-tamer's costume when he flew and took his pet lion, Gilmore, along in the second cockpit. An amused public paid to see this bizarre partnership, providing Turner with the money he needed for his plane, its fuel, and its maintenance.

Speaking to reporters, Bessie now began to draw upon everything at her command—her good looks, her sense of theater, and her eloquence—to put her own campaign of self-dramatization into high gear. Some of the things she said were contradictory or even close to pure fabrication. She recognized that the press was her path to the public. But if the reporters were going to create her public image, she was going to sculpt its details. Everything she told them was purposefully selected to enhance the image of a new, exciting, adventurous personality.

Bessie told the *New York Times* reporter, for example, that she had learned to fly after going to France with a Red Cross unit during the war. Brigaded to a French aviation group, she persuaded the officers to give her flight lessons. She also told him that she made a series of flights in a Dornier seaplane at the German base of Friedrichshafen, flights requiring "unusual aeronautical skill."

In Germany, she told the same reporter truthfully, she flew over the Kaiser's palace at Potsdam. And at Staaken, the former German military airfield, she said, she piloted, alone and without previous instruction, the largest airplane ever flown by a woman. There is no verification for this.

In another interview she said she was applauded in both Berlin and Munich for her "daredevil skills" and met a number of German celebrities. Baron Puttkamer of Berlin invited her to dinner, and she was presented to the Lord-Mayor of Baden-Baden. No doubt some of her accounts were true but the confirmation that might have been in German newspaper and magazine articles was lost when the Allies destroyed German archives in the course of the saturation bombing of World War II.

Certainly some of Bessie's claims were falsehoods. A week after she arrived in New York she told a *New Age* reporter that she was awaiting delivery of a dozen Fokker planes she had ordered for her aviation school. But the planes never appeared. In later interviews she added England, Belgium, and Switzerland to her list of countries visited although her passport bears no visa stamps from any of them. In every story she gave her age as 24 although she was by now 30.

The widespread news coverage Bessie now began receiving brought her increasing recognition. And it added distinction to her old friend, Robert Abbott. Convinced she would be useful in his aggressive campaign to increase circulation, he gave Bessie a desk in the New York office of the *Chicago Defender* at 2352 Seventh Avenue and told his staff there to arrange an air show for her. To build a large turnout for the show, Abbott's staff publicized it as being ostensibly in honor of the Fifteenth New York Infantry, veterans of the all-black 369th American Expeditionary Force of World War I. The show would star Bessie as the "world's greatest woman flyer," protégée of the "world's greatest weekly," the *Chicago Defender*. It was scheduled for August 27 at Glenn Curtiss Field in Garden City, Long Island, with a program that included parade-ground drills by doughboys of the Fifteenth, music by the Fifteenth's band, and eight other "sensational flights [by] American Aces" in addition to Bessie's performance. For spectators who wanted to take sightseeing tours after the show, giant passenger planes would be provided.

That Sunday morning when Bessie looked out the window of her room at the Hotel Pennsylvania, she saw rain. It rained all day. She hurried to the *Defender* office where she was assured the show would not be canceled, only postponed until the following Sunday.

During the week it was again publicized by the *Defender*, this time to take place on September 3 at 3:30 P.M., when the "wonderful little woman" would do "heart thrilling stunts." Bessie had been booked for an exhibition flight in Chicago on Labor Day but

Above: Commemorative buttons struck before the official record of Bessie's birthdate was uncovered. She was actually born in 1892, not 1893. (Courtesy of Marion Coleman.) *Below:* Bessie with an unidentified friend. (Courtesy of Arthur W. Freeman.)

Above: Bessie's sisters, Elois and Nilus, in the 1930s. (Courtesy of Arthur W. Freeman.) *Below:* Pilot Arthur W. Freeman, Bessie's nephew, whose decision to become an aviator was made as a young boy after watching his aunt give an exhibition in Chicago. (Courtesy of Arthur W. Freeman.)

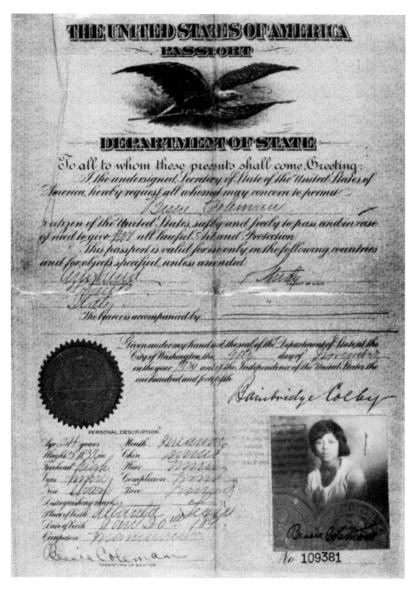

The front panel of Bessie's passport, issued in November 1920. (Courtesy of Thomas D. and Vera Jean Ramey.)

Above: Detail from passport. (Courtesy of
Thomas D. and Vera Jean Ramey.) *Right:*
Portrait of Bessie taken during her first
stay in Paris in 1921. (Courtesy of Marion Coleman.)

Bessie standing in front of the Nieuport plane in which she learned to fly. The heavy flying suit protected her against the cold in the plane's open cockpit. (The Bettmann Archive.)

The picture used on Bessie's license issued by the Fédération Aéronautique Internationale. She was the first African American to receive one. (Courtesy of Arthur W. Freeman.)

Fédération Aéronautique Internationale

FRANCE

Nous soussignés pouvoir sportif reconnu par la Fédération Aéronautique Internationale pour la France certifions que:

Mme Bessie Coleman né a Atlanta, Texas le 20 Janvier 1896 ayant rempli toutes les conditions imposées par la F.A.I. a été breveté:

Pilote-Aviateur

a la date du 15 Juin 1921

Commission Sportive Aéronautique Le Président:

Signature du Titulaire

Bessie Coleman

Bessie's flying license issued by the Fédération Aéronautique Internationale in Paris on June 15, 1921. (Courtesy of Marion Coleman.)

Above: Bessie with her pilot instructor, Robert Thelen, at Aldershof, near Berlin, where she took advanced aerobatic and flying lessons in 1921. (Courtesy of John Underwood.) *Below:* Bessie's mother, Susan Coleman, holding the silver cup given to her daughter in 1922 by the company of the Broadway musical *Shuffle Along.* (Courtesy of Arthur W. Freeman.)

G ernor Dona. y and Mayor Thomas
Welcome Bessie Colen.an to Columbus

GOVERNOR'S LETTER.
STATE OF OHIO
EXECUTIVE DEPARTMENT
COLUMBUS

Miss Bessie Coleman,   August 27, 1923.
    Chicago, Illinois.
Dear Miss Coleman:—

The committee in charge of the Labor Day Celebration in the City of Columbus inform me that you are to be present and assist in the proper observance of that day.

Therefore I extend to you a hearty welcome to the State of Ohio and to the City of Columbus and trust your stay may be both pleasant and profitable.

                Very truly you...
                VIC DONAHEY, Gov.

MAYOR'S LTTER.
CITY OF COLUMBUS
OFFICE OF MAYOR

Miss Bessie Coleman,   August 27, 1923.
    Chicago, Illinois.
My Dear Miss Coleman:—

Word comes to me that you are to pay us a visit on next Monda to aid in the observance of Labor Day.

Being f miliar with your career and the skill, daring and courage you have exhibited on so many occasions, and knowing how your efforts have been recognized by the heads of many European governments, I deem it an honor an a pr ilege to welcome you to the City of Columbus.

With the hope that your stay in our city will be pleasant and profitable, and again expressing my gratifi ion that you are to visit us, I am,

Letters of welcome to Bessie from the governor of Ohio and the mayor of Columbus where she appeared in an air show in 1923. (Courtesy of Arthur W. Freeman.)

Above: The Curtiss JN-4 Bessie bought at Rockwell Field in Coronado, California, late in January 1923. Her ownership ended on February 4 when the aircraft stalled after takeoff. The plane was completely demolished and Bessie spent three months in a hospital. (Courtesy of Arthur W. Freeman.) *Left:* Bessie and an unidentified friend. (Courtesy of Arthur W. Freeman.)

Pictured with her Curtiss JN-4, Bessie wears the handsome leather coat and cap that she had made in France. (Courtesy of Arthur W. Freeman.)

Above: An advertisement for one of Bessie's aerial exhibitions in the African American weekly *Houston Informer* in July 1925. (Courtesy of State Historical Society of Wisconsin.) *Below:* Bessie in the tailored officer's uniform designed for her aerial exhibitions. (Courtesy of Fredia Delacoeur.)

The Savannah Tribune.

VOLUME XLI.   SAVANNAH, GEORGIA, THURSDAY,   JANUARY 7, 1926.   NUMBER 15

N. A. A. C. P. Reports Most Successful Year In History

LOSES LIFE OVER CHITTERLINGS

Own Lawyer Wins His Case

Tales From Florida

Another Enterprise For West Broad

New Undertaking Establishment to Open

International Alliance Of Negroes Appoints V. D. Jenkins In Georgia

WINS $150 PRIZE FOR MOST COTTON

Dr. And Mrs. Sweet And W. White On Speaking Tour

EMANCIPATION CELEBRATION WAS GENERALLY OBSERVED

But Few Organizations In The Procession—Rev. E. W. Rakestraw Delivers Forceful Address—Church Packed to the Gallery

REV. E. W. RAKESTRAW
Financial Speaker at Emancipation Celebration

FORMER BOSSES BACK ACCUSED

DARING COLORED FEMALE AVIATRIX

Spit Fire

She Is The Only Colored Authorized Flier In The Country

Dr. Alexander Reports Mississippi Roused Over Lynching

WOULD RATHER BE A SLAVE

Gives Away Pot Of Gold

CONGRESS STUDIES LABOR MEASURES

STEAL YEAR'S SAVINGS FROM AGED WOMAN

NEGRO COMPANY PLANS BIG ENTERPRISE

PURCHASE VALUABLE WEST BROAD STREET SITE

"COLOUR BAR" VOTED DOWN IN SOUTH AFRICA

MRS. HENRY LINCOLN JOHNSON JOINS U. S. LABOR STAFF

$50,000 Gift For Colored Hospital In Greensboro

18 LYNCHINGS IN 1925 GAIN 2 OVER LAST YEAR

MISSISSIPPI LEADS WITH SIX, FLORIDA SECOND WITH THREE

Bessie's appearance in Savannah, Georgia, was the subject of a front-page story in the African American weekly *Savannah Tribune* when she arrived there in January 1926. (Courtesy of State Historical Society of Wisconsin.)

Left: Bessie in her show uniform wearing an officer's cap with wings. (Courtesy of Arthur W. Freeman.) *Below:* Extract from a letter by Bessie to African American film producer R. E. Norman. (Courtesy of the Lilly Library, Indiana University–Bloomington.)

you may know what a Real
Film of 6 reels would mean
you only have to ask the Mgr.
at some of the Theatre in Fla.
Tampa was not advertized
"at all" But in St. Petersburg
it was imposit to Show that
Have a chance to return.
The picture that I want Film
maybe we could get together on
it, yesterday, To day and tomorrow
It would be better if we jointly
put out the photo, as I am not
able to produce it independently
now if you are enerested let
Me know also give my a view
on directing 6 reels and let you know
Bessie Coleman 510 First St.

W. Palm Beach Fla
Feb. 23, 1926

My Dear Mr. Norman
I just rec' your
letter it was Not fowared to
me as it should have been.
yes Mr Norman I am More
than Sure my picture will
go big in Colored houses
I know this, as a Fast as
my two News reels have
drawn in house more so
than some colored Drama.

The Elite Circle and Girls DeLuxe Club
expect you and your friends to enjoy
"An Aerial Frolic"
honoring
Miss Bessie Coleman
Sat May 1, 8:30 to 12 P. M.   Pythian Auditorium
Subscription 75c
Music by the Imperial Jazz Orchestra

Invitation to the dance that was to be held in Bessie's honor on
May 1, 1926, following her air exhibition in Jacksonville, Florida. She died in a crash the day before, while rehearsing for the
show. (Courtesy of the Eartha White Collection, Thomas G.
Carpenter Library, University of North Florida–Jacksonville.)

Above: Members of the Bessie Coleman Aviatrix Charity Club, formed in Chicago in 1928 to honor her memory, with Bessie's mother, Susan Coleman (*back row, center*). (Courtesy of Arthur W. Freeman.) *Below:* The author at the memorial stone at Bessie's grave in Lincoln Cemetery on Kedzie Avenue in Blue Island, Illinois (outside Chicago).

Above: Portrait of an aviator.
(Courtesy of Arthur W. Free-
man.) *Right:* Bessie Coleman.
(Courtesy of Bettman Archives.)

the paper rescheduled it in favor of the New York show "to please the thousands of persons of both races [who were] disappointed by the cancellation on August 27."

On September 3 Bessie made the "first public flight of a black woman in this country." The Fifteenth's men paraded and the band played jazz until Bessie walked out on the field. She wore the tailored uniform her nephew Arthur recalled so vividly years later, the Sam Browne belt, puttees, shiny boots, and leather coat, and she pushed up her goggles over her pilot's helmet so the crowd could see her face. The uniform was a combination of show business and Paris haute couture. The spectators loved it.

A writer from the Kansas City weekly, *The Call*, thought the crowd of from 1,000 to 3,000, depending on which reporter was counting, was "not as large as the occasion warranted." He expressed satisfaction, however, that it included important professional and business people, and added that the only disappointment was that Miss Coleman did not use her own plane. She flew a borrowed Curtiss JN-4 with a company pilot along with her on her first takeoff. Only after he had checked her out and passed her as satisfactory was she allowed to fly alone. This time she took off to loud applause and a rousing rendition of "The Star Spangled Banner." Though flying alone, she performed no stunts. Stunting was forbidden by Curtiss. Nor did any of the eight "American Aces" promised by the show's organizers appear.

Also on the program that day was Hubert Fauntleroy Julian, Bessie's equal in the art of self-dramatization. A flamboyant black aviator and parachutist from Trinidad, Julian made a 2,000-foot jump off the wing of a plane. Although he seemed to know no fear, he could not always be trusted to perform. At a show in Wilmington, Delaware, Julian disappointed Louis Purnell, a young boy who had come to see him play "The Flight of the Bumble Bee" on a saxophone while descending in a parachute, as advertised. Before the jump Julian rode in a limousine past the crowded grandstand. A few minutes later, and at a considerable distance, Purnell saw a figure leap from a plane and descend under an open chute, but he heard no saxophone. Purnell watched as the chutist landed and was picked up by a truck. The boy separated himself from the crowd and ran in the truck's direction. When he caught up with the vehicle pulling up to a stop behind an isolated hangar, he saw a man with a parachute in the truck bed, but the man was not Julian. Shortly afterward Julian reappeared in front of the grandstand, in the limousine, a parachute wrapped around him.

If competition from this tall (six feet, two inches), handsome man, who claimed to have been in the Canadian air force and prefixed his name with "Lieutenant," "Colonel" or "Doctor," annoyed Bessie, she gave no indication of it. The pilot of Julian's plane was Capt. Edison C. McVey, who frequently worked with the more colorful Julian. A year later McVey suffered a broken jaw and breaks of both legs and one arm in a flying accident but was back on the air-show circuit in less than twelve months. At the close of the show, McVey presented Bessie with a bouquet as the band played loudly and the audience cheered in approval. Julian, generally referred to in the press as "The Black Eagle of Harlem," had not detracted from Bessie's starring role.

New York newspaperman J. A. Jackson, a black columnist for the white-owned entertainment trade paper *Billboard*, wrote that Bessie was a conservative flier but a very good one. He quoted W. H. McMullen, assistant to the Curtiss Field chief pilot, as saying her "quick handling" of the Curtiss plane, an aircraft she had never flown before, "was very good." Jackson also noted that at Bessie's show a large passenger plane that took paying passengers on brief sightseeing flights over the area "did a big business all afternoon. Probably more people of color," he wrote, "went up that day than had ever flown since planes were invented."

Jackson was not only a very influential theater reviewer but was also busy organizing an association of black state fair promoters. Such fairs were gaining popularity in the rural areas of the South where entertainment for black audiences by black performers was severely limited by white authorities. On Jackson's strong recommendation, Bessie was engaged to appear at the Negro Tri-State Fair in Memphis, Tennessee.

In Memphis, Bessie's performance was billed as "the principal thrill" of the fair's opening day, October 12. Describing the occasion months later, Bessie said 20,000 spectators watched that first-day performance. (Twenty thousand was actually the attendance for the full three days of the fair.) The *Memphis Commercial Appeal* praised Bessie's performance as "nervy flying." And Jackson said Bessie proved "so great a draw that she should have been obtained for a dozen [shows]." Now, with two successful appearances behind her, Bessie Coleman, "the world's greatest woman flyer," was all fired up to return to Chicago and show her hometown what she could do.

CHAPTER 5

Pleasing the
Crowds, Alienating
the Critics

L ate in September of 1922 a triumphant Bessie Coleman, only three months shy of 31, came home to Chicago. In seven years she had engineered a personal transformation from penniless Southern immigrant to South Side celebrity. The former manicurist who as a young woman had observed and admired the prominent of her community along the Stroll was now herself one of the observed and admired.

For the first time her name appeared in the social notices of the *Defender* when a Sunday luncheon in her honor was given by a Miss Anthea Robinson of Forty-fifth Place and attended by a married couple and four bachelors. That same afternoon, September 30, as if to remind her of the fragile and ephemeral basis of her newly established reputation, a plane crashed in the middle of Main Street in Mount Vernon, Ohio, killing both its occupants.

"Queen Bess, Daredevil Aviatrix" was undeterred by the almost daily reports of fatal accidents in the high-risk occupation she had chosen. Two weeks after she came home she gave her first air show in Chicago, the one that had been originally set for Labor Day but which publisher Robert Abbott had to reschedule when rain delayed Bessie's first New York appearance.

In its pre-show publicity the *Defender* declared Bessie had "amazed continental Europe and been applauded in Paris, Berlin and Munich." For this show she would do four flights, starting at 3 P.M. at Checkerboard Airdrome, at Roosevelt Road and Fifth Street. More enthusiastic than accurate, the publicity could have been written by Bessie herself. One typical blurb ran:

> Her flight will be patterned after American, French, Spanish and German methods. The French Nungesser start will be made. The climb will be after the Spanish form of Berta Costa and the turn that of McMullen in the American Curtiss. She will straighten out in the manner of Eddie Rickenbacker and execute glides after the style of the German Richtofen. Landings of the Ralph C. Diggins type will be made.

The *Defender*'s "Berta Costa" was Bertrand B. Acosta, an American, not Spanish test pilot who would be one of Adm. Richard Byrd's crew to fly the Atlantic in 1927. The other Americans were Curtis McMullen, Ralph C. Diggins, and Edward Vernon Rickenbacker, all American aces of World War I. Charles Eugene J. M. Nungesser was a French ace and Baron Manfred von Richtofen was Germany's famous "Red Baron."

The same reporter wrote that on a second flight Bessie would cut a figure eight in honor of the Eighth Illinois Infantry and afterwards, presumably with Bessie at the controls, Jack Cope, veteran balloonist, wing-walker, and rope-ladder expert, would perform. On the fourth and last flight, Bessie's sister Georgia, younger by six years, was to do a "drop of death" as the parachutist. "No one has ever attempted this leap," the *Defender* trumpeted.

As far as Georgia was concerned it would be best if no one ever did. Certainly *she* would not. The night that Bessie came home with the costume she planned for Georgia to wear and blithely began to give her instructions on how to jump, Georgia shouted, "I will not, absolutely not, jump!"

"You'll do what I tell you!" Bessie shouted back.

"Who in the hell do you think you're talking to?" Georgia snapped.

They stood, toe to toe, glaring at each other until Georgia grinned and began chanting "Unh, unh, not *me!*" It was a chant the Coleman youngsters had used time and time again whenever Susan wanted them to do some distasteful chore. Bessie laughed, surrendered, and arranged for Cope to do the jump.

On October 7 and again a week later, the *Defender* ran a two-column advertisement stating that "The Race's only aviatrix" would make her initial flight at Checkerboard Airdrome on Sunday, October 15. The admission was one dollar for adults and twenty-five cents for children.

The *Defender* gave Bessie the publicity but the field and plane were provided by David L. Behncke, a white man. Behncke was a rare combination of shrewd businessman and aviation enthusiast who pushed his planes to maximum performance and encouraged other pilots to do the same. An Army Air Service instructor at 19, he was five years Bessie's junior but nevertheless owned Checkerboard, where he operated his own air express and charter service, fueling and repair station, and a sales office for new aircraft.

Because of this young entrepreneur, Sundays in Chicago had become a time for pilots to show the public their skills in racing, stunting, wing walking, and parachuting. Behncke, who would later become a commercial airlines captain, and president of the Air Line Pilots' Association for thirty years, had no reservations about Bessie's race or gender. He himself had replaced her in the Labor Day show at Checkerboard that conflicted with her New York appearance and won a speed derby by flying fifty-five miles in forty-five minutes.

On the Sunday of Bessie's Chicago debut the spectators, black and white, numbered about 2,000. Among them were her mother

Susan; sisters Georgia, Elois, and Nilus; nieces Marion, Eulah B., and Vera, and Nilus's son, eight-year-old Arthur Freeman. Arthur had always marveled at Bessie's stationery with its picture of an airplane on every sheet. Now, actually watching her perform, he was ecstatic. " 'My aunt's a flier!,' I thought, 'and she's just beautiful wearing that long leather coat over her uniform and the leather helmet with aviator goggles! That's my aunt! A real live aviator!' "

Bessie performed in one of Behncke's planes with considerably more dash and daring than at her first show in New York. She finished the first act in ten minutes before taking off again to do a figure eight in honor of the Eighth Infantry, turning and twisting the aircraft as if she had lost control, then soaring up again.

Sixteen months earlier Bessie may well have thought that nothing would ever surpass the joy of her first solo flight in France. If she later changed her mind and transferred pride of place to her Curtiss Field experience, that would hardly have been surprising. Her New York debut was charged with the double-barreled thrill of being her first chance to display her skills to her own countrymen and the first American performance ever by a black woman flier. Still, nothing brought as much elation to Bessie as her show at Checkerboard. Her solo flight had been for foreigners; her New York show for mostly strangers. Only Checkerboard could embody the unique satisfaction of performing in her hometown, flaunting her skills before her own people, her own friends, her family.

After the show passengers lined up for rides in one of five two-seater planes on the field. Bessie piloted one while Behncke and his assistants flew the others. They continued giving rides until dark.

Among the passengers that day was only one woman, Elizabeth Reynolds, who was so delighted with her brief flight that she asked Bessie for lessons. Applying with Reynolds was an unnamed male friend, a manufacturer of "extracts" and proprietor of a cigar store at East Forty-third Street. "Extracts" were often the ingredients used in the manufacture of alcoholic drinks forbidden by law. Most of the illegal liquor on the South Side was locally produced. But for wealthy clients who were willing to pay more for imported spirits, bootleggers were beginning to use airplanes as a means of eluding the law. Bessie could have made enough money from only a dozen smuggling runs to finance her flying school. But if she ever had any such offers she turned them down.

Only three weeks after her triumphant homecoming show at Checkerboard Airdrome, Bessie put her entire career as an aviator in jeopardy. Through the assistance of *Billboard* critic-columnist J. A. Jackson, who had given her rave reviews for both her New York and Memphis air shows, she had signed a contract to star in a full-length feature film. In a long *Billboard* article Jackson wrote that the projected venture was to be an eight-reel movie financed by the African American Seminole Film Producing Company and was tentatively titled *Shadow and Sunshine*. Bessie, he said, would be supported by twelve experienced actors, including Leon Williams, "one of the few race members of the movie branch of [Actor's] Equity."

One hundred extras would be hired for the film, which was to be jointly directed by script-writer Jesse Shipp and director Leigh Whipper. A Chicago businessman, Trueman Bell, was financial manager. Shooting was scheduled to start on October 18 in the hope of a Christmas release date.

But in another *Billboard* story scarcely three weeks later an angry Jackson wrote that Bessie "threw it up and quit cold." She did so after being told she would have to appear in the first scene dressed in tattered clothing and with a walking stick and a pack on her back, to portray an ignorant girl just arriving in New York. "No Uncle Tom stuff for me!" was her parting shot.

Jackson returned Bessie's blast with a caustic stab at Bessie's own humble beginnings in Waxahachie. "Miss Coleman," he wrote, "is originally from Texas and some of her southern dialect and mannerisms still cling to her." He followed this story with another in which he said the woman who would replace Bessie as the lead was an experienced actress with "unmistakable culture and social status which will be an asset to the company."

In an interview with Jackson, Peter Jones, president of the Seminole Company, said *Shadow and Sunshine*, originally intended to feature Miss Coleman, was delayed in production "because of the temperament of that young lady, who, after coming to New York at the expense of the company, changed her mind and abruptly left New York without notice to the director."

"Six autos filled with a cast of thirty people, two photographers and the directors waited in vain for two hours for the lady," Jackson wrote, "after which time Mr. Jones called upon her and was advised that she was too ill to accompany him to Curtis[s] Field for the few hours of outdoor stuff that was scheduled. That day she departed for Baltimore."

More than a broken film contract, one that he himself had had a hand in arranging, was behind Jackson's bitter denunciations of Bessie. In addition to his role as journalist and critic, Jackson aspired to be an impresario. At a meeting of the National Negro Business League he and Dr. J. H. Love, manager of the Colored State Fair of Raleigh (North Carolina), had founded the National Association of Colored Fairs. Both men were keenly aware of the desperate need for coordination in booking entertainers for the black state fairs that were just beginning to burgeon in popularity and they hoped their new alliance would end the existing state of booking chaos and confusion. But in that endeavor Bessie was to be of no help. Even before breaking her contract with the Seminole Company, Bessie had, in fact, managed to alienate both Love and Jackson as well as organizers of African American fairs in Virginia and North Carolina.

While still in Germany Bessie, acting on her own or through some unknown agent, communicated with officials of the Norfolk Colored Agricultural and Industrial Fair Association and agreed to appear at their Virginia fair on September 16. Bessie at the time was completely unaware of the existence of either Jackson or Love and could have had no way of knowing that her Norfolk booking had been arranged through their newly formed National Association of Colored Fairs. As early as July the black weekly *Norfolk Journal and Guide* enthusiastically reported, "Negotiations are underway to secure the wonderful colored aviatrix." And as late as September 8 the same newspaper featured a photograph of Bessie taken at her Curtiss Field exhibition in New York along with a statement that "Miss Coleman may appear at the Norfolk Colored Fair next week." But Bessie did not appear.

Since he had been instrumental in publicizing Bessie in *Billboard* and in setting up the Norfolk show, Jackson took Bessie's failure to appear as a personal affront. And now his new business partner, Love, was beginning to be concerned. For, when the Norfolk appearance was being negotiated, Bessie also assured Jackson she would play a date in Love's particular bailiwick of Raleigh, North Carolina. She promised she would send her final terms directly to Love but she never did get in touch with him. And, in the end, Bessie disappointed Love just as she did Jackson, flying instead at the twelfth annual Negro Tri-State Fair show on October 12 in Memphis, Tennessee.

Jackson now was publicly referring to Bessie as "eccentric," a common show-business euphemism for "unreliable." She had already had three managers in five months, he wrote. The first was William White, New York manager of the *Chicago Defender*. The second was Alderman Harris, owner of the black weekly *New York News*. And the third was "a white man whom she brought into the *Billboard* office. The lady," Jackson fumed, "seems to want to capitalize her publicity without being willing to work."

By this time Bessie had managed to offend men who already were or soon would be among the most powerful in the black entertainment world. Within a year entertainment entrepreneur and movie producer Peter Jones would be the manager of airshow fliers Edison C. McVey and Hubert Fauntleroy Julian and, in fact, owner of the airplane they used. Not only would Jackson continue his influential *Billboard* column until 1925, but he was also on the East Coast staff of the Chicago-based Associated Negro Press, a wire service used by most of the African American weeklies in the country. And Doctor Love continued his association with Jackson as promoters-bookers-organizers of black fairs throughout the United States.

When Bessie chose to do battle with Jackson, Love, and Jones, she was fighting on two fronts, in defense of equal rights for blacks in a white-dominated society and equal rights for black women in a male-dominated black society. Clearly her walking off the movie set was a statement of principle. Opportunist though she was about her career, she was never an opportunist about race. She had no intention of perpetuating the derogatory image most whites had of most blacks, an image that had already been confirmed by the Chicago Race Commission's report on the 1919 riots.

Bessie had refused to reinforce these white stereotype misconceptions or to further reduce the self-esteem of her own people by acting out on screen the role of an ignorant Southern black woman. As she herself had put it so bluntly when she stormed off the *Shadow and Sunshine* set, "No Uncle Tom stuff for me." Bessie was determined to bolster black pride. She was supported in this by a black press that seized on her image as both role model and grist for editorials promoting black equality.

The *Norfolk Journal and Guide* chastised the *New York Evening Journal* for stating in a long article that Bessie was from Europe and "took to flying naturally without any teaching." Bessie, the Norfolk editorial pointed out, not only was an American but was forced to take flying lessons in France because no one would teach her in her own country. The Norfolk paper further took the *Evening Journal* to task for stating that "the colored race should supply many excellent flyers" because blacks have a natural physical balance superior to that of whites and that blacks "usually ride a bicycle the first time they try." This latter claim, the Norfolk paper said in angry rebuttal, is a typical "Arthur Brisbane thought which runs through the editorial columns of all the Hearst newspapers . . . If we had more 'balance' along some other very important lines," the *Journal and Guide* concluded, "we would be hitting a greater stride in the race of races."

As to the balance Bessie sought—equal opportunity both for her race and her gender—the latter was far harder to attain. The majority of the nation's men, black or white, regarded women as the "weaker sex." Men headed households, governments, and businesses. Women were not even allowed to vote until the passage of the Nineteenth Amendment in August 1920. And women who aspired to careers were often mocked and even feared by men who saw them as a threat to their hitherto unshakeable faith in male superiority. The *Norfolk Journal and Guide*, for example, the same paper that had so readily praised Bessie as a black, was notably less ready to praise her as a black woman. In a miscellania column titled "Stray Thoughts" it included the following:

> **Blivens:** I see Miss Headstrong is taking aviation lessons.
>
> **Givens:** That so? Always had an idea that girl was flighty.

Bessie was ahead of her time as an aviator and as an advocate of equal rights for African American women. It would be three years before the rising tide of protest from black men was reflected by a blistering editorial in the *Amsterdam News* headed "Colored Women Venturing Too Far From Children, Kitchen, Clothes and Church."

The popular black weekly noted that "the biological function of the female is to bear and rear children," and said Kaiser Wilhelm of Germany had the right idea with his slogan of *"Kinder, Küche, Kleider, Kirche."* Calling black Washington, D.C., "a city of bachelors and old maids," the editorial focused on the fifty-five female teachers of Howard University, pointing out that whereas they had been raised in families totaling 363 children, or 6.5 children per family, the teachers themselves had only 37 children, or 0.9 per family. Worse still, it said, only 22 of the 55 were married, of whom four had one child and four had none. That the average age of the unmarried teachers was over 32 disqualified them as "good breeders," the editorial said, denouncing as "race suicide" what was happening in the nation's capital. "Liberalization of women," the editorial concluded, "must always be kept within the boundary fixed by nature."

Like the Howard teachers Bessie had crossed that boundary. She married late and was not living with her husband. She had no children and had become a pilot at a time when black male fliers could be counted on the fingers of one hand. She had a mind of her own. She was neither apologetic nor ashamed of being a so-called threat to the race or an affront to its men.

After breaking her film contract, Bessie left New York for Baltimore, where she gave an "interesting outline of her work" at the monthly meeting of the "Link of Twelve" at Trinity A.M.E. Church. She also went to Logan Field on the outskirts of that city to look over a number of airplanes but was reported to have found "none to her liking in which to take a spin."

Bessie's movie career was over before it began and her future as a stunt flier seemed in jeopardy as well. A good agent might have persuaded the Seminole Company's writer and director to reshape their script into a story stressing black pride and prowess, characteristics that would have had Bessie's approval. A good agent certainly could have eliminated her confusion over bookings for air shows. But none of the men Jackson named were really agents. White was a newspaperman put in charge of Bessie by Robert Abbott. Harris was a politician with a newspaper he used to promote his own interests. The unidentified white man is nowhere on record and may have been invented by Jackson.

Bessie returned to Chicago with little to sustain her beyond pride and obstinacy. There she launched into a search for new backers. If show-business people on the East Coast would not give her a break, she would look elsewhere.

CHAPTER 6

Forced Landing

In December of 1922 Bessie walked off that New York movie set without a contract, a sponsor, or a plane. Anyone else might have given up. She could not. Instead, at her next stop in Baltimore, she told a reporter that she was opening a school for aviators at 628 Indiana Avenue in Chicago.

From the moment she received her pilot's license in the summer of 1921 she had repeatedly talked of teaching the members of her race to fly. The school was not a publicity stunt, it was an obsession. An office at 628 Indiana was a start. The next step was to renew her contacts on the Stroll and at Checkerboard Airdrome, where David Behncke, Checkerboard's owner, would lend her a plane. But teaching in a downtown office was a far cry from founding a school. In addition to that office, she needed at least one plane of her own and a hangar for her aircraft and its maintenance, and she needed capital.

A giant step toward getting some came through one of her students, Robert Paul Sachs. Sachs was an African American, the Midwestern advertising manager for a California firm, the Coast

Tire and Rubber Company of Oakland. Soon after Sachs began taking lessons, Bessie proposed that for a fee she would go to California to "drop literature from the clouds" advertising the superiority of Coast Tires. The money would finance the purchase of a plane on the West Coast.

Bessie left freezing, windswept Chicago in late January on a train bound for sunny California. That state had already become a Mecca for the country's growing number of young fliers. In their fragile aircraft of wood and fabric and without radio or radar, they found Southern California's climate a godsend. For stunt fliers such as Bessie, California provided a built-in audience of unabashed admirers for the stars of the nation's newest pioneering industries—aviation and motion pictures.

A number of women were already benefiting from the recent trend of using women and airplanes in advertising, a fad that brought columns of free newspaper advertising. Two wing-walker/ parachutists, Gladys Ingle and Gladys Roy, had found sponsors in real estate developers. Their free air shows over property up for sale drew crowds of potential customers. Six months before Bessie arrived in California, two young women pilots, Amelia Earhart and André Peyre, were flying for publicity and profit. In a printed testimonial, 22-year-old Earhart endorsed the airplane she flew for its builder, Bert Kinner. Peyre, a French actress, appeared with Earhart in exhibition flights to boost her own career in Hollywood.

Bessie knew none of these women. In the 1920s whites and blacks met socially only on the rarest of occasions. Whites generally ignored blacks but often held them up to ridicule and contempt, as in an article that appeared in the *Los Angeles Times* a month after Bessie arrived in that city. "Using his razor for social purposes cost Frank P. Hodge, a negro, $100 yesterday," the article said. Explaining that Hodge was courting a woman who objected to one of his remarks, the writer continued, "Hodge pulled out his razor and flourished it in such a manner as to abrade the lady's epidermis."

Yet in the face of such racism and without help from fellow woman aviators, Bessie could still match any competitor in skillful press agentry. She truly belonged in this land of flamboyant aviators and film stars.

For her first stop Bessie went straight to Oakland, to the Coast Tire and Rubber Company, to watch tires being made. In an interview after this inspection she "pronounced it the best, most modern and scientifically equipped plant ever seen in all her travels," although it is unlikely that she had ever seen a tire factory before. She declared herself so impressed she again volunteered, as she had to Sachs in Chicago, to "distribute advertising material from the clouds."

If the offer by this time was something less than spontaneous, the reporter nevertheless took Bessie at her word and wrote that, with every fifth person in California a car owner, "this method of advertising should be quite effective, stimulating the popularity of Coast tires and tubes." He also wrote that Bessie intended to open an aviation school in Oakland and would be giving exhibition flights in California and throughout the Pacific states.

Having paid a courtesy call on her new Oakland employer, and leaving a flurry of publicity for Coast Tires in her wake, Bessie headed south for Los Angeles to buy an airplane, most likely with Coast Tire Company money. The military kept a stockpile of surplus planes at the Rockwell Army Intermediate Depot on North Island at Coronado. By the time she arrived, there were only fifty left, many still in their original crates and not yet fully assembled. They were going for $400 each—or $300 each if you were rich enough to buy ten. Bessie settled for one—an early, almost obsolete Curtiss JN-4.

While at Coronado she was interviewed by a reporter for the *Air Service News Letter,* an influential professional publication that was widely read, especially by pilots. Acknowledging Bessie as "probably the only colored woman in the world who can pilot an airplane," he seemed more interested in her appearance than her accomplishments.

> Miss Coleman is a neat-appearing young woman who has discarded the shirt waist and short skirt for O.D. [olive-drab] breeches, leather leggings, Sam Browne belt and coat cut on the lines of Canadian officers. She says she went to France for two purposes, to drink wine and learn to fly. It goes without saying that she has been successful in flying, but we don't know yet her capacity or ability to drink wine.

Bessie told the *Newsletter* reporter that she had bought three planes, which she would arrange to have flown to San Francisco, where she would supervise their assembly and testing. Like many of his colleagues before (and after), he took her words for gospel and quoted her accordingly.

When Bessie spoke as a pilot she became an actress on stage, uttering fictional lines with total conviction. In her personal life she did not lie to friends and family, either telling the truth or remaining silent. At first she had simply embellished her stories in her search for recognition as a pilot. With that obtained, she now tended to even greater overstatement, saying whatever she thought might help her achieve her goal of founding an aviation school for her fellow African Americans.

She also called at the office of the African American weekly, the *California Eagle*, which proved as cooperative and obliging as the *Newsletter*. Taking eight years off her life, the paper described Bessie as an "enthusiastic, charming girl, only 23 years old, apparently unspoiled by the honors, social and professional, that have been showered upon her in the capitals of Europe."

Bessie's lying about her age was hardly unusual. Americans at the time considered a woman of 40 to be middle-aged. If any woman, especially a celebrity, could deceive the eye of the beholder, she might spend two decades claiming she was in her twenties. And Bessie, with her smooth face unblemished by any sign of aging, could deceive the eye of any beholder.

Never less than zealous in her persistent pursuit of the press, Bessie had prepared and always carried with her a press handout summarizing her accomplishments. Obviously she gave a copy to the *Eagle* correspondent. According to his story, Bessie had now flown in six European countries (up from three) and held German as well as French flying credentials. She again displayed the purported testimonial letter signed by the German Keller and told of being the guest of Anthony Fokker in Holland and of Baron Puttkamer and the Lord Mayor of Baden-Baden in Germany. She also claimed to be the only woman to hold an international flying license recognized "all over the world." Bessie was, in fact, the only American woman to receive her license directly from the Fédération Aéronautique Internationale in Paris. But there were other women, both American and foreign, who held licenses issued by aero clubs affiliated with the FAI.

In yet another interview, this one with the black weekly *Dallas Express*, she said she had come to California intending to open an aeronautical school in Oakland but the good people of Los Angeles had persuaded her to remain there instead. Their warm welcome included a dinner given by Mrs. S. E. Bramlett, of 1409 East Eighteenth Street, on February 1. All the guests were women, whom Bessie had begun to recognize as the dominant force for change in the black community and therefore the people most important to her plans.

The only hint of criticism in the otherwise laudatory press coverage of Bessie's visit up to this point appeared in an unattributed article in the Baltimore *Afro-American*. The story noted that "while Miss Coleman is an expert at the wheel of an airplane, she has made no long distance flights, confining herself mainly to exhibitions." This, it said, was the reason "the public is watching her effort" to break the flying record from Los Angeles to San Francisco. At the time, pilots were expected to attempt record-breaking flights, but Bessie had never said anything about wanting to break this or any other record. Quite possibly, "the public" was simply a euphemistic reference to the writer himself, who could well have been J. A. Jackson, the *Billboard* critic Bessie had so enraged by walking out on the Seminole movie contract in New York. Whoever wrote it, the implied criticism was certainly vitiated by the author's next statement that the distance between the two cities was about 80 miles and could be covered in an hour. A casual reader in Baltimore or almost any other part of the country might have overlooked that, but a Southern California reader, knowing that the distance was closer to 425 miles, would certainly have been brought up sharply by the glaring error. Breaking records was not on Bessie's agenda; opening a flying school for members of her race was.

As soon as her newly purchased plane was ready Bessie tried to arrange an exhibition flight at Los Angeles's Rogers Field, where Amelia Earhart had taken her very first airplane ride (as a passenger) just two years before. This plan never materialized, but Bessie found backers for another exhibition, on February 4, to celebrate the opening of a new fairgrounds at Palomar Park near Slauson Avenue.

As she left her room that Sunday at the Young Women's Christian Association's Twelfth Street Center for Colored Girls, she must have thought the worst of her struggles—learning to fly, soloing, getting her license, working for three different managers, and the ever-present pursuit of financial backing—was over. Now, finally, she had a plane of her own. And she had the use of a park for a show where she wouldn't just be one small part of a larger flying circus but the sole attraction. This time 10,000 people had gathered to see her, Bessie Coleman, just to "see a woman handle a plane."

But they never did see her. Moments after she took off from

Santa Monica for the short twenty-five-mile flight to the fairgrounds her motor stalled at 300 feet. Her newly purchased Jenny nose-dived, smashing into the ground. Airdrome workers ran to the demolished machine to pull Bessie's apparently lifeless body from the wreckage. She was alive but unconscious. A doctor who gave her emergency care at the crash site said she had a broken leg, fractured ribs, multiple cuts around her eyes and chin, and possible internal injuries.

On regaining consciousness, Bessie begged the doctor to "patch her up," just enough to enable her to fly the twenty-five miles to Palomar Park so that her waiting fans would not be disappointed. He refused and called an ambulance. En route to St. Catherine's Hospital on Pacific Avenue in Santa Monica, Bessie sent a message to her field manager, a Mrs. Bass, saying she would be along later—as soon as she could be "patched up."

Much more than a patch job was required. Bessie's broken leg had to be set in a cast from ankle to hip. She had three broken ribs, which made it painful for her to breathe. Many of her multiple cuts required stitches and caused massive swelling, especially around her eyes and chin. But far worse to Bessie than the pain and disfigurement was the loss of her one and only plane and of the chance to parade her skills before thousands of fans and ticket-buyers.

The accident itself was not unusual. In the preceding sixteen days one Los Angeles newspaper alone reported three similar crashes caused by motor failure right after takeoff, killing a total of five pilots. And two years later the Army Air Service's most distinguished aviator, Col. William F. "Billy" Mitchell, would crash just as Bessie had when his motor went dead at eighty feet.

What was, perhaps, at least somewhat unusual was the reaction of the crowd that had paid to see her perform and was waiting to see her at Palomar Park. Instead of the sympathy that might have been expected, many of the fans reacted in outrage, making strident demands for refunds "then and there." Bessie and her managers were accused of promoting a "bunco game." The charges were so widespread and virulent that one journalist, Dora Mitchell, a reporter for the *California Eagle*, wrote that it was "with shame for our own people" that there was a "most deplorable dearth of chivalry among our men and an utter lack of womenly feeling and sympathy among those of Miss Coleman's own sex." The story went on:

> a brave little Race girl was condemned without a hearing while she lay on a bed of pain, unable even to send a message . . . although such a message would doubtless have been received with sneers and incredulity . . . Certain people on Sunday night even declared this poor girl's injuries to be a punishment from on high for the sin of attempting to fly on Sunday.

Battered and bandaged but unwilling to admit defeat, Bessie sent this telegram from her hospital bed to friends and well-wishers:

> TELL THEM ALL THAT AS SOON AS I CAN WALK I'M GOING TO FLY! AND MY FAITH IN AVIATION AND THE USEFUL OF IT WILL SERVE IN FULFILLING THE DESTINY OF MY PEOPLE ISN'T SHAKEN AT ALL.

What Bessie lacked in grammar she made up for in ambition. And she insisted she would carry on her plans for a school where "our boys may acquire the mastery of the air."

A week later, a reporter who described her as "undaunted by a broken leg" recalled that in a lecture before the accident she had said, "I am anxious to teach some of you to fly, for accidents may happen. I may drift out and there would be someone to take my place." This was not all self-dramatization for the sake of publicity. Like all circus and stunt fliers Bessie knew she was engaged in a high-risk occupation. And she accepted those risks because she loved flying and regarded her participation in it as a way of generating self-respect for her race.

When news of the accident reached the family, Elois said, "Well, that's the end of it." Susan, who once claimed she "had thirteen children, raised up nine and one of them was crazy," knew her Bessie. "Oh, no," she told Elois. "That's only the beginning."

Bessie's leg took far longer to heal than she had anticipated, but she got some badly needed support from her former Chicago student and Oakland business sponsor Robert Sachs. Sachs wrote a long letter on her behalf, which was published in the *Eagle*. He, too, parroted all the background information Bessie had given him, including her age as 23, and added:

> Her race should be proud of her accomplishments, as she is the only colored woman pilot in the world—and there are no colored men flyers at all. And with all her hard work and her great record of accomplishments, she has always borne her race in mind and worked for its advantage. One of her plans for the future is the establishment of a school of flying especially for colored people.

Her work, Sachs continued, had been delayed "as much by lack of funds as by her personal injuries" and "worry over financial problems" was likely to delay her recovery.

A subscription had been taken up at the company's office, he said, and he was offering the African American people of California "the opportunity to really come to the assistance of the race." Sachs said black churches and other organizations might raise funds from their members, and those who wanted to contribute to Bessie directly could send money to her at the hospital, to a Mrs. Melba Stafford at 939 Willow Street in Oakland, or to the office of the *Eagle*.

Bessie was down, but far from out. Without capital or a single airplane, she placed a paid notice in the *Eagle*, a contract for the Coleman School of Aeronautics. Its terms included a student tuition fee of $400 in advance, $25 on signing the contract, and the remainder to be individually agreed upon in monthly installments. In return she promised to provide competent instructors and to put all the money paid in escrow until the lessons started. The student would be responsible for all injuries to his or her person, and Bessie would be the judge of his or her ability to operate an aircraft. She would give the student every available means of instruction and equipment to demonstrate that ability.

Her advertisement of the contract came to the attention of the *New Age*, a black newspaper clear across the country in New York, that published an editorial headlined "Queen Bess Opens School." The rules of the contract, the editorial said, could be interpreted as meaning that "broken bones and other dangers are at the risk and expense of the flyer in embryo. There is nothing like having such details understood in advance."

Bessie's contract was a conventional one for the times. In fact it was virtually identical to the "Conditions de l'Apprentissage," the contract given student pilots at the famous Caudron School of Aeronautics at Le Crotoy. Bessie may have hoped that the contributions received from the collections by Sachs, together with advance payments from a few students, would be enough for a start. It was not.

It was early May, nearly three months after the accident, before Bessie was able to leave the hospital, her leg still in a cast. "Although hardly able to put the weight of her body on the injured member," said one newspaper account, "the gritty and progressive woman has, during the past week, completed a motion picture started on the day before her unfortunate fall." Sachs too had mentioned Bessie's making a motion picture in Los Angeles but neither report went into further detail and there is no confirmation that a movie was ever released, made or, indeed, even begun.

For the remainder of May Bessie stayed at the Los Angeles home of a Mrs. S. E. Jones on West Fifty-seventh Street. Still striving to raise capital for her school, she arranged to give five nightly lectures between May 5 and 12 at the Ninth Street branch of the

Young Men's Christian Association. At these she showed films of her flights in Europe and the United States. Wendell Gladden of the *Eagle* wrote that one of the reels, from Pathé News, had already been shown at a popular theater in Los Angeles. It showed Bessie standing beside her plane in Germany and later airborne, "soaring high above the Kaiser's palace." The admission charge for her combined lecture/picture show was twenty-five cents for children and thirty-five or fifty cents for adults. Bessie, of course, did not get all the money. Some of it went to the YMCA.

Bessie left Los Angeles for Chicago in June without a plane, without a job, and with very little money. Once again she would just have to start over.

CHAPTER 7

Grounded

When Bessie Coleman returned to Chicago in late June 1923 she was determined to halt the downward course of a career that had lost its momentum during the months she had spent recuperating from the crash in Santa Monica. It was not going to be easy. The *Chicago Defender* appeared to have lost interest in her. After a brief notice about her return, the paper failed to publicize her first post-crash air show, scheduled for September 3, Labor Day, in Columbus, Ohio. This time a white daily, the *Columbus Evening Dispatch*, gave her the advance publicity she needed. The *Dispatch* billed her as "the only colored aviatrix in the world," who would stage a number of flying stunts at Driving Park, a racetrack, on Labor Day.

Whether owing to an unusually credulous reporter or a copy-short editor, the *Dispatch* story used almost all of Bessie's prepared publicity handout, claiming she had given exhibition flights in Europe and that she was raised on a ranch in Texas where she was "born twenty-three years ago." The *Dispatch* reported that the show would include airplane, automobile, and motorcycle races

75

and stunts by a man billed as "Daredevil Erwin" and by two women, Iona McCarthy and Bessie Coleman. Erwin's repertoire included hanging by his teeth from a strap suspended beneath an airplane. McCarthy, who did triple parachute leaps, said she would try to land on the wing of another plane. The pre-show publicity did not say what Bessie would do. Presumably just being black, female, and the first of her kind to fly a plane was thought to be sufficiently intriguing.

But the Labor Day show that had promised to be a bonanza for Bessie's flagging career was rained out. She sat all day at Driving Park, along with her colleagues, waiting for the weather to clear. While she waited, more than 2,000 people—almost as many women as men—gathered at the state fairgrounds a few miles away for an all-day celebration of the Ku Klux Klan. Their enthusiasm undampened by the rain that was washing Bessie's show off the calendar, it was primarily an initiation festival, men being inducted into the Klan itself and women into the Ladies Auxiliary. Called "naturalization," the initiation preceded the marriage of two Klan members by a Klan minister. Music for the festivities was provided by an orchestra of Columbus Klan members. Almost everyone, including the musicians, dressed in the ghost-like white hoods and robes that inspired fear in so many African Americans.

Leaving rain-soaked Driving Park and the state park full of Klan members behind, Bessie returned to Chicago, where she called at the *Defender*'s office to announce that her Columbus air show had been rained out but her plans to open an aviation school in Chicago were "well underway." According to the *Defender*, Bessie also stated that she would give her first flight in Chicago "as soon as her machine arrives in the city" and the plane itself would be put on exhibit at the Eighth Regiment Armory. The paper affirmed that Chicago's favorite aviator had completely recovered from her accident in Los Angeles and said "rumors were afloat at the time that some mechanics tampered with the steering apparatus in an effort to keep her from gaining the recognition due her." The *Defender* was the only paper to publish such rumors.

On September 9 Bessie returned to Columbus for the air show that had been rained out a week earlier and performed before an enthusiastic crowd of 10,000. Back in Chicago once more she made another announcement to the *Defender*, saying she would give "a farewell flight" in the city before leaving for "a tour of the south." The show was advertised for Sunday, September 23, at the Chicago Air Park at 63rd Street and 48th Avenue. Yet the following week's *Defender* had no story about it. It may have been canceled because the plane she had offered to put on exhibit never arrived. Worse yet, there would be no Southern tour for Bessie, who had run head-on into a series of changing managers and broken contracts.

Bessie's contract troubles became public in the spring of 1924 when the *Afro-American* ran an unbylined story headlined "Aviatrix Loses Another Manager." In a style strongly resembling that of her enemy, J. A. Jackson, it described Bessie as "the Colored girl who has been presenting herself as an aviatrix for the last two seasons" and accused her of accumulating "a long list of incomplete contracts and an almost as lengthy list of managers and agents." Naming the managers and agents as D. Ireland Thomas, M. C. Washington, and, the most recent, Raymond Daley, the writer asserted, "we understand that Washington has not added to his prestige in Ohio by his experience with the temperamental aviatrix."

By that spring, Bessie had already had five known managers in twenty-eight months—*Defender* editor William White, New York alderman George W. Harris, and the three named in the *Afro-American*'s article. Her sister Elois added a sixth, David Behncke, to the list, although he could do little more for Bessie than book her in local air shows at Checkerboard, along with another black pilot, Edward Young, from Iowa.

Whether or not she was as "temperamental" as the *Afro-American* writer claimed, Bessie certainly was strong-willed and independent enough to insist on having a say in her bookings. Her African American agent/managers, on the other hand, were equally determined, raised in a business tradition where the manager led and the client followed, especially if the manager was male and the client female. That she and they should clash was probably inevitable.

Purposeful though she was, Bessie now found herself in Chicago at the end of a long series of misadventures with no employment in sight. The loss of her only plane in California, her long hospitalization, and the alienation of her managers—first on the East Coast and now in the Midwest—seemed to be having their cumulative effect. She appeared ready to give up, telling her sister Elois that she had decided to take "a good long rest." Bessie rented an apartment at Forty-second and South Parkway (now Dr. Martin Luther King Jr. Drive) and, according to Elois, "took her furniture out of storage, and clad in coveralls, went about cleaning the place up."

Despite her being a celebrity, there was no niche in black Chicago into which Bessie could fit and feel really at home. She was not a businesswoman like Madame C. J. Walker or an educator like Dr. Mary McLeod Bethune. She had none of the editorial skills of Ida Wells Barnett. Even as an entertainer she had nothing in common with the Chicago celebrities of the time—Ethel Waters, Bill "Bojangles" Robinson, Alberta Hunter, Bessie Smith, or Lil Hardin, who had just married Louis Armstrong.

Neither was Bessie interested in the activities of the Pilgrim Baptist Church, which were so essential to her mother's social life. However, she did accompany Susan there occasionally and once, in a lower-floor lecture room, was introduced to Marjorie Stewart Joyner, friend and employee of Madame Walker. An observant and gregarious businesswoman, journalist, and community activist, Dr. Joyner remembered Bessie as "middle-class, like me and most of the people there that night. She dressed nicely—nothing too showy—and was well-spoken. She was trying to *be* somebody. Not like people who say they want to be somebody but don't really try. We tried, Bessie and I, and we succeeded."

Dr. Joyner was right. Bessie was already a South Side celebrity. So was Joyner, who was active in the Cosmopolitan church, Bethune-Cookman College, the National Council of Negro Women, and the United Beauty Schools Owners and Teachers Association, as well as in raising funds for any number of Chicago charities. (In 1990 Chicago's mayor Richard M. Daley, Jr., declared October 24 to be Dr. Marjorie Stewart Joyner Day in honor of her ninety-fourth birthday.)

"But we weren't upper class," Dr. Joyner added. "True society people were all doctors, professional people or very wealthy business men and their families. That upper crust didn't want to help Bessie. Robert Abbott helped because he was interested in news and Bessie was news."

There was no place for a manicurist-turned-pilot in Chicago's black society, not when some black clubs and churches would not accept members unless one could see "the blueness of their veins through the skin of their wrists." "The right shade of black is yellow" was the general rule. So Bessie turned to her friends and family.

At home, her experiences as pilot and world traveler had affirmed Bessie's pivotal role in the family. Her apartment was where all the Coleman women met. A center for both happy reunions and fierce arguments, it was open to all family members. In their eyes she was famous and because of that all the Colemans had gained status. Her Southern, black, rural heritage of allegiance to kin was expressed in the support and loyalty she gave all of them.

The wife of Bessie's brother, Isaiah, who had gone to Canada, received a warm welcome when she came to Chicago on a visit. Bessie was charming, she said, amusing her one evening by giving a series of impressions of Asians, Caucasians, and blacks, all done with the skillful use of theatrical makeup, wigs, and a wardrobe that included French frocks and a beautiful silk negligee. "Your Aunt Bessie is very pretty and very clever," she said. "Somehow she can even make herself up to look like different nationalities."

Bessie's apartment became a second home for her nieces and nephews—a home she herself had never had with hardworking Susan who had neither the time nor money to give her children anything except the bare necessities. Bessie hugged and kissed her sisters' children and let them play her records on the wind-up Victrola. A marvelous cook, she made them meals, ate with them, and taught them table manners without scolding. She bought them clothes and toys and gave them spending money. Bessie gave Elois's daughter Vera, who was eleven, something Bessie herself must have longed for as a child, a room of her own, in Bessie's apartment. "Back when I was little, when she was still doing manicures," Vera said, "she'd give me money for the movies and walk down to the corner of Thirty-fifth to watch when I crossed the street. Then, when she came back from California, I got to live with her."

Vera's brother Dean recalled that "Vera had a little dog, a white Spitz and she loved that little dog so much she named it Bessie."

Their favorite aunt also paid the tuition for Georgia's daughter, Marion, to attend kindergarten at the Little Shepherd School on a site now occupied by a public school at Forty-ninth Street and Indiana Avenue. And it was Aunt Bessie who rescued Marion from her grandmother's insistence that the child remain at the Pilgrim Baptist Church every Sunday for the entire day. "I was so restless and bored," Marion said, "until Bessie got me excused for the afternoons. She gave me money for the movies and two dimes to ride the El to and from the Peerless Theater. I saw a movie almost every Sunday afternoon."

In contests of will Bessie prevailed over her mother and sisters—except for Georgia. Georgia was a fun-loving "flapper" of the Twenties, a devotee of the Stroll's nightlife. She coveted Bessie's elegant clothes and often took things from her closet without permission. Bessie's wardrobe was mostly conservative, but it did include a glittering, red, beaded designer creation from Paris. One night when Susan was at Bessie's apartment, she saw Georgia taking the red dress out of the closet. "Bessie will kill you!" Susan shouted. Georgia just grinned. She was in it and out and dancing at Dreamland before Bessie could catch her.

Bessie went out, too, to many of the same clubs on the Stroll, but her evenings were more sedate than impetuous Georgia's and her escorts were older and richer. One was Kojo Takalou Housemou, the handsome visiting prince of Dahomey, whose company was much sought after by African American society throughout the United States. Kojo had been invited by the National Association for the Advancement of Colored People to visit the United States under the direction of the association's secretary, Walter F. White. The invitation was the NAACP's way of showing its disapproval of Marcus Garvey's Universal Negro Improvement Association, the back-to-Africa movement that claimed thousands of ardent followers. Kojo, who said he had a doctorate degree in philosophy from the University of Paris, headed an organization of his own in French Colonial Africa known as the League for Defense of the Black Race.

Bessie gave a tea in the prince's honor, charming him and his retinue by speaking French. Other of Kojo's evenings were far less pleasant. One night at a downtown Chicago restaurant, he was refused service by the waiters because he was black. When the prince objected, the restaurant summoned the police and, in the ensuing melee, the guest of honor lost two gold teeth.

Bessie had other men friends but aside from "Uncle Claude," Bessie's husband, who still called on her, Bessie's nieces did not know their names. One is thought to have been her old acquaintance from her manicurist days, Jesse Binga. By this time Binga, real estate magnate, owned the lot at the corner of State and Thirty-fifth on which he was to build "the most modern bank and office building owned by colored people in the world."

As for Robert Abbott, the publisher seems to have distanced himself from his one-time protégée. Possibly he was concerned that Bessie's alienation of so many managers and agents would cut into the *Defender*'s advertising revenue, a large share of which came from African American theatrical notices. Between October 1923 and May 1925, his newspaper published only one item with Bessie's name in it. Datelined New York, May 12, 1924, it was an announcement by Maj. Lorillard Spender that Dutch plane designer Anthony H. G. Fokker had leased the former Witteman Aircraft factory in Hasbrough Heights, New Jersey, to build planes in the United States. Although Bessie was not at all integral to the story, inserting her name would most assuredly increase local interest in Chicago. The *Defender* identified Fokker as the man who had taught Bessie to fly and again quoted Bessie as saying that Fokker had promised, "after considerable persuasion on her part," to establish an aviation school for men and women, regardless of race, creed, or color, which would charge the least possible tuition. A year later Spender and Fokker formed an Atlantic Aircraft Corporation partnership but nothing ever came of the school.

Languishing in Chicago with no job and no capital, Bessie was not the only pilot whose luck had run out. Hubert Julian was placed on probation for six months for parachuting inside the city limits of New York. When the probation period was over he lectured to paying audiences such as one at the Attucks Theater in Norfolk, Virginia, in a struggle to raise money for a "world flight."

The following July Julian was ready—or thought he was. His takeoff was delayed for hours because the builders of his seaplane, the Chamberlin Roe Aircraft Corporation, demanded the $1,400 he still owed on the $8,000 airplane. Julian stood on top of a taxi pleading for donations from the crowd gathered to see him leave from the Harlem River at 125th Street. Assisting Julian were several friends who sold photographs of him and his airplane to onlookers until two admirers, a minister and an undertaker, pledged the final $500.

Meanwhile, the plane had become mired in mud as the tide receded. Mechanics dragged it into deeper water where three attempts to start the motor failed. On the fourth try it started but the plane hit the rocks, now exposed at low tide at 140th Street. From there the aircraft was hauled by boat down the Harlem River to the East River. Immediately after takeoff a pontoon dropped off and fell into the water. The plane tilted, then crashed into Flushing Bay, six miles from where it had started. The whole affair was a fiasco, but one that easily could have happened to any flier in that era of primitive planes and capital-starved pilots.

Just as brave, as daring, as determined, and perhaps as foolish as Julian or any other of these aviation pioneers, Bessie could not give up flying. She kept looking for a part in an air circus—anywhere. After eighteen months of searching she finally succeeded, lining up a series of lectures and exhibition flights in Texas.

CHAPTER 8

Texas Triumph

In May of 1925, twenty-eight months after her plane was shattered on a California airfield, and with it all of her plans, Bessie arrived in Texas. Leaving behind her the bad press of the East Coast and the bad luck of the West, she returned to her first home—Texas. For her home base she chose Houston and for her first performance a lecture at the Independent Order of Oddfellows (I.O.O.F.) temple on May 9 with films of her flights, a format she had developed during her lecture series at the YMCA in Los Angeles.

Bessie had lost none of her ability to glean good advance press notices, one of which, in the *Houston Post-Dispatch*, a white daily, quoted the "widely-known daredevil pilot" as saying "the Negro race is the only race without aviators and I want to interest the Negro in flying and thus help the best way I'm equipped in to uplift the colored race." Her great ambition, she went on, was "to make Uncle Tom's cabin into a hangar by establishing a flying school."

She also said she was 23 instead of 33, a college graduate who had received flight training in Amsterdam, Berlin, Paris, and "elsewhere," and had recently returned from her latest trip to Europe where she had given exhibition flights.

Five weeks later Bessie made her first flight in Texas, on June 19 in Houston. Her Chicago sponsor, David Behncke, may have had a hand in securing the Texas engagements, but well-known local flier Capt. R. W. Mackie was in charge of the Houston show, which was promoted by O. P. DeWalt, owner of the Lincoln Theater, and J. M. Barr, secretary of the Negro Amateur Baseball League. Mackie himself probably provided the postwar Jenny powered by an OX-5 engine that Bessie flew. Because of the extensive advance publicity and because Bessie's appeal was so widespread that it cut across all prevailing color lines, DeWalt, Barr, and Mackie had to provide "special reservations" (that is, separate seating) for white patrons.

An extra drawing card for Bessie's first Texas air appearance was that her show was scheduled for the same day that African American Texans celebrated as the most important holiday of the year—Juneteenth, the anniversary of the day blacks in Texas achieved their freedom. Although President Lincoln had issued his Emancipation Proclamation on New Year's Day, 1863, it didn't take effect in Confederate-governed Texas until more than two and a half years later—June 19, 1865—when Union troops under Gen. Gordon Granger marched into that state from the North. Also, just a week before Bessie's Houston show, the U.S. House of Representatives, after a three-hour floor fight led by Representative Thomas Lindsay Blanton of Texas, defeated a proposal to make Lincoln's birthday a legal holiday. Thus Bessie's Houston appearance came as a thrilling defiance of such bigotry.

The air-circus heroine, born in a dirt-floored cabin in Atlanta and raised by a single parent in Waxahachie, was as much a celebrant as her audience. That Juneteenth the one-time manicurist did not *tell* her people what they could do if only they were determined enough, she showed them. Ascending high above the clouds, she stalled her motor and dove to within a few feet of the ground before pulling up as the crowd let out a roar of relief and admiration. Interspersed with dives were barrel rolls, figure eights, and loop-the-loops. Her aeronautic pyrotechnics sent a direct message to the spectators below: If I can do it, so can you!

The *Post-Dispatch* ran two advance stories on the show. The first told of the city's black ministers making flights over Houston along with supervisor Mackie. It further reported that Bessie was "said to be daring as well as skillful" and a huge crowd was expected to witness her flights. The second story, the day before the show, noted that Bessie was one of the big attractions and that "thousands of spectators of both races expect to be on hand to see stunt flying promised by aerial daredevils." Citing numerous clippings from other newspapers all over the country, it further lionized Bessie and propagated her message:

> This negro woman attracted attention all over the country for her efforts to interest the negro race in aviation. She is well educated and spent a number of years abroad. She says the negro race is the only one in the world that has not yet developed aviation and she is trying to get negroes flying.

The fact that she could wrest such laudatory and extensive coverage from a newspaper generally regarded as a spokesman for the white Texas establishment, and which steadfastly refused to capitalize the word "Negro," was more than just a major triumph for Bessie. It was sure evidence of the revolutionary change, however slow, subtle, or imperceptible, she was effecting on white attitudes toward blacks, at least in her home state of Texas. However, almost as if embarrassed by its coverage of a story that many of its readers probably considered of primary interest only to blacks, the next day the *Post-Dispatch* ignored Bessie's show completely and featured a "shooting and cutting rampage in the city" in its reports on Houston's Juneteenth black community celebrations.

Bessie's show was not as professional as those presented by entrepreneurs who could afford better equipment or by high-priced crowd-pleasers such as Mabel Cody, who climbed from a speeding boat up a ladder suspended from a plane overhead, and Gladys Roy, who danced the Charleston on the upper wing of an airborne biplane. Still, the combination of an opportunity both to see the world's only black woman flier and to ride in an airplane oneself proved an irresistible rousing success. Thousands came to the old speedway auto racetrack at the end of Main Street, renamed Houston Aerial Transport Field in honor of the occasion, to see Bessie perform.

Juneteenth 1925, Bessie's flying debut in Texas, was more than just a big day for Bessie and for Houston. It also marked an event of historic significance for her country. Although overlooked or unnoticed by virtually everyone else at the time, the city's leading black newspaper, the *Houston Informer*, recognized and reported that it was "the first time colored public of the South ha[d] been given the opportunity to fly." Calling it "the biggest thrill of the evening and of a lifetime for that matter," the *Informer* noted that "about 75 of our fearless citizens, most of whom were women," climbed aboard one of the five small passenger planes available to get "a birds-eye view of Houston from the sky." The paper later identified a Reverend Harrison of the Antioch Baptist Church as "the first colored person in the city to hop off for a ride. Rev. Woolfolk of Trinity Methodist Episcopal, not to be outdone, took little Miss Woolfolk with him." Not everyone at the show, however, was quite so courageous. "One of our leading ministers reneged on his wife when she requested him to escort her for a ride."

The Houston African Americans who dared that day did indeed need courage, not so much to fly as to make themselves conspicuous in a mixed crowd of Texans. Texas blacks at the time were leaving the state in record numbers to escape the worst Jim Crow laws in the country. A contributor to the Baltimore *Afro-American* wrote that although there were two trains a day going from Houston to San Antonio, black passengers were not allowed to ride the morning train, which had Pullman cars. They had to wait twelve hours for the night train and were confined to a partitioned half of the men's smoking car, which had only one toilet for both sexes. When the reporter went into the dining car a green curtain was thrown up around him to separate him from the rest of the passengers, and, at the end of his meal, he was handed not the regular dining check but a special one identifying him as a Negro.

The Baltimore writer's experience notwithstanding, Houston African Americans that day not only dared to fly but, according to the *Informer*, were left "clamering for another [chance to do so] soon." And, no doubt speaking for itself as well as for its source, the *Informer* quoted "the management" of Bessie's show as saying it "was more than glad to see [Houston African Americans] disprove the assertion that the Negro is afraid to fly." The paper concluded that the day had been an unmitigated success, "the only difficulty experienced" being how to accommodate in the few aircraft available everyone who wanted to go up and to fly them "as fast as they wanted to go."

Bessie's Houston debut was so successful, in fact, that she was immediately booked for three more appearances, one at Richmond, Texas, on July 10 at a meeting of the Southeast Central Baptist Association, a second on the following day at a picnic of the Southern Pacific railway's African American employees in Houston, and a third on July 12, also in Houston.

Promoter O. P. DeWalt predicted that thousands would attend the Houston show on July 12, and they did. The *Chicago Defender*'s reporter wrote that "she took off with a perfect start in the OX5 plane [a JN-4 with an OX-5 engine], circled the field several times, then 'gave it the gun' and spiraled and looped to the satisfaction of the roaring crowd." Mackie closed the exhibition with his own "daredevil act" of aerobatics and a parachute jump, but Bessie was the undisputed star of the show.

Although the *Houston Informer* reports of Bessie's Juneteenth show did not specifically say so, they strongly suggested (by their style and their occasional judicious use of the word "volunteer") that the sightseeing flights for Houston's leading black clergymen had been offered free as an inducement to the rest of the spectators. This time, however, with the management making an extra sixth passenger plane available, the crowds lined up to pay twenty-five or fifty cents apiece for their flights over the city.

In Houston, Bessie appealed to African American women to take an interest in aviation. As she had begun to do in California, Bessie would continue to focus her attention more on women than on men, convinced that black women would be more effective activists than men in furthering the participation of African Americans in the political and educational systems and in the struggle for equal rights. DeWalt agreed, claiming only women truly "appreciated the fact that it took a woman to put the Race on the map." In spite of her pride of race, Bessie was pragmatic in appealing to African American women rather than men. She knew that the white, male-dominated society found it easier to accept assertive black females than to accept the same characteristic in black males.

The next recorded show for Bessie was on August 9 in San Antonio, where she appeared again with Mackie at the old San Antonio Speedway. This time the parachutist was a local African American woman, Liza Dilworth, of 1417 Crockett Street. She would be the first black woman to make a jump, Bessie said. Later Bessie told her sister Elois that Dilworth had been afraid. She had cause to be. In an aerial circus at the same speedway in April of the previous year a woman pilot, Bertha Horchem, was killed during an aerial circus when her plane crumpled in midair and fell 1,200 feet, landing perilously close to a group of spectators. Dilworth might also have heard of the death only the day before of Dallas pilot H. C. Foster in a plane collision near Breckenridge, Texas. In spite of her fear, and undoubtedly with considerable persuasion from Bessie, she made a successful leap off the wing of Bessie's plane at 1,000 feet. Following San Antonio, Bessie did another show at Galveston before moving to Dallas to give a lecture.

Just as Bessie had begun to realize that African American women should be her primary audience, she began to wonder if perhaps the lecture platform might not be more effective than the airfield as a pulpit for preaching her gospel of flight. Consequently, she now lectured more and more frequently, accompanying her talks with clips of the 2,000 feet of film of her performances in Europe and the United States, which she had judiciously pre-served and taken with her. Finding that the lectures brought in more income and audiences and were far less expensive than air shows, Bessie spoke at both theaters and schools. But she never charged admission for students, whom she hoped to inspire to become future pilots.

Bessie delivered her first Dallas lecture at the Ella B. Moore Theater, which was managed by Mrs. Moore's husband, Chintz Moore. A local newspaper, describing her as a "very vivacious and loquacious little beauty," also said she was staying in the city with a relative, Lilly V. Beale. None of Bessie's living relatives knew of the Beales. But the city's 1925 city directory lists Mrs. Beale and her husband LeRoy, a janitor, as living at the rear of 208 South Ewing Street.

The hospitality of the Beales in a city with almost no public accommodations for blacks was typical of that shown Bessie in her travels. Wherever she went, fellow African Americans welcomed her, fed her, and housed her. Many of the men were janitors, mailmen, or Pullman porters who, despite their poverty, had found a way to start churches, baseball leagues, and social clubs. Their wives and daughters, often employed as maids and cooks, likewise formed organizations to nurture and protect their communities. Clearly they admired Bessie's accomplishments and supported her aspirations.

Bessie said she would give an aerial exhibition in Dallas on August 27, weather permitting. If she did, there is no evidence of it. If she didn't, it might well have been because she could not borrow a plane. In fact, her main purpose in coming to Dallas had been to buy a plane of her own at nearby Love Field, a veritable shopping mall of used aircraft. Located only seven miles from the present city center, Love Field was then surrounded by cotton fields and ranches. A depot for the storage, dismantling, and sale of surplus aircraft, it was a good place to buy a Jenny for $400 or less. The runway, bordered by a line of wooden hangars, was unpaved and the only housing in view a former Army barracks. To get there from Dallas one took a twenty-five-cent ride in the topless jitney bus that met incoming trains at the Elm and Pacific railway station.

To Bessie, who was accustomed to being regarded as a curiosity at airfields because of her race and sex, Love Field must have been a pleasant surprise. Behncke at Checkerboard Airdrome in Chicago had been kind to her, but she had never before experienced the casual acceptance she found at Love from men who were more interested in piloting and aircraft than in Bessie's gender or color. In fact, one black man, Louis Manning, was already virtually a permanent resident of Love Field. Identified in the city directory as a "yard man," Manning was accepted as a full-fledged member of Love Field's easy-going fraternity. A former delivery man for a drug store, he had been befriended by Byron Good, who serviced and sold planes with his partner H. C. Foster until the latter's death on the day before Bessie's show in San Antonio. When Bessie met Manning, he was Good's right-hand man, acting as office secretary and receptionist, servicing barnstormers' airplanes, controlling crowds at weekend air shows, and directing the parking of cars. And he answered the office telephone with a cheerful "This is Louis Manning, Good and Foster's colored mechanic!"

Like most of the other young men who worked at the field, Manning lived there, sleeping in a hangar or outdoors under the wing of a plane when the weather was good. One of the group, Arthur Spaulding, said,

> [Manning] joined the social life of the field. Everyone met at Ma and Pa Vencill's home, in the old Officers' Quarters on

the field. They sat around the kitchen table, eating Myrtle Vencill's food. Occasionally a stranger would visit the group and gape at Louis in the crowd. When that happened, someone would invariably yell out, "We're all black people here!"

Bessie did not stay long enough to join the Vencill crowd but Spaulding recalled seeing her talking with the pilots in a little café near the edge of the field. "She was a pretty woman," he said, "wearing a fancy uniform. I never did get a chance to talk to her but the boys who did wouldn't have cared if she was colored or not. It was the same as with Louis [Manning]."

Bessie looked over all the planes for sale before she decided on one at the Curtiss Southwestern Airplane and Motor Company. It was probably another JN-4 with an OX-5 engine. However, she could not take it with her. It needed servicing, and the company wanted full payment for it, more money than she had.

Her next stop was at Wharton, eighty miles southwest of Houston, to give two shows—on September 5 and 6. The *Wharton Spectator*, a white-owned newspaper, declared "there was some fancy flying by Bessie Saturday [for] the gathering of a great number of the colored folks at the rodeo grounds." There was also a jazz band from Houston for a dance on Saturday night. However, the report added, "the crowd was disappointed in the failure of Eliza to make her parachute jump." She had been "taken ill," the paper said. But Bessie told Elois that Dilworth had refused to jump. Perhaps that first leap in San Antonio convinced her that a second would be unwise.

The next day Bessie wired Houston for a pilot to come and fly her plane. The patrons had paid to see a jump and she would do it herself. As much an actress as she was a pilot, Bessie was not letting someone else ring down the curtain on her show. On Sunday, after she had finished her stunt flying, Bessie strapped on the jumper's harness and climbed into the front cockpit of her plane while the pilot from Houston took over the controls in the rear.

Straps on the sides of the harness were attached to a parachute inside a canvas laundry bag tied to the outer first strut of the plane. At 3,000 feet Bessie climbed out of the cockpit and walked out on the wing's catwalk to the leading edge, then along the front of the wing to the parachute. When she jumped, her harness pulled the chute out of the bag and she did a free fall until well away from the plane before pulling the release cord. Being Bessie, she landed in the center of the crowd. "The ovation given her was calculated to make her quit flying to do the more spectacular feat of parachute jumping." This was part of a rave review by a white reporter for a white newspaper about a show attended by a mixed audience.

A few days later, the *Houston Post-Dispatch* reported that the people of Wharton had subscribed money for Bessie to buy a plane of her own that would arrive in Dallas on September 13. The source for the story was obviously two of the show's three managers, Hubert Taylor of Wharton and G. R. M. Newman of Houston, who claimed to be in charge of the fundraising campaign. Speaking to the same paper twenty-four hours later, however, Bessie gave a contradictory version. It was she who was buying the plane, with money derived from her air-show earnings and from "additional flights and parachute jumps" she planned to give. Bessie was not letting Newman, Taylor, or Taylor's brother Emmet, all of whom had an interest in the show, take credit for or make any claim on the plane she meant to get from Love Field.

Bessie gave another show in Waxahachie. Elois did not know or say whether her sister went to look at the old house on Palmer Road, a house that her mother still owned and that Bessie had left more than ten years ago. Presumably she did. If so, she would have walked along red dust roads, passing not only the small homes of Wyatt Street blacks but also the Victorian mansions of white Waxahachie, whose front doors were still barred to her. The owners of those mansions undoubtedly had a say in the arrangements for Bessie's show, which was to be before a mixed audience, blacks in one area, whites in another, with separate admission gates for each race. But Bessie drew the line at the two gates. There would be one entrance only, she said, or she would not perform. The organizers finally agreed so long as the audience was segregated after they came in. This time Bessie had to give in. She needed the money from the show and the passenger rides that would follow to pay for her new airplane at Love Field.

Although neither the white nor the black press took note of it, Elois claimed that Bessie also gave a show in the state capital of Austin, after which she was entertained by the governor, Miriam A. "Ma" Ferguson. A colorful figure from Temple, Texas, "Ma" Ferguson succeeded her husband in office after he was impeached. She was an opponent of prohibition and an advocate of tenant farmers. More important to Bessie, she was a Democrat who neither feared nor collaborated with the powerful Ku Klux Klan. And although blacks had almost no voting clout in Texas, one of Ferguson's first acts in office was to pardon a large number of black prisoners. It is entirely possible that the governor asked Bessie to dinner.

Bessie gave other shows in Texas, often in places too small to support a newspaper and thus unrecorded by the press. But she described them to Elois, who provided a good composite picture of her sister's typical performance. Usually it took place at a fairgrounds or racetrack. At the beginning Bessie appeared in her uniform and stood silently while a prayer was said for her safety. Then she walked to her plane and posed for photographers before climbing into the cockpit and taking off for a series of aerobatics. Upon landing, she climbed out of the cockpit and waved her thanks to the cheering crowd that surged forward, pressing against the rope barriers set up to protect her and her plane.

On one occasion a determined small boy slipped under the ropes and through the hands of her assistants. Bessie beckoned to him to approach her. Once he was in her arms, he said, "Lady, didn't your plane stop up there for a little while?" It had, she told him, in a stall. As he walked away, he smiled smugly at the crowd, now the confidant of a celebrity.

Among Bessie's other admirers was a young white man, Paul McCully, who later became National Governor of the OX-5 Pioneers—pilots who flew planes with the OX-5 engine. "I credit this lady with helping launch my flying career," he said. McCully recalled a group of black aviators landing in the river bottom near Fort Worth. With them was "a beautiful black lady in charge of the group, and between selling ride tickets and helping gas the air-planes, she didn't have much time to talk. She drew almost as much crowd as the fliers," McCully recalled.

> She told us of the difficulties a black person had learning to fly. White women were not considered pilot material and black women—no way. She told us she had to leave the coun-try to find an instructor . . . and that she and her friends were trying to make enough money to open a flying school for blacks. I went back to the field early next morning to find out more about the group but they had departed and I never saw them again.
>
> My father was "dead set" against my flying, until he saw these people perform. On the way home papa was puffing and blowing up the long dusty road out of the river bottom. He stopped for air and, wiping his forehead, said, "Well, son, I guess if they can do it you can too." I soloed one month later.

McCully's account, delivered almost seventy years later, is testi-mony to the vivid impression Bessie left on the people of Texas who saw her, the majority of them African Americans. That summer was the most successful she had ever had. There were no crashes, no canceled engagements, no irate managers or crowds. Her career was no longer under attack from critic J. A. Jackson, who had left the *Billboard* within days of her first appearance in Houston be-cause his column no longer attracted enough paid advertising to justify it. Another good season on the road might well assure her enough capital to open her aviation school.

CHAPTER 9

Nearing the Goal

From Texas Bessie went home to Chicago for three months—long enough to resolve whatever differences she had had with her sometime agent, D. Ireland Thomas. Through the Theater Owners Booking Agency, Thomas now arranged a series of lectures for Bessie in black theaters in Georgia and Florida. There is no recorded evidence of it, but it is possible he may also have had a hand in setting up some of the exhibitions Bessie was booked to give during her speaking tour of the South.

On the night before Christmas Bessie appeared at Elois's door, her bags packed, ready to leave for the first lecture in Savannah, Georgia. "We had a joyful Christmas Eve," Elois recalled. "We made and wrapped Christmas presents and I hemmed a black taffeta dress of hers. We cooked, tasted, and drank coffee and chatted the whole night through. All of a sudden it was broad daylight and time for Bessie to leave."

Bessie arrived in Savannah during the first week of the new year. "The city is thrilled," the black weekly *Savannah Tribune* proclaimed on its front page. She appeared at a local theater where the paper's reviewer was impressed by her "unlimited confidence" in herself.

Whether due to that unlimited confidence or the reporter's ineptness, the story contained three errors: that Bessie went to France in 1919, was licensed by the French government, and was "the only accredited black flyer in the country." Bessie went to France in 1920, got her license from the nongovernmental Fédération Aéronautique Internationale, and, although the only black woman aviator in the world, was not its only accredited black pilot. Hubert Julian had received his flying license from the National Aeronautic Association in the United States in February 1925. The story was correct, though, in saying Bessie's "main anxiety is to see the number of black aviators increased."

Booked for another lecture in Augusta on January 7, Bessie promised she would return to Savannah in time to give an exhibition flight on Sunday, January 10, at Daffin Park Flying Field, where she would "walk the wings of an airplane and do some other stunts." Bessie did not give that show. All available evidence indicates she could not because she had no plane. There was a show that Sunday at Daffin Park Field but it was given by the nationally known Gates Flying Circus. When Bessie originally promised a performance there, she may have assumed she could borrow a plane from them.

From Atlanta Bessie moved on to Florida, appearing at the Liberty Theater in St. Petersburg first, then at theaters in Tampa and West Palm Beach. The popularity of her lectures soon convinced her that her road to future fame might be paved with celluloid. In pursuit of that possibility she wrote to an African American film producer in Arlington, Florida, as follows:

Tampa, Florida.
Feb. 3 1926

Norman Studios
Arlington, Florida

Dear Sirs:—

I was given your address by Mr. Trumbull, owner of the Liberty Theater in St. Petersburg after I appeared there in person and on the screen with 2 reels showing my flights in Europe and America I have my life work that I want put into pictures I know I have been a success in every house I have played in Chicago and other cities. And with only 2 reel as an added attraction I have titled my play Yesterday—Today & Tomorrow.

If you are enerested which I am sure a few remarks from Mr. Trumbull will give you the correct idea what I will mean to you. and as any intelegent person knows that I am the world's first Col Flyer. man or woman. We have one man now a pilot 9 months in Tulsa if you are inerested I will be willing to go father into the matter with you. I am, and know it, the Most Known Colored person (woman alive) other than the Jazz singers)

Write me what you would like to do about this matter.

Bessie Coleman
c/o 1313 Mairon
Street.
Tampa, Fla.

Obviously interested, film company manager R. E. Norman replied almost immediately in a letter dated February 8:

February 8, 1926

Miss Bessie Coleman
Tampa, Fla.

Dear Miss Coleman:—

In your letter of the 3rd., we note that you wish your life work put into a film entitled, "YESTERDAY, TODAY and TOMORROW." There is no doubt that with a picture of five or six reels, properly acted and full of action with you in the leading role, would be a good drawing card in the colored theatres.

Mr. D. Ireland Thomas wrote us several weeks ago along the lines of your letter, but we didn't hear from him further.

You may go further into the matter and tell us what your plans are for making such a picture.

Yours very truly,
Norman Film Manufacturing Company
(signed) R.E. Norman

Bessie's writing ability was limited but her business acumen was sharp enough to realize that Norman's "tell us what your plans are" meant "Are you going to pay for it?" Her reply suggested some sort of coproduction:

W. Palm Beach Fla
Feb. 23, 1926

My Dear Mr. Norman

I just rec' your letter it was *NOT* forwared to me as it should have been. Yes Mr. Norman I am more than *SURE* my picture will go big in Colored houses I *know* this, as a Fact as my two News reels have drawn in house more so than some Colored Dramas

You may know what a Real Film of 5 reels would mean You only have to ask the Mgr. at some of the Theatres in Fla. Tampa was not advertized "at all" But in St. Petersburg it was impossible to Show to all. Have a chance to return.

The picture that I want filmed maybe *we* could get together on it, yesterday Today and tomorrow It would be better if we jointly put out the photo, as I am not able to produce it independty Now if you are enerested let me know also give me a prise on directing 5 reels and est.

Yours

Bessie Coleman

530 First Street.

Apparently Bessie heard nothing more from Norman and all correspondence stopped.

At some time during those first two months in Florida Bessie met an Orlando couple, the Reverend Hezakiah Keith Hill and his wife, Viola Tillinghast Hill. Both respected community activists, the Hills thought Bessie's message of real importance to African Americans in Orlando. To ensure that it would be heard, Mrs. Hill invited Bessie to stay with them at the parsonage of her husband's church, the Mount Zion Missionary Baptist Institutional Church, west of downtown Orlando in a neighborhood known as the Callahan section.

For Bessie the spacious house on shady, tree-lined Washington Street became the center of her life, the idyllic home she had never had. And Viola Hill became friend, sponsor, and surrogate mother, presiding over a household that radiated generosity and warmth. Behind the parsonage Viola kept a large vegetable garden, several cows, a flock of chickens, and fruit trees, all of them providing food for a table where members of her husband's congregation were welcomed. There Bessie met many of Orlando's church workers and school-board members, devout, hardworking people who shared whatever they had. Within this bastion of faith Bessie rediscovered hers, and with the help of the Hills, whom she called "Mother" and "Daddy," she became a "born-again" Christian, professing her belief in public.

Bessie had always liked children but, with the exception of her nieces and nephews, had known few of them. In Orlando several children lived near the Hills, all of them in delighted awe over the presence of the woman who could fly. Six of them—Aretha Hill Norwood, Jessie Lee Green Griffin, Edyth Lindsey Crooms, Bernice Cox Wheeler, Guyretha Thomas Courtney, and Felix Cosby— can still recall Bessie's affectionate manner with all the children of the neighborhood.

Aretha Hill, the Reverend Hill's niece, who looked after the Hills' cows and chickens, thought Bessie brought excitement and joy to the household and to the whole neighborhood, and especially to the children. "I would always run to take a quick look at that beautiful lady," she said. Aretha was only one of Bessie's many small admirers. At first, Edyth Lindsey, who lived down the block on Washington Street, used to hide under the elevated porch, "to see her come out of the parsonage. She was a very beautiful woman," Edith recalled. Bernice Cox, who lived next door, agreed.

Before long Bessie's shy admirers were no longer hiding under the porch. Jessie Lee Green said Bessie talked to them frequently.

> She was an important lady, famous in air races, but she gave us her time. I lived the second door down but I had a lot of time and every time I saw Bessie Coleman, I was right there. She was a friendly lady, talking to everyone, even us children. She couldn't get rid of us. We admired everything about her, especially her driving airplanes. She promised us she would give us a ride in the airplane. The last time I saw her she had on her uniform. She was smiling. She had beautiful short hair and a beautiful personality.

Among the boys who were awed by Bessie was Felix Cosby, whose father owned a bar and ice-delivery company. Felix and some of his cronies planned to take flying lessons.

> I was so inspired by her talk and her flying that I immediately decided that when I was old enough I'd become a pilot for the

U.S. Army. Of course, growing boys always change their minds if they want to. When World War II began I changed my mind and decided I wanted to be something else. But you know, I will always, always remember seeing and hearing Bessie Coleman.

Not only boys aspired to take lessons. Guyretha Thomas's father suggested that she learn how to fly from Bessie but her mother, she said, "was one of those kind . . . She didn't believe in things like that . . . So she [Bessie] couldn't give me lessons and I cried and I cried. I couldn't get my mother to gee or haw." Bessie's ability to charm the children stretched beyond the Hills' neighborhood. Soon she was a frequent speaker at black schools in Orlando.

For a small fee or passing the hat, depending where she was, she spoke to adults, too, at churches, theaters, and once in a pool hall (with an escort), "so determined was she to have Negro men become air-minded," said Elois.

Viola Hill soon realized that these collections were never going to be enough to fund an aviation school. However, she disapproved strongly of Bessie's raising money by stunt flying and wanted her to confine herself to lecturing and teaching students. To replace the fluctuating income from air-circus flying, Bessie's new friend suggested that she open a beauty shop in Orlando and helped her to do so. Soon after, Bessie wrote to Elois, "I am right on the threshold of opening a school."

Years later, Viola Hill's sister-in-law Audrey Tillinghast, while cleaning out a family dwelling that her husband Sidney had inherited, discarded a "raggedy rattan chair." Her mother-in-law told her, "That was Bessie Coleman's. Haven't you seen her picture on Viola's piano? That lovely lady standing by her airplane? That was her chair."

"That chair was meant for Bessie's beauty shop," Mrs. Tillinghast said, "and I was sorry I didn't know that before I threw it out."

Neither speaking engagements nor part-time beautician's work put a stop to Bessie's drive to fly and her determination to teach others of her race. Orlando resident Edyth Crooms recalled that once, after making a speech at the Mount Zion church, Bessie took many of her audience to a field where she gave the Reverend and Mrs. Hill a ride "in her plane."

There is no evidence that Bessie owned the plane she used in Orlando. It could have been borrowed or rented from the white pilots who lived there. Pilots tended to be a breed apart, far less racist than the general population, and if a woman as charming as Bessie showed her license and a few press notices to a local owner he might well have loaned or rented her his plane. In Savannah she did not give the show she had promised, probably because she couldn't get a plane. In Tampa and St. Petersburg she appeared in theaters but did not fly. She did give an exhibition in West Palm Beach but the plane she used may not necessarily have been hers. The fact that she had her picture taken in Orlando in front of an airplane does not mean that she owned it.

According to Viola Hill, Bessie had been booked for a parachute jump at the annual flower show of the Orlando Chamber of Commerce but threatened to send her plane back to Texas and withdraw from the show when she learned that it would be for whites only and that African Americans would not be allowed to attend. Principles aside, it was a gutsy and even dangerous stand for her to take. Racial segregation in Florida at the time was rigorously enforced, sometimes sanctioned by law but often not. In Daytona Beach, blacks could go nowhere after dark unless they carried a special pass. In Tampa, vigilante-type "night riders," instigated by real-estate operators, engaged in wholesale attempts to frighten black property owners into selling their property and leaving the city.

Bessie's threat to withdraw from the show worked. Orlando's white businessmen backed down, the Chamber of Commerce agreed to allow blacks, and Bessie performed her jump as promised. But her threat to send her plane back to Texas was typical

Bessie bluster. She had no plane yet and was still trying to raise money for the final payment and delivery of the plane on which she had made partial payment the previous summer and which was still sitting in Dallas at Love Field.

Bessie needed a benefactor badly and finally found one in the person of Edwin M. Beeman, son and sole heir of Harry L. Beeman, the millionaire manufacturer of Beeman Chewing Gum. How they met is not known, but they could easily have done so at a three-day flying show that the *Orlando Evening Reporter Star* sponsored in February, or at a new airfield then under construction at a racetrack on East Winter Park Road. Beeman's father had built a mansion across the road from the racetrack, an estate known as Beeman Park. He also built the San Juan Hotel, which Edwin, widely remembered as a short, pleasant-looking, and well-liked man, managed.

It is hardly surprising that the Beeman-Coleman relationship became a matter of much speculation and gossip. Edwin was rich and married and a member of Orlando's white aristocracy. Bessie was poor and black and, while popular within her own community, was hardly welcome in Beeman Park. Beeman's interest in Bessie was probably platonic, that of a rich young man fascinated by aviation and aviators. Whatever his relationship with her, it was he who gave Bessie the money she needed in order to make final payment on her plane in Dallas and have it flown to Jacksonville, the site of her next engagement, scheduled for May 1.

Before Bessie left she promised the Hills that she would return to make her home in Orlando and that she would give up exhibition flying to lecture and teach. She also told them that the time she had spent with them was the happiest she had ever known. For the first time in her life, she said, she was enjoying the comforts of religion. She left Orlando April 27, traveling by train to Jacksonville to await the arrival of her plane from Dallas. With plane and money problems now a thing of the past and with her career in an upswing, Bessie was ready to fly again.

CHAPTER 10

The Sky Has
a Limit

When Bessie stepped off the train in Jacksonville, she was met by the publicity chairman of the city's Negro Welfare League, John Thomas Betsch. The 21-year-old Betsch was an attractive fellow, an aviation enthusiast who had, in spite of his youth, convinced League members that Bessie Coleman should be the star attraction at the organization's annual Field Day on May 1.

The charming, gregarious Betsch, son of a German brick mason and an African American woman, was a Howard University graduate just starting out in what would be a lifetime as a community activist, businessman, agent for the Atlanta Life Insurance Company, and Thirty-second-Degree Mason. More than a decade after he met Bessie he would have a daughter, Johnnetta Betsch Cole, the present-day president of Spelman College in Atlanta.

Despite the fact that the May 1 show was being put on by a black organization in a strictly segregated city, Betsch wrangled an advance story from a white newspaper. Having recently performed charity shows in Orlando and Palm Beach, the article said, Bessie that coming Saturday would give an exhibition of her flying skills and do a parachute jump from a speeding plane at 2,500 feet. It pointed out that she had 350 career solo flights to her credit and was "one of the two colored persons [in the world] licensed to fly"—the other being Hubert Julian. And, in what could have been a crude attempt to increase attendance, it contained the first published reference in Florida to the California crash in which she was seriously injured.

As her sponsor, young Betsch had also booked a number of speaking engagements for Bessie at churches and theaters. On the night of April 29 she spoke at the Strand Theater after spending the entire day visiting every African American public school in Jacksonville where she found her most fervent, indeed awe-stricken admirers.

"We thought it was just great," said Marian Johnson, then a 15-year-old student at Stanton High School, the only one for African Americans in Jacksonville. "Imagine, a woman flying a plane!" Schoolmate Frances Tyson readily agreed, adding that Bessie was "small and very pretty in her uniform." Hettie Thompson heard Bessie at the Davis Street Elementary School and said she felt "thrilled that a real flier had come to talk to us," pointing out that "in those days we almost never saw planes." And, almost seventy years after the event, Eugenia Mathews, 11 at the time, recalled how impressive Bessie looked when she appeared at the Darnell-Cookman Elementary School "wearing that smart uniform."

Having earlier in Texas switched her focus of attention from audiences of black men to black women, Bessie now, in Orlando and Jacksonville, was concentrating on children as potential aviators of the future.

While she was preaching her doctrine to the people of Jacksonville, 24-year-old white mechanic William D. Wills was taking off from Love Field in Dallas in the Jenny Bessie had finally purchased. One of the earlier models, it acted accordingly, malfunctioning twice during the twenty-one-hour flight and requiring Wills to make forced landings at the Mississippi towns of Meridian and

Farmingdale for a total of five instead of the three originally scheduled landings.

Wills reached Jacksonville on April 28, landing at Paxon Field,
a sixty-acre private field at Edgewood Avenue and Lake City Road
(now Enterprise and Broadway). Local pilots W. H. Alexander and
Laurie Yonge met him at the field and said later that Wills seemed
an excellent pilot but they couldn't understand "how he could get
that plane to fly all the way from Dallas" since its OX-5 engine,
rated at 90 horsepower, was so worn and so poorly maintained that
it could really develop only 60 horsepower at most.

Because of the rigid segregation in Jacksonville, Betsch made
arrangements for Wills to room at the Church Street home of another white man, Karl Westerfield, who judged him a "fine young
fellow." Wills told his host he had worked his way up at Love Field
from helper to mechanic to pilot and was currently employed by
the Curtiss Southwestern Airplane and Motor Company, a firm
managed by his brother-in-law Capt. S. C. Coons, who was also in
charge of surplus sales at Love Field.

The day after the young Dallas man arrived with her plane,
Bessie walked into a restaurant where Robert Abbott was dining
with friends. The *Defender* editor had been visiting his mother,
Flora Abbott Sengstacke, in Savannah the previous week and had
continued south to Jacksonville to see friends there. Bessie rushed
over to his table, kissed him on both cheeks, and told her friends,
"This is the man who gave me my chance. I shall never forget
him." Whether the two men had met earlier or whether Wills was
actually with Bessie at the time, the Chicago publisher "didn't like
the looks of the Texan who was to fly with her, and [told Bessie so]
in no unmistakable terms," suggesting that she change her plans
and find another partner. Bessie ignored Abbott's advice, however,
and decided to go up with Wills the next day for a preliminary
flight over the racetrack where the exhibition was to be held and to
choose a suitable landing site for her jump on Saturday.

At 6:30 Friday morning Bessie called Wills at Westerfield's and told him that she and Betsch, who would be driving them both to Paxon Field, would come by to pick him up in a few minutes. Wills, Betsch, and Bessie arrived at the field at 7:15. As soon as Wills declared the plane ready, Bessie knelt in prayer beside it. Then she asked Wills to take over the controls in the front cockpit while she sat in the rear so she could study the field for a good jump site. She didn't put on her seat belt because to look down on the field required peering over the edge of the cockpit and she was too short to do so if the belt was fastened.

With Wills at the controls, the JN-4 headed for the racetrack and circled for five minutes at 2,000 feet before climbing to 3,500 feet and turning back to Paxon Field. Wills was cruising at 80 miles an hour when the plane suddenly accelerated to 110 miles an hour, then nose-dived. Eyewitness accounts of the altitude to which the plane descended varied but aviators who were watching said it went into a tailspin at 1,000 feet, then flipped upside-down at 500 feet. Bessie fell out, somersaulting end over end until she hit the ground "with a sickening thud, crushing nearly every bone in her body," 100 yards north of Broadway Avenue, a block from Edgewood Street.

Still strapped in his seat, Wills struggled to regain control of the aircraft but failed. Shearing off the top of a pine tree near the edge of the field, the plane crashed on the adjacent farmland owned by Mrs. W. L. Meadows, more than 1,000 feet from where Bessie's body lay. While Mrs. Meadows's son Raymond called the police, she and a neighbor tried to lift the plane off Wills's body but it was too heavy. John Betsch and several police officers reached the scene at just about the same time. Distraught, his hands shaking with shock, Betsch lit a cigarette to calm his nerves. The match ignited gasoline fumes from the plane, which was immediately engulfed in flames. After the trousers of one of the policemen who was trying to pull Wills's body from the plane also caught fire, Betsch was arrested and taken to the Jacksonville jail, where he was held for several hours.

Even before souvenir hunters stripped the barely cooled skeleton of Bessie's plane of all but its charred engine, officials discovered the cause of the accident. A wrench that either had been left in the plane or worked loose from its fitting slid into the control gears and jammed them. Field pilots Alexander and Yonge said Bessie's plane was "an old wreck" to begin with and noted that the gears of a newer plane would have had protective covering. Their view was supported by J. W. Price, himself the owner of four planes, who said, "We've been taking people up for four months and have not had a single accident" and claimed Bessie's accident "wouldn't have happened" had she been using one of his new-model aircraft.

Before noon that Friday several members of the Jacksonville Negro Welfare League took over for the jailed Betsch, arranging for Bessie's terribly mutilated body to be brought from the field to Lawton L. Pratt's funeral home, an African American establishment. In accordance with Jacksonville's rigid segregation practices, Wills's charred body was taken to the Marcus Conant Company's white mortuary pending the arrival of his brother Connor from Tallahassee.

The next day, Saturday, May 1, with souvenir hunters still picking at the skeleton of the plane, Connor Wills boarded the noon train for Terrill, Texas, with his brother's body. Mourning Bessie, the black community of Jacksonville was plunged into gloom. The fairgrounds were deserted, the lemonade and hot dog stands empty, the sawdust-covered midway silent, the newly painted ferris wheel motionless in the burning Florida sun.

Also canceled was the celebration dance that the young women of the Elite Circle and Girls' DeLuxe Club had scheduled to follow the exhibition. Planned especially in Bessie's honor, the dance had been advertised by cards bearing her picture and inviting guests to "An Aerial Frolic" from 8:00 to midnight for an admission fee of seventy-five cents. That night, instead of attending a dance Bessie's admirers came to the funeral home, which was kept open late for people wishing to pay their respects. The line filed past the closed casket from sundown until long after midnight.

On Sunday Bessie's coffin was moved to the Bethel Baptist Institutional Church to lie in state until 3 P.M., when a memorial service was held under the auspices of the Negro Welfare League and the Pride of Maceo, Daughters of Elks, of which Bessie was a member.

More than 5,000 people attended the service, among them hundreds of schoolchildren who had heard Bessie speak the day before she was killed. Only 2,000 mourners were able to crowd into the Greek revival and Romanesque church at Caroline and Hogan streets. The rest had to stand outside the handsome brick and marble building that would have suited even Bessie's grandest aspirations.

The printed programs bore Bessie's picture, and after soloist Daisy Harding reduced the mourners to tears with Bessie's favorite hymn, "I've Done My Work," three ministers—the Rev. John E. Ford, pastor of the church from 1907 to 1943, the Rev. T. H. B. Walker, and the Rev. Scott Bartley—all gave eulogies. At 7:30 that night, a second memorial service was held at St. Phillip Episcopal Church at Cedar and Union streets. Then Bessie's coffin was put aboard a train to Orlando, where her friends the Hills were arranging still another service.

The Hills were heartbroken. They had already received the last letter she wrote before her death, which contained her promise that she would give up flying and confine herself to the lectern. When Bessie's remains reached Orlando, Viola and Hezakiah brought the casket to their church for a Monday service that was one of the largest ever attended at Mount Zion Missionary Baptist.

On that Monday morning the church was banked with flowers from the Mount Zion's Women's Home Missionary Society, the Mount Pleasant Baptist Church, and the leading black citizens of Orlando. The Reverend Hill spoke of Bessie's newly found faith and her kindness to all who knew her. Ministers from every black church in Orlando read resolutions honoring her, and the voices of the church choir and a quartet rang out with "Lead Kindly Light" and "I've Done My Work."

At the close of the service the mourners—including Viola Hill, who would ride with Bessie on her last journey—accompanied the casket to the station to see it off for Chicago. In the words of one report, "As the body of Miss Coleman was being raised into the baggage car, en route to its final resting place, more than five hundred voices, representing the colored population of the city, hummed sweetly, 'My Country 'Tis of Thee.' "

While the African Americans of Orlando praised and prayed for Bessie, William D. Wills was buried in Terrill, Texas, after services attended by his wife and infant daughter; his parents, Mr. and Mrs. David Wills of Dallas; his sister, Mrs. S. C. Coons, and her husband Captain Coons, also of Dallas; and his brothers Connor and Glenn. There is no record of how many others were there.

Bessie's remains arrived in Chicago the morning of Wednesday, May 5. Already several thousand people were crowded into the Forty-third Street Station to get a glimpse of the casket bearing "The Daring Manicure Girl." Also waiting was a military escort from the African American Eighth Infantry Regiment of the Illinois National Guard, which took the coffin to the South Side funeral home of Kersey, Morsell and McGowan. An estimated 10,000 people filed past the coffin that night and all day Thursday to pay their final respects before it was moved Friday morning to the Pilgrim Baptist Church, scarcely a block away at the corner of Thirty-third Street and Indiana Avenue.

At eleven o'clock that morning a trembling Susan, leaning on the arm of her son John and followed by Bessie's sisters Elois, Nilus, and Georgia, was escorted up the aisle and seated in front. After them came six uniformed pallbearers, overseas veterans of the Eighth Regiment, carrying the flag-draped casket. Fifteen hundred mourners filled the church. Another 3,500 people stood outside on the sidewalk and adjacent streets.

Among the twenty-two pallbearers were Congressman Oscar DePriest and attorney Earl B. Dickerson, who had won a major case before the U.S. Supreme Court which upset real-estate covenants barring blacks from residential areas. Ida Wells Barnett, well-known proponent of and passionate speaker for equal rights, was mistress of ceremonies. And Viola Hill delivered a eulogy, speaking proudly of Bessie's refusal to perform any place where African Americans were not admitted.

Pilgrim Baptist pastor Junius C. Austin delivered the funeral oration, saying of Bessie, "This girl was one hundred years ahead of the Race she loved so well, and by whom she was least appreciated." In answer to complaints he had heard that Bessie accepted too much help from white people, he replied, "Heaven help her if white people had not helped her!"

Bessie was buried that day in Lot 580, Section 9, of Chicago's Lincoln Cemetery at Kedzie Avenue and 123rd Street. One year later the Reverend Austin unveiled a memorial stone at the grave, which read:

> In memory of Bessie Coleman, one of the first American
> women to enter the field of aviation.
>
> Remembered for her courage and accomplishments.
> She fell 5,300 feet while flying in Jacksonville, Florida,
> April 30, 1926.
> Presented by the Cooperative Business Men's League, Cook
> County, and Florida friends.

The donors were wrong about the length of her fall but not about her courage and accomplishments.

Mourning her sister, Georgia sobbed, "Oh, Bessie, you tried so hard." Bessie, who had said during her lifetime, "If I can create the minimum of my plans and desires there shall be no regrets," might well have responded with the words of her favorite hymn:

I've done my work. I've sung my song.
I've done some good.
I've done some wrong.
And I shall go
Where I belong.
The Lord has willed it so.

Or perhaps the strong-willed, warm-hearted, uninhibited Bessie
would have preferred the ringing declaration of African American
poet Mari Evans's "The Rebel:"

When I
die
I'm sure
I will have a
Big funeral.

Curiosity
seekers

coming to see
if I
am really
Dead

or just
trying to make
Trouble.

Epilogue

Ever since Bessie Coleman died more than sixty years ago, a tenuous but persistent undercurrent of speculation has rippled through African American communities as to the real cause of her death. Some suggest sabotage: envious white fliers who couldn't bear the thought of being outdone by a black woman left the wrench loose in her plane on purpose so it might jam the controls. Others hint at outright murder, picturing Wills as either a spurned lover or jealous colleague who intentionally flipped the Jenny so that Bessie would fall but then was not able to get it back under control.

Bessie wasn't the first nor would she be the last pilot to fall out of an airplane. Ten months earlier Carter Leach, a 19-year-old pilot from Waxahachie, was killed at Love Field when his plane overturned and he slipped from the cockpit, falling 500 feet. Like Bessie, Leach had not fastened his safety belt. Soon after that, pilot Al Johnson fell 1,000 feet from his plane while flying over Mineola, New York. Johnson had loosened his belt to peer over the side when the aircraft hit an air pocket and he bounced out.

Nor was an unsecured wrench unique as the cause of a plane crash. As recently as September 1991 a five-inch wrench left inside the front wheel compartment of a U.S. Navy jet caused the $33-million aircraft to crash off the coast of Virginia Beach. It had come loose on takeoff and lodged in the left engine, jamming it and severing the hydraulic lines.

Bessie's sister Elois may not have started the rumors of sabotage but she certainly gave them soil to grow in by writing that Robert Abbott did not trust Wills. And she enriched that soil by adding, "I cannot reconcile myself to . . . Bessie not wearing her parachute [since] safety first has always been one of her strongest mottoes."

The press, too, may have contributed its share by reporting almost unanimously that the wrench "probably" caused Bessie's accident. The wrench did cause it. Whether it was there accidentally or on purpose is the question. There is not a shred of evidence to support insinuations of sabotage or murder.

At the same time, however, it is not hard to see how the tenor of racial relationships that prevailed at the time (and continues to this day) could give birth to and sustain rumors of chicanery and evil. Throughout her career, the white press, with only rare exceptions, either ignored Bessie or treated her with contempt. No more glaring example of this can be found than the way in which Jacksonville's two leading white newspapers covered her death. Both focused total attention on Bessie's flying companion, white mechanic-pilot Wills. The *Florida Times Union* printed Wills's picture, went into detail about his family life and his having recently become a father, and exalted him as the person who was "teaching Bessie how to fly!" Bessie's name appears only twice—in a photo caption and in the lead, and then merely as identification. Thereafter, and only far down in the story, does the paper refer to her three times as "the Coleman woman" and once simply as "the woman." For its part, the *Jacksonville Journal* did concede that Bessie was "said to be the only negro woman aviatrix in the world." But Wills's name came first and for the rest of the story Bessie was never referred to by name, only as "the woman."

Bessie fared no better in the white press outside of Jacksonville. The *Chicago Tribune* buried the crash story on page 10 in a one-paragraph story from the United Press wire service. The paper never even thought to identify her as a resident, let alone a celebrity of Chicago. And back in Florida, the *Orlando Reporter-Star* used the same United Press story and made no mention of Bessie's recent air show and many speaking appearances in that city.

If the white world ignored Bessie, several black newspapers commented when she died on how even her own people had tended to overlook her and failed to appreciate her importance. A *Dallas Express* editorial upon her death remarked that "there is reason to believe that the general public did not completely sense the size of her contribution to the achievements of the race as such." The *Norfolk Journal and Guide* reminded its readers, "Whether they take to it or not, Miss Coleman has taught our women that they can navigate the air and, like all pioneers, she has built her own monument." It was the only monument she would have, other than her tombstone.

In 1927, when the *New York News* sponsored naming an apartment building at 140th Street in Harlem "Coleman Manor," it simultaneously delivered the following scolding to its readers: "It is regretted that colored Americans have not before now had the gratitude and the vision to raise some enduring monument to this black 'Joan of Arc.'"

In 1931 a group of black pilots from the Chicago area, the Challenger Pilots' Association, led by Cornelius R. Coffey, flew their planes over Lincoln Cemetery and dropped flowers on Bessie's grave. The flyover continued as an annual event but eventually lapsed as the originating pilots retired from flying.

In 1934 Lt. William J. Powell, who had already founded Bessie Coleman Aero Clubs on the West Coast, dedicated his book, *Black*

Wings, to her memory. "Because of Bessie Coleman," he wrote, "we have overcome that which was much worse than racial barriers. We have overcome the barriers within ourselves and dared to dream."

In 1977 a group of black women student pilots from the Chicago-Gary, Indiana, area formed the Bessie Coleman Aviators Club, and in 1980 aviation historian and pilot Rufus A. Hunt revived the flyover of her grave. A decade later, on April 28, 1990, pilots from the Chicago American Pilots Association, along with members of the Detroit and New York affiliates of the Negro Airmen International, joined in the annual flyover. And that same year Mayor Richard M. Daley renamed Old Mannheim Road at O'Hare Airport "Bessie Coleman Drive."

On February 26, 1992, during Black History Month, the Chicago City Council passed a resolution presented by Alderman Arenda Troutman and signed by Mayor Daley which requested that the U.S. Postal Service issue a stamp "commemorating Bessie Coleman and her singular accomplishment in becoming the world's first African American pilot and, by definition, an American legend." The resolution noted in part that even sixty-five years after her death "Bessie Coleman continues to inspire untold thousands, even millions of young persons with her sense of adventure, her positive attitude, and her determination to succeed."

Finally, Mayor Daley issued a proclamation on April 27, 1992, urging "all citizens to be cognizant of the accomplishments of the first African American aviatrix of record." Declaring May 2 to be "Bessie Coleman Day in Chicago" and coming sixty-six years after her death, the proclamation represented the first truly official recognition of her many achievements. On that day, aviation historian Hunt again led the annual memorial flight from Gary (Indiana) Municipal Airport to Lincoln Cemetery, where flowers were dropped on her grave.

Notes

Notes appearing in abbreviated form at first reference appear in full in the bibliography.

Chapter 1 • The Reluctant Cotton Picker

3   "SHE HAD THIRTEEN CHILDREN": U.S. Census of 1880 and 1900, provided by Jean Albright Gilley.

"NEITHER SUSAN NOR HER HUSBAND": Marion Coleman, interviews with author, October 24 and 25, 1990 (hereafter Coleman interviews); and Patterson, *Memoirs.*

BESSIE'S PARENTS: Gilley, from U.S. Census.

"ATLANTA WAS A PLACE": Frank X. Tolbert, *Tolbert's Texas Scrapbook,* Vols. 3, 18, 19. Unpublished. Barker Texas History Center, Austin.

4   MEMPHIS LYNCHING: Giddings, *When and Where I Enter,* 17–20.

PARIS LYNCHING: Ibid., 89.

TEXAS VIOLENCE: *Norton's Union Weekly Intelligencer,* January 10, 1892, pp. 1–3.

5   GEORGE COLEMAN'S LAND PURCHASE: Bernice Hamilton, per-

sonal communication with author, Ellis Associates, Surveyors, Waxahachie, Tex., May 13, 1991.

"WAXAHACHIE WAS A TEEMING HUB": Historic Waxahachie Inc. Driving Tour Map. Historic Waxahachie, Inc., Waxahachie, Texas, 1987.

ELLIS COUNTY RECORDS: Beulah Florence, personal communication with author, from the land deeds of Ellis County, Texas, May 13, 1991; Hamilton, communication with author; Historic Waxahachie, Inc., Driving Tour Map.

"ON HIS QUARTER ACRE": Historic Waxahachie Inc. Driving Tour Map; Annie Pruitt, interview with author, May 14, 1991.

6   BESSIE'S EARLY CHILDHOOD: Patterson, *Memoirs;* and Coleman interviews.

"IN 1894": Gilley, from U.S. Census, 1900.

BESSIE'S FAMILY RESPONSIBILITIES: Patterson, *Memoirs.*

FAMILY MEMBERS: Gilley, from U.S. Census, 1900.

7   BESSIE'S SCHOOL: Pruitt interview.

"AS LATE AS 1922": *The Call* (Kansas City, Mo.), August 26, 1922, p. 1.

"STAR PUPIL": Patterson, *Memoirs.*

"IF HE PROTESTED": Meier and Rudwick, *From Plantation to Ghetto,* 196–214.

"115 LYNCHINGS": Allen, *The Big Change.*

8   "SAVAGE AND TREACHEROUS": *A Memorial and Biographical History of Ellis County, Texas* (Chicago: Lewis Publishing, 1892), 61.

"IN OKLAHOMA": Greenberg, *Staking a Claim,* 21–28.

"NEITHER PIONEER NOR SQUAW": Patterson, *Memoirs;* Coleman interviews.

BESSIE'S BROTHERS: Alberta Lipscombe, interview with author, February 25, 1991 (hereafter Lipscombe interview).

"SHE HAD NO KINFOLK": Jones, *Labor of Love, Labor of Sorrow,* 77–78.

WORKING FOR THE JONESES: Coleman interviews; Patterson, *Memoirs.*

9   BESSIE'S HOMELIFE: Patterson, *Memoirs;* Coleman interviews.

10   "DEPENDING ON THE WEATHER": Jones, *Labor of Love*, 17.
      PICKING COTTON: Jones, *Labor of Love*; Patterson, *Memoirs*;
      Coleman interviews; Taulbert, *Once Upon a Time When We
      Were Colored*, 28.

11   "WE LAID OUR COTTON SACKS ASIDE": Patterson, *Memoirs*.
      BESSIE AS FAMILY ENTERTAINER: Ibid.; Coleman interviews.

12   FUNDRAISING: Patterson, *Memoirs*.
      WORKING AS LAUNDRESS: Coleman interviews.
      UNIVERSITY ENROLLMENT: Jilda Stallworth Motley, interviews
      with author, September 16, 1990, and February 18, 1992. The
      school is now Langston University, from which Bessie's great-
      niece, Jilda Stallworth Motley, the granddaughter of Elois, was
      graduated in 1967.
      LANGSTON: Rampersad, *The Life of Langston Hughes*, 1:9; Mot-
      ley interviews.

13   "CONQUERING HERO": Patterson, *Memoirs*.
      "KEEPING HER PLACE": Jones, *Labor of Love*, 125.
      "ELOIS LATER SAID": Patterson, *Memoirs*.
      "THINKING": Ibid.

14   "LEAVE THAT BENIGHTED LAND": Smith, *Chicago*, 388.
      "IN DALLAS": *Dallas Morning News*, May 17, 1912, pp. 10, 17.
      "AMOUNT TO SOMETHING": Coleman interviews.

Chapter 2 • That Wonderful Town

15   CONDITIONS ON THE TRAIN: Taulbert, *Once Upon a Time When
      We Were Colored*, 141.

16   *BIRTH OF A NATION:* Perrett, *America in the Twenties*, 85.
      "SHUNNED BY ALL OTHER GROUPS": Meier and Rudwick, *From
      Plantation to Ghetto*, 237.
      "A VIRTUAL SAVAGE": Smith, *Chicago*, 392.

17   BESSIE MOVES IN WITH WALTER: *Lakeside Directories, 1912–
      1922*; Coleman interviews.
      PULLMAN CONDITIONS: Travis, *An Autobiography of Black Chi-
      cago*, xviii–xix.
      BESSIE'S BROTHERS: Coleman interviews.
      "BLACK BELT" OF CHICAGO: Meier and Rudwick, *From Planta-*

tion to Ghetto, 235; Frazier, *The Negro Family in the United States*, 227–34.

18 BESSIE'S SISTER-IN-LAW: Coleman interviews.
"IN 1915": Giddings, *When and Where I Enter*, 101.
"ONE SUCH CRITIC": Ibid., 115.
HAIR AND SKIN PREPARATIONS: *Chicago Defender*, June 6, 1925; *Guinness Book of Records* (New York: Facts on File, 1977), 496.

19 "THE DEBATE CONTINUED": *Chicago Defender*, July 4, 1925, pt. 2, p. 1.
"SHE WAS GOING TO BE A BEAUTICIAN": Coleman interviews; Vera Stallworth Buntin, interview with author, September 6, 1991 (hereafter Buntin interview).

20 "BEST AND FASTEST MANICURIST": Patterson, *Memoirs*.
"AS A MANICURIST": *Lakeside Directories*.
"THE STROLL": Travis, *An Autobiography of Black Chicago*, 35–37.
NIGHTCLUBS: Ibid., 36; Travis, *An Autobiography of Black Politics*, 42–44; Coleman interviews.

21 PEKIN CAFÉ: *Birmingham* (Ala.) *Reporter*, September 4, 1920, p. 1, and December 4, 1920, p. 1.
"BESSIE'S BROTHER JOHN": Coleman interviews.
ROBERT ABBOTT: Coleman interviews; Patterson, *Memoirs*; Travis, *An Autobiography of Black Chicago*, 25, 39.

22 JESSE BINGA: Travis, *An Autobiography of Black Chicago*, 25, 37, 39, 209, 219; *Norfolk Journal and Guide*, September 3, 1921, p. 8; *Afro-American* (Baltimore), February 14, 1925, p. 13.
OSCAR DEPRIEST: Travis, *An Autobiography of Black Politics*, 42–44, 61.
"IN ADDITION TO BINGA": Penderhughes, *Race and Ethnicity in Chicago Politics*, 20.

23 BESSIE'S MARRIAGE: Coleman interviews; Buntin interview; Marriage License No. 750728, County Clerk, Bureau of Statistics, Cook County, Illinois; *Lakeside Directories*.

24 "I COULDN'T STAND HIM": Coleman interviews.

"In 1918": Coleman interviews; Dean Stallworth, interview with author, September 12, 1991 (hereafter Stallworth interview); Arthur Freeman, interview with author, September 25, 1991 (hereafter Freeman interview).

Blacks in the service: Quarles, *The Negro in the Making of America*, 181–87.

25   "Twenty percent": Travis, *An Autobiography of Black Chicago*, 18.

"Still others had been hired": Quarles, *The Negro in the Making of America*, 185–87.

Mayor Thompson: Travis, *Black Politics*, 55–61, 70–71.

26   Chicago race riot: Quarles, *The Negro in the Making of America*, 390–99; Smith, *Chicago*, 390–91.

"As Joe and his mother fled": Travis, *An Autobiography of Black Chicago*, 25.

"The riots left 38 dead": Ibid., 20.

27   "That's it!": Coleman interviews.

Chapter 3 • Mlle. Bessie Coleman—Pilote Aviateur

29   Trying to find a teacher: Patterson, *Memoirs*.

Robert Abbott: *Chicago Defender*, September 6, 1924, p. 2.

30   "Taking Abbott's advice": Patterson, *Memoirs*; Coleman interviews; *Lakeside Directories*.

31   "As the first African American woman pilot": Dr. Marjorie Stewart Joyner, interviews with author, October 26 and 28, 1990 (hereafter Joyner interviews).

"An unnamed Spaniard": *Pittsburgh Courier*, May 8, 1926.

"A lot of men callers": Coleman interviews.

Passport application: Patterson, *Memoirs*; State Department Files, Group 59, National Archives, Washington, D.C.

32   Visas: State Department Files; U.S. Passport No. 109381, dated November 9, 1920 (courtesy Vera Jean and Thomas D. Ramey, Harvey, Illinois).

"By the English": *Chicago Defender*, October 8, 1921, p. 2. The reference is to Le Crotoy in the Somme. The mistake may have been the error of a *Defender* reporter.

ACCOUNTS OF BESSIE'S STAY: *Chicago Defender*, October 8, 1921, and numerous subsequent interviews with the U.S. press; Marie-Josèphe de Beauregard, Fédération des Pilotes Européennes, letter dated December 10, 1992.

33   THE NIEUPORT: *All the World's Aircraft*, 172–73.

"STRUCTURAL FAILURE": Tallman, *Flying Old Planes*, 56–61; Ethell, "Wings of the Great War," 64.

"EACH TIME SHE TOOK A LESSON": Tallman, *Flying Old Planes*, 56–61; Ethell, "Wings of the Great War," 64.

34   FAI TEST: Coleman File, National Air and Space Museum Archives, Washington, D.C.; Young and Callahan, *Fill the Heavens with Commerce*, 67.

FAI LICENSE: Coleman file, NASM Archives; *L'Aerophile*, Bulletin Officiel de L'Aero-Club de France, p. 287.

"IN FRANCE": Dr. Wilberforce Williams, *Chicago Defender*, December 10, 1921, p. 4.

35   "BESSIE LEFT FRANCE": Passport No. 109381; *Air Service News*, November 1, 1921, p. 11; *Aerial Age Weekly*, October 17, 1921, p. 125; *New York Tribune*, September 26, 1921.

"USING INDIA AS AN EXAMPLE": *Dallas Express*, October 8, 1921, p. 1.

SHUFFLE ALONG: Chicago Defender, August 27, 1921, p. 5; Haney, *Naked at the Feast*, 34–35.

DEFENDER INTERVIEW: *Chicago Defender*, October 8, 1921, p. 2.

37   "THE GOLIATHS": "L'Aeronautique Marchande aux Répertoire Commercial de l'Industrie Aeronautique Francez" (Paris: Gauthier Villars, 1922), 13, 60.

38   "THIS IS MY MOTHER": Coleman interviews.

STUNTS BY WOMEN: Kathleen Brooks-Pazmany, *United States Women in Aviation* (Washington, D.C.: Smithsonian Institution Press, 1991).

39   MORE TRAINING: *Chicago Defender*, October 18, 1921, p. 2.

Chapter 4 • Second Time Around

41   "SHE WAS THE GUEST": *Chicago Defender*, February 25, 1922, p. 3.

"THIS WAS THE HEART OF HARLEM": *Pittsburgh Courier,* October 11, 1924, p. 13.

42 CHARLES GILPIN: *Afro-American* (Baltimore), March 17, 1922, p. 1.

CHICAGO RACE COMMISSION REPORT: *Chicago Daily News,* October 5, 1922, p. 30.

43 GEORGE HARRIS: *Chicago Defender,* February 25, 1922, p. 3.

SPEAKING IN NEW YORK: Ibid.

44 CENSURING HARRIS AND MOORE: *Washington Bee,* August 20, 1921, p. 6; *Afro-American* (Baltimore), July 20, 1923, p. 7.

NIEUPORT PLANE: *Afro-American* (Baltimore), March 10, 1922, p. 10; Patterson, *Memoirs.*

"WHITE AMERICANS": *Cleveland Gazette,* August 4, 1922, p. 1.

"ONE EXPATRIATE SPOKESMAN": *Los Angeles Times,* January 2, 1923, p. 2.

45 HOLLAND VISA: Passport No. 109381.

FOKKER: Hegener, *Fokker,* 54–57; *Chicago Defender,* May 21, 1924, p. 1; *Aerial Age,* June 5, 1922, p. 292.

"OBVIOUSLY DETERMINED": Passport No. 109381.

46 NEWSREEL: *California Eagle,* April 29, 1923, p. 11.

ROBERT THELEN: Correspondence from aviation historian John Underwood to Arthur W. Freeman. Undated.

LETTER FROM CAPTAIN KELLER: *Chicago Defender,* September 9, 1922, p. 3.

47 REPORTERS: Ibid.

NEW YORK TIMES REPORTER: *New York Times,* August 14, 1922, p. 4.

FEATS IN GERMANY: Ibid.; research by Herr Wolfram Müller in Hamburg and East and West Berlin;

48 DOZEN PLANES ORDERED: *New Age,* August 26, 1922, p. 6.

COUNTRIES BESSIE SAID SHE VISITED: Passport No. 109381.

"THE WORLD'S GREATEST WOMAN FLYER": *Chicago Defender,* September 2, 1922, p. 3.

"HEART-THRILLING STUNTS": *Chicago Defender,* September 9, 1922, p. 3.

49   REPORTS ON THE SHOW: *The Call* (Kansas City), September 15, 1922, p. 6; *New York Times*, September 4, 1922, p. 9. HUBERT JULIAN: *Chicago Defender*, September 9, 1922, p. 3; Louis B. Purnell, "The Flight of the Bumble Bee," *Air and Space Magazine* (October/November 1989): 33–34.

50   EDISON MCVEY: *Amsterdam News*, August 1, 1923. *BILLBOARD:* Afro-American (Baltimore), September 15, 1922, p. 9.

51   "IN MEMPHIS": *The Commercial Appeal* (Memphis, Tenn.), October 12, 1972, and October 12, 1922; *Afro-American* (Baltimore), December 15, 1922, p. 3.

Chapter 5 • Pleasing the Crowds, Alienating the Critics

53   "SOCIAL NOTICES": *Chicago Defender*, September 30, 1922, p. 4. PLANE CRASH: *Cleveland Plain Dealer*, September 1, 1922, p. 1.

54   PRE-SHOW PUBLICITY: *Chicago Defender*, October 2, 1922, p. 2. "AS FAR AS GEORGIA WAS CONCERNED": Coleman interviews; and Patterson, *Memoirs*.

55   TWO-COLUMN ADVERTISEMENT: *Chicago Defender*, September 30, 1922, p. 2. "BECAUSE OF THIS YOUNG ENTREPRENEUR": *Chicago Aviation News*, April 27, 1922, p. 9. SPEED DERBY: *New York Times*, September 6, 1922, p. 17.

56   "MY AUNT'S A FLIER": Freeman interview. "BESSIE PERFORMED": *Chicago Defender*, October 21, 1922, p. 2. "AFTER THE SHOW": *Chicago Defender*, September 30, 1922, p. 4.

57   *BILLBOARD* ARTICLE ON FILM: *Afro-American* (Baltimore), December 1, 1922, p. 13. WALKING OUT ON FILM: *Afro-American* (Baltimore), November 10, 1922, p. 3, and December 1, 1922, p. 13. "JACKSON RETURNED BESSIE'S BLAST": Ibid., November 10, 1922, p. 3, and December 1, 1922, p. 13.

58   "IN AN INTERVIEW WITH JACKSON": Ibid. NATIONAL ASSOCIATION OF COLORED FAIRS: *Norfolk Journal and Guide*, August 26, 1922, p. 8.

59 "As early as July": Ibid., July 8, 1922, p. 1, and September 9, 1922, p. 1.

"eccentric": *Afro-American* (Baltimore), December 1, 1922, p. 13.

Most powerful men: *The Call* (Kansas City), July 10, 1923, p. 1; *Norfolk Journal and Guide*, June 27, 1925, p. 9; *Afro-American* (Baltimore), December 15, 1922, p. 3.

60 Chicago Race Commission report: *Chicago Daily News*, October 5, 1922, p. 30.

"took to flying naturally": *Norfolk Journal and Guide*, December 16, 1922, p. 4.

61 "Blivens": *Norfolk Journal and Guide*, January 14, 1922.

62 "the biological function": *Amsterdam News*, February 4, 1925.

Bessie in Baltimore: *Afro-American* (Baltimore), November 17, 1922, p. 2, and November 10, 1922, p. 3.

Chapter 6 • Forced Landing

63 school for aviators: *Afro-American* (Baltimore), November 12, 1922, p. 3.

64 "drop literature from the clouds": *California Eagle*, March 4, 1923, editorial page.

Free air shows: *Los Angeles Evening Herald*, February 17, 1923, Real Estate section, p. 8; *Los Angeles Times*, May 6, 1923, part L, p. 8.

André Peyre: Hatfield, *Los Angeles Aeronautics*, 23, 53, 57.

Peyre and Earhart: Hatfield, *Los Angeles Aeronautics*, 23, 53, 57, 80.

Frank Hodge: *Los Angeles Times*, February 26, 1923.

65 Coast Tire and Rubber Company tour: *Los Angeles Evening Herald*, February 3, 1923, p. 1.

Rockwell Army Intermediate Depot: Edward L. Leiser, letter to author, September 21, 1991.

66 "Miss Coleman is a neat-appearing young woman": *Air Service Newsletter*, February 20, 1922.

"ENTHUSIASTIC, CHARMING GIRL": *California Eagle*, January 27, 1923, p. 3.

INTERNATIONAL FLYING LICENSE: Ibid.

IN LOS ANGELES: *Dallas Express*, February 10, 1923, p. 1; *California Eagle*, February 3, 1923, p. 9.

67   "THE ONLY HINT OF CRITICISM": *Afro-American* (Baltimore), February 2, 1923, p. 3.

68   AIR SHOW: *Afro-American* (Baltimore), February 11, 1923, p. 1.

69   CRASH: Ibid.

OTHER CRASHES: *Los Angeles Times*, pt. 2, January 19, 1923, p. 6; pt. 5, January 21, 1923, p. 14; and January 31, p. 6; *Houston Post-Dispatch*, September 1, 1925, p. 1.

CROWD REACTION: *California Eagle*, February 10, 1923, p. 1.

70   TELEGRAMS: Ibid.

LECTURE: *Afro-American* (Baltimore), February 18, 1923, p. 1.

71   FAMILY REACTION: Coleman interview, August 1, 1992.

SACHS'S LETTER: *California Eagle*, March 4, 1923, p. 4.

PAID NOTICE: *California Eagle*, March 18, 1923, p. 4.

72   "QUEEN BESS OPENS SCHOOL": *The New Age*, May 8, 1923, p. 4.

CONTRACT: *L'Aviation* magazine, Editeur à Saint-Valery-sur-Somme. F. Paillart à Abbeville, 1988, p. 170.

FILM POSSIBILITY: *Afro-American* (Baltimore), May 11, 1923, p. 1; *California Eagle*, March 4, 1923, p. 4.

73   YMCA LECTURES: *California Eagle*, April 29, 1923, p. 11, and May 5, 1923, p. 5.

Chapter 7 • Grounded

75   POST-CRASH AIR SHOW: *Chicago Defender*, June 23, 1923, p. 2, and September 2, 1923, p. 2; *Columbus Evening Dispatch*, September 2, 1923, p. 9, and September 9, 1923, p. 9.

76   KLAN MEETING: *Columbus Evening Dispatch*, September 2, 1923, p. 10, and September 4, 1923, p. 18.

VISIT TO *DEFENDER* OFFICE: *Chicago Defender*, September 8, 1923, p. 2.

77   AIR SHOW: Ibid., September 22, 1923, p. 2.

UNBYLINED STORY: *Afro-American* (Baltimore), May 2, 1924, p. 4.

MANAGERS: Patterson, *Memoirs; Afro-American* (Baltimore), July 7, 1923, p. 6.

78   "A GOOD LONG REST": Patterson, *Memoirs.*

"NO NICHE IN BLACK CHICAGO": *Chicago Defender,* February 9, 1924.

DR. MARJORIE STEWART JOYNER: Joyner interviews.

79   "THE RIGHT SHADE OF BLACK": Travis, *An Autobiography of Black Chicago,* 72, 73.

80   "YOUR AUNT BESSIE IS VERY PRETTY": Coleman interviews and Lipscombe interview.

NIECES AND NEPHEWS: Coleman interviews; Buntin interview; Stallworth interview.

GEORGIA'S DAUGHTER MARION: Coleman interviews. Marion was taught by nuns from Corpus Christi Church at Forty-ninth and Dr. Martin Luther King Jr. Drive.

81   THE RED DRESS: Ibid.

PRINCE OF DAHOMEY: Patterson, *Memoirs;* Coleman interviews; *Dallas Express,* November 29, 1924; *Afro-American* (Baltimore), March 7, 1922, p. 5; *Fort Worth Star-Telegram,* June 9, 1925, p. 8.

"OTHER MEN FRIENDS": Coleman interviews.

JESSE BINGA: *Afro-American* (Baltimore), February 14, 1925, p. 13.

82   ARTICLE ON FOKKER: *Chicago Defender,* May 27, 1924, p. 1.

HUBERT JULIAN: *Norfolk Journal and Guide,* April 19, 1924, p. 1; *Chicago Defender,* July 12, 1924, p. 1.

Chapter 8 • Texas Triumph

85   NEW HOME BASE: *Houston Post-Dispatch,* May 7, 1925, p. 4.

86   FIRST TEXAS FLIGHT: *Houston Informer,* June 27, 1925, p. 1.

JUNETEENTH: *Afro-American* (Baltimore), June 12, 1925, p. 1.

87   ADVANCE STORIES ON AIR SHOW: *Houston Post-Dispatch,* June 14, 1925, p. 9, and June 18, 1925, p. 1.

"SHOOTING AND CUTTING RAMPAGE": *Houston Post-Dispatch,* June 20, 1925, p. 1.

88 MABEL CODY AND GLADYS ROY: Lomax, *Women of the Air*, 35.
"BIGGEST THRILL": *Houston Informer*, June 27, 1925, p. 1, and
July 11, 1925, p. 6.
JIM CROW LAWS: *Afro-American* (Baltimore), August 17, 1923,
p. 1.

89 "THE BALTIMORE WRITER'S EXPERIENCE": *Houston Informer*,
July 11, 1925, p. 6.
SHOWS IN TEXAS: *Houston Post-Dispatch*, July 5, 1925, p. 7;
Chicago Defender, July 25, 1925, p. 2.

90 INDUCEMENT TO SPECTATORS: *Afro-American* (Baltimore), July
18, 1925, p. 4.
"PUT THE RACE ON THE MAP": *Houston Informer*, July 11,
1925, p. 6.
SHOW AT SAN ANTONIO SPEEDWAY: *San Antonio Express*, August 9, 1925, p. 12; *Dallas Express*, September 29, 1925, p.
1; Patterson, *Memoirs*.
OTHER ACCIDENTS: *Pittsburgh* (Texas) *Gazette*, March 7, 1924,
p. 3; *Fort Worth Star Telegram*, August 9, 1925, p. 1.
DILWORTH'S JUMP: *Dallas Express*, August 29, 1925, p. 1.

91 SCHEDULE OF LECTURES: Ibid.
THE BEALES: Ibid.; City Directory of Dallas, Texas (Dallas:
John F. Worley Directory Co., 1925), 1385, 1974.
AERIAL EXHIBITION IN DALLAS: *Dallas Express*, August 28,
1925.

92 LOVE FIELD: Al Harting, "A Love Affair," *Westward Magazine*
(July 15, 1984): 14–21; *Dallas Times-Herald*, July 15, 1984,
pp. 14–21.
LOUIS MANNING: Arthur Spaulding, interviews with author,
May 15 and 16, 1991 (hereafter Spaulding interviews).

93 DILWORTH REFUSES TO JUMP: *Wharton Spectator*, September
11, 1925, p. 6.
BESSIE'S JUMP: Ibid.; Spaulding interviews.

94 MONEY FOR PLANE: *Houston Post-Dispatch*, September 9,
1925, p. 11, and September 10, 1925, p. 5.
SHOW IN WAXAHACHIE: *Waxahachie Daily Light*, April 30,
1926, p. 1; Patterson, *Memoirs*.

95   GOVERNOR FERGUSON: Jordan and Heron, *A Self-Portrait*, 21.
      "ON ONE OCCASION": Patterson, *Memoirs*.
96   PAUL MCCULLY: Paul McCully, letter to author, December 28,
      1991.
      J. A. JACKSON: *Norfolk Journal and Guide*, June 27, 1925, p. 9.

Chapter 9 • Nearing the Goal

97   SERIES OF LECTURES: R. E. Norman, correspondence with Bes-
      sie Coleman (February 8, 1926), Manuscript Department, Lilly
      Library, Indiana University–Bloomington.
      VISIT WITH ELOIS: Patterson, *Memoirs*.
      "THE CITY IS THRILLED": *Savannah Tribune*, January 7, 1926,
      p. 1.
98   JULIAN'S LICENSE: *Pittsburgh Courier*, March 14, 1925, p. 9.
      "WALK THE WINGS OF AN AIRPLANE": *Savannah Tribune*, Janu-
      ary 7, 1926, p. 1.
      GATES FLYING CIRCUS: *Savannah Morning News*, January 10,
      1926, p. 10.
99   LETTERS: Norman-Coleman correspondence.
101   THE HILLS: Audrey Tillinghast, interviews with author, April
      21 and 22, 1991.
102   NEIGHBORHOOD CHILDREN: Tape-recorded interviews provided
      by Audrey Tillinghast. Subsequent interviews, by the author
      with Mrs. Tillinghast, were conducted April 21–22, 1991
      (hereafter Tillinghast interviews).
103   "SO DETERMINED WAS SHE": Patterson, *Memoirs*.
      BESSIE'S CHAIR: Tillinghast interviews.
104   EDYTH CROOMS: Ibid.
      THREAT TO WITHDRAW: *Chicago Defender*, May 15, 1926, p. 2.
      SEGREGATION IN FLORIDA: *Pittsburgh Courier*, April 10, 1926, p.
      2; *Norfolk Journal and Guide*, January 16, 1926, p. 1.
105   EDWIN BEEMAN: *Evening Reporter Star*, February 29, 1926,
      p. 1.
      FINAL PAYMENT ON PLANE: *Chicago Defender*, May 8, 1926,
      p. 1.
      COMFORTS OF RELIGION: Patterson, *Memoirs*.

Chapter 10 • The Sky Has a Limit

107   JOHN THOMAS BETSCH: *Florida Times Union*, May 1, 1926,
       sect. 2, p. 1; *Chicago Defender*, May 8, 1926, p. 1; Dr.
       Johnnetta Betsch Cole, letter to author, November 27, 1991.
       ADVANCE STORY: *Jacksonville Journal*, April 28, 1926, p.
       11.

108   STRAND THEATER: *Atlanta Independent*, May 6, 1926, p. 1.
       SCHOOLCHILDREN'S IMPRESSIONS: Interviews with author: Mar-
       ian Johnson Jeffers, Hettie Thompson Mills, and Eugenia
       Mathews Brown, all October 17, 1991; Frances Tyson John-
       son, interviews with author, October 17 and 24, 1991.
       WILLS FLYING BESSIE'S PLANE: *Jacksonville Journal*, April 30,
       1926, p. A24.

109   WILLS STAYING WITH WESTERFIELD: *Florida Times Union*, May
       1, 1926, sect. 2, p. 1; Spaulding interviews.
       MEETING ABBOTT: Patterson, *Memoirs*.
       PRELIMINARY FLIGHT: *Jacksonville Journal*, April 30, 1926,
       p. A24.

110   FLIGHT AND ACCIDENT: *Florida Times Union*, May 1, 1926,
       sect. 2, p. 1; *Chicago Defender*, May 8, 1926, p. 1; *Dallas
       Express*, May 15, 1926, p. 1; unidentified newspaper clipping
       on microfilm, FSN SC #000386-1, Schomburg Center for Re-
       search in Black Culture, New York Public Library; *Amsterdam
       News*, May 8, 1926, p. 1.
       BETSCH'S ARREST: *Florida Times Union*, May 1, 1926, sect. 2,
       p. 1; *Jacksonville Journal*, April 30, 1926, p. A24.

111   CAUSE OF CRASH: *Jacksonville Journal*, April 30, 1926, p.
       A24, and May 1, 1926, p. 1.
       WILLS'S BODY: *Florida Times Union*, May 2, 1926, p. 2;
       Jacksonville Journal, May 1, 1926, p. 1.
       FAIRGROUNDS: *Chicago Defender*, May 8, 1936, p. 1.
       DANCE CANCELED: Eartha M. M. White Collection, Thomas
       G. Carpenter Library, University of North Florida, Jackson-
       ville.
       VIEWING: *Florida Sunday Times Union*, May 2, 1926, p. 2.

112   MEMORIAL SERVICE: White Collection; *Florida Sunday Times-*

Union, May 2, 1926, p. 2; *Chicago Defender*, May 8, 1926, p. 1; *Dallas Express*, May 15, 1926, p. 1; Camilla Thompson, interview and tour with author, April 25, 1991.

ORLANDO SERVICE: *Jacksonville Journal*, May 1, 1926, p. 1; *Pittsburgh Courier*, May 15, 1926, p. 1; *Atlanta Independent*, May 13, 1926.

113   WILLS'S FUNERAL: *Dallas Journal*, May 1, 1926, p. 6.
VIEWING AND SERVICE IN CHICAGO: *Norfolk Journal and Guide*, May 8, 1926, p. 1; *Dallas Express*, May 15, 1926, p. 1; *Chicago Defender*, May 15, 1926, p. 1; Julia Whitfield, *A Brief History of the Pilgrim Baptist Church;* Jacqueline Smith, interview with author, October 31, 1990.

114   MEMORIAL STONE: Patterson, *Memoirs*.
"OH, BESSIE": *Chicago Defender*, May 15, 1926, p. 1.
"I'VE DONE MY WORK": White Collection

115   "THE REBEL": Adoff, *The Poetry of Black America*.

Epilogue

117   OTHER ACCIDENTS: *Houston Post-Dispatch*, June 17, 1925, sect. 2, p. 1, and September 13, 1925, p. 10; *Washington Post*, September 2, 1991, p. B3.
"I CANNOT RECONCILE MYSELF": Patterson, *Memoirs*.

118   FOCUS ON WILLS: *Florida Times Union*, May 1, 1926, sect. 2, p. 1; *Jacksonville Journal*, April 30, 1926, p. A24.
"THERE IS REASON TO BELIEVE": *Dallas Express*, May 15, 1926, editorial page.

119   "WHETHER THEY TAKE TO IT OR NOT": *Norfolk Journal and Guide*, May 15, 1926, p. 14.
"BLACK 'JOAN OF ARC' ": *New York News*, July 30, 1927.
FLYOVER: *Chicago Defender*, March 8, 1990, pp. 1, 34–35.

120   "BECAUSE OF BESSIE COLEMAN": Powell, *Black Wings*.
BESSIE COLEMAN AVIATORS CLUB: "They Take to the Sky," *Ebony* (May 1977): 89–97.
REVIVAL OF FLYOVER: *Chicago Defender*, March 8, 1990, pp. 1, 34, 35.
BESSIE COLEMAN DRIVE: Reports and Communications from

city officers, *Journal-City-Council-Chicago*, February 28, 1990, p. 11794.

"BESSIE COLEMAN CONTINUES TO INSPIRE": Resolution, the City Council of the City of Chicago, February 26, 1992.

BESSIE COLEMAN DAY: Office of the Mayor, City of Chicago, proclamation dated April 27, 1992; announcement, Bessie Coleman Annual Memorial Flight, distributed by Rufus Hunt.

Bibliography

Adoff, Arnold. *The Poetry of Black America: Anthology of the Twentieth Century*. Edited by Arnold Adoff. Introduction by Gwendolyn Brooks. New York: Harper and Row, 1973.

Alford, Sterling G. *Famous First Blacks*. New York: Vantage Press, 1974.

All the World's Aircraft. Founded by the late Fred T. Jane. Edited and compiled by C. G. Gray. London: Sampson Low, Marston and Co., 1918; rpt: New York: Arno Press, 1968.

Allen, Frederick Lewis. *The Big Change*. New York: Harper and Row, 1952.

————. *Only Yesterday*. New York: Harper and Row, 1957.

Anderson, Jervis. *This Was Harlem: A Cultural Portrait, 1900–1950*. New York: Farrar, Straus, and Giroux, 1982.

Bacon, Eve. *Orlando: A Centennial History*. Chuluota, Fla.: Mickler House, 1975.

"Bethel Baptist Institutional Church, 1838–1988." Booklet issued in 1988 on the 150th anniversary of the church in Jacksonville, Florida.

Bilstein, Roger, and Jay Miller. *Aviation in Texas*. Austin: Texas Monthly Press, 1985.

Birmingham, Stephen. *Certain People: America's Black Elite*. Boston: Little, Brown, 1977.

"Black, Brave and Flying." *Ebony, Jr.*, February 1979, pp. 16–18.

Bulletin Officiel de L'Aero-Club de France, Brevets de Pilots, *L'Aerophile*, September 1–15, 1921.

Bundles, A'Lelia. *Madame C. J. Walker—Entrepreneur*. New York: Chelsea House, 1991.

Caiden, Martin. *Barnstorming*. New York: Duell, Sloan, and Pearce, 1965.

Carisella, P. J., and James W. Ryan. *The Black Swallow of Death*. Boston: Marlborough House, 1972.

"Chicago Colored Girl Learns to Fly." *Aerial Age Weekly*, October 17, 1921.

Clark, Kenneth B. *Dark Ghetto: Dilemmas of Social Power*. New York: Harper and Row, 1967.

Coles, Robert. *Children of Crisis: A Study of Courage and Fear*. Boston: Little, Brown, 1964.

"Colored Aviatrix Bobs Up Again." *Air Service Newsletter*, February 20, 1923.

Dedmon, Emmett. *Fabulous Chicago*. New York: Random House, 1953.

Ethell, Jeffrey L. "Wings of the Great War." *Air and Space* (October/November 1991): 64.

Evans, Sara M. *Born for Liberty: A History of Women in America*. New York: Free Press, 1989.

Ferenbach, T. R. *Lone Star: A History of Texas and Texans*. New York: Macmillan, 1968.

Frantz, Joe B. *Texas: A History*. New York: W. W. Norton, 1976.

Frazier, E. Franklin. *The Negro Family in the United States*. Rev. and abridged. Chicago: University of Chicago Press, 1948.

Giddings, Paula. *When and Where I Enter: The Impact of Black Women on Race and Sex in America*. New York: William Morrow, 1984.

Greenberg, Jonathan. *Staking a Claim: Jake Simmons, Jr. and the Making of an African-American Oil Dynasty*. New York: Atheneum, 1990.

A Guide to Orlando's Afro American Heritage. A special project of the Central Florida Society of Afro American Heritage, Inc. No date.

Haney, Lynn. *Naked at the Feast: A Biography of Josephine Baker.* New York: Dodd, Mead, 1981.

Hardesty, Von, and Dominic Pisano. *Black Wings: The American Black in Aviation.* Washington: Smithsonian Institution Press, 1987.

Hatfield, D. D. *Los Angeles Aeronautics: 1920–1929.* Los Angeles: Hatfield History of Aeronautics Alumni Library, Northrop Institute of Aeronautics, 1973.

Hegener, Henry. *Fokker: The Man and the Aircraft.* Garden City, N.Y.: Garden City Press, Ltd., 1961.

Hetzel, Gary. "A Flight into the Past." *Orange County Historical Quarterly,* June 1986.

Hughes, Langston, Milton Meltzer, and C. Eric Lincoln. *A Pictorial History of Black America.* 5th rev. ed. New York: Crown, 1983.

Ingle, John P. *Aviation's Earliest Years in Jacksonville (1878–1935).* Jacksonville, Fla.: Jacksonville Historical Society, 1977.

Johnson, James Weldon, ed. *The Books of American Negro Poetry.* Rev. ed. San Diego: Harcourt Brace Jovanovich, 1959.

Jones, Jacqueline. *Labor of Love, Labor of Sorrow: Black Women, Work and the Family from Slavery to the Present.* New York: Basic Books, 1985.

Jordan, Barbara, and Shelby Hearon. *A Self-Portrait.* Garden City, N.Y.: Doubleday, 1979.

King, Anita. "Family Tree: Brave Bessie, First Black Pilot." *Essence:* pt. 1, May 1976; pt. 2, June 1976.

Lakeside Directories of the City of Chicago. Chicago: Chicago Directory Company, 1912–22.

"License for Air Pilots." *Flying,* June 1921.

Lomax, Judy. *Women of the Air.* New York: Dodd, Mead, 1986.

Mayer, Harold M., and Richard C. Wade. *Chicago: Growth of a Metropolis.* Chicago: University of Chicago Press, 1964.

Meier, August, and Elliott Rudwick. *From Plantation to Ghetto.* 3d ed. New York: Hill and Wang, 1976.

Meltzer, Melton. *The Truth about the Ku Klux Klan.* New York: Franklin Watts, 1982.

The Negro Almanac: A Reference Work on the African American. 5th ed. Edited by Harry A. Ploski and James Williams. Detroit: Gale Research, 1989.

"Negro Aviatrix to Tour Country." *Air Service Newsletter*, November 1, 1921.

O'Neil, Paul, and the editors of Time-Life Books. *Barnstormers and Speed Kings*. Alexandria, Va.: Time-Life Books, 1981.

Patterson, Elois. *Memoirs of the Late Bessie Coleman, Aviatrix*. Privately published by Elois Patterson, 1969.

Penderhughes, Dianne M. *Race and Ethnicity in Chicago Politics*. Urbana: University of Illinois Press, 1987.

Perrett, Geoffrey. *America in the Twenties: A History*. New York: Simon and Schuster, 1982.

Peters, Raymond Eugene and Clinton M. Arnold. *Black Americans in Aviation*. San Diego: New World Aviation Academy, Inc., and Clinton M. Arnold, 1975.

Powell, Lt. William J. *Black Wings*. Los Angeles: Ivan Deach, Jr., 1934.

Quarles, Benjamin. *The Negro in the Making of America*. 2d rev. ed. New York: Collier Books, Macmillan, 1969.

Rampersad, Arnold. *The Life of Langston Hughes*. Vol. 1: *1902–41. I, Too, Sing America*. New York: Oxford University Press, 1986.

Smith, Henry Justin. *Chicago: The History of Its Reputation*. Part 2. New York: Harcourt, Brace, 1929.

Stowe, Harriet Beecher. *Uncle Tom's Cabin*. New York: Viking Press, 1982.

Tallman, Frank. *Flying Old Planes*. Garden City, N.Y.: Doubleday, 1973.

Taulbert, Clifton Lemoure. *Once Upon a Time When We Were Colored*. Tulsa, Okla.: Council Oak Books, 1989.

Terrell, Mary Church. *A Colored Woman in a White World*. Washington, D.C.: National Association of Colored Women's Clubs, Inc., 1968.

"They Take to the Sky." *Ebony*, May 1977, pp. 16–18.

Travis, Dempsey J. *An Autobiography of Black Chicago*. Chicago: Urban Research Institute, 1981.

——. *An Autobiography of Black Politics*. Chicago: Urban Research Press, 1987.

Wallace, Michelle. *Black Macho: The Myth of the Super-Woman*. New York: Dial Press, 1978.

Waters, Enoch. *American Diary: A Personal History of the Black Press.* Chicago: Path Press, 1987.

Woodson, Carter G. *The Negro in Our History.* 9th ed. Washington, D.C. Associated Publishers, 1947.

Young, David, and Neal Callahan. *Fill the Heavens with Commerce: Chicago Aviation, 1825–1926.* Chicago. Chicago Review Press, 1981.

AFTERWORD

Knowledge of aviator Bessie Coleman fills me with sunshine, enthusiasm, daring, courage, sadness, hope, joy, triumph, and a firm grip on reality. Ms. Coleman's zeal for life and dedication to her own vision strengthens for each one of us our own tenuous grip on reality simply because she was—because she existed. To me she is that ephemeral daydream of adventure, strength, audacity, and beauty that we all seek, hope, and somehow know must be present in the world. That daydream truly exists as every day we see people making small strides in overcoming obstacles of gender, birthright, race, ethnicity, economics, illness, poor technology, education, societal condemnation, and fear. In our hearts we know there is someone out there who leaps over conventional wisdom to reality in a single bound! Bessie Coleman's leap nourishes the heart and spirit of each of us who "meets" her.

I did not have the pleasure of meeting up with Ms. Coleman until I was already an astronaut. But when I did, it was as if I had known this best friend all my life. I met her one cold winter afternoon at the Du Sable Museum of African American History in Chicago in 1992. Artifacts from her life were in a glass display case. There staring at me from the picture on her aviation license was a young woman who at first glance looked startlingly like me. I read of her travels to France to get instruction in flying airplanes, since no one in her own country, the United States, would teach her because she was a woman and black. I was warmed by the optimism and self-confidence she must have had to determine her own life at a time when most people and especially black women were supposed to take the lot they were given in life and accept the limitations others imposed on them.

I grew up in Chicago and as the first and only black woman astronaut in the world and an African American history aficionado, I was embarrassed and saddened that I did not learn of her until my space flight beckoned on the horizon. In fact, I felt cheated. At the time I was thirty-five, a year older than Bessie when she died. I wished I had known her while I was growing up, but then again I think she was there with me all the time. You see I am convinced that the fall from the airplane that killed her did not kill her spirit. And the calculated arrogance and disdain of the country's major newspapers, magazines, and history books that ignored her did not diminish her existence either. Because what Bessie Coleman affirms is the *life* in each of us.

I could have written about how aviation technology has progressed over the last seventy years and that now space is the new frontier. I could have emphasized the need to acknowledge that every group of people in the world has had scientists, explorers, and adventurers and that to make the best of space exploration all people must be given the chance to participate. I could have stood on my soapbox and said, "Make sure that we do not lose over 75 percent of the talent available to aerospace technology in this country because of ignorance, sexism, and racism." But no words from me could touch the elegance of these demands made manifest by the life of Bessie Coleman.

It is tempting to draw parallels between me and Ms. Coleman. On the surface folks could say we were both young black women born in the South who lived in Chicago and became involved in science and technology fields—aviation and aerospace—when women and especially black women whether through commission or omission were traditionally kept from participation. One could speak of the lack of role models and facing adversity. It is easy to get caught up in the airplane, mechanical, flyboy stuff. But what I hope is common to me and Bessie is the smile of adventure, self-determination, and dogged will to see beauty in the world even as ugly things happen around us and to us.

I point to Bessie Coleman and say without hesitation that here is a woman, a being, who exemplifies and serves as a model to all humanity: the very definition of strength, dignity, courage, integrity, and beauty.

It looks like a good day for flying.

Mae Jemison, M.D.

Index

ROBERT LOUIS SHEPARD was the first Black Ph.D. scientist in the organization where he started his research career. The book details his life, the influences—personally and professionally—that molded him into a successful scientist, scholar, entrepreneur, and role model for hundreds of individuals in the United States and others abroad.

Shepard provides a look at the historical context for his life experiences growing up in a small town in North Carolina during the *Jim Crow era,* and how these experiences interfaced with the history of African Americans in America to shape the man he became. Shepard reflects on his amazing life in *Fulfilling My Destiny, Step By Step,* crediting the people that helped him on his journey, starting with his parents and community. While the book provides a look at how Shepard parlayed his own life lessons into a successful business in the science and engineering fields, it has a broader purpose of laying down a blueprint for future generations across the broad spectrum of social, economic, historical and political arenas in the United States.

Most especially, Shepard wrote this book for his children, grandchildren and his future descendants, providing a glimpse of how he stood on a foundation of faith and courage, belief in the best of the human spirit, and care and love for all that entered his world looking for nothing in return, to create a life of immeasurable fulfillment for himself, his family and for countless others in the United States and others abroad.

Fulfilling My Destiny, Step By Step, is the story of how Shepard discovered his purpose then nurtured it to become a contributor to the global community. The book is certain to inspire whoever reads it—no matter their race or position in life.

them to my library as I looked to find the style and structure of what I believed would be the approach for my book. I owe a big debt of gratitude to Dr. Green for pushing and strongly encouraging me to write this book.

This book embodies the spirit of The Shepard Institute (TSI). However, it could not have been written without the early focus and insights provided by my friend Dr. Gwendolyn S. Bethea, the director of the Expository Writing Program and manager of the Preparing Future Faculty Program at Howard University Graduate School. She has been a tremendous and loyal friend and has earned my heartfelt gratitude and thanks.

This is my autobiography. However, others were there to cheer me on: my son, Shawn, and daughter-in-law, Bridget; my daughters, Dr. Robin Shepard Broughton and Pamela Shepard; my grandchildren, Ryan and Stefani Shepard and Nina Grace Broughton; my brothers, Roger (who wrote the foreword to this book) and Anthony; my sisters, Maxine, Janice, and Celestine (who rechecked the edited version for errors), and the late Ruth. I am grateful for their support and love from the start to the finish of this project.

I am indebted to my publishing team at AuthorHouse,™ including Project Coordinator Sarah Smith, Lead Editor Amanda Carmichael, Lead Designer Bob DeGroff and Marketing Consultant Jeff Reuter. They all were outstanding professionals, and I owe the team a big thank-you for their expert guidance and support in the production and marketing of this book.

Finally, this book could not have been attempted without the support of the beautiful young girl I write about in chapter 9—my wife, Alzonia—who took time from her own schedule to read and edit the original manuscript. Her discussions with me on various crucial issues at important junctures helped clarify the thinking and writing of this book. In addition, she encouraged me in her quiet but powerful way to tell my story. We thank God for the production of this book and hope that we have done our family and friends proud.

INTRODUCTION

Over the years, many have sought my advice on matters pertaining to their lives. The range of inquiry has been vast. Young people in the early stages of their lives have come to me wanting to know what school they should attend, what subjects they should pursue, and even at times what career paths and options they should consider. There have been occasions when parents have brought their children to me with the entire family requesting my insight and guidance on matters of tremendous importance to a life struggling to find its place in the universe.

Others in the more seasoned phases of their lives have asked my advice to help them find their roles and purposes in life. The inquiries from this group have dealt with trying to find solutions to family issues, including financial concerns; focus for their careers and work options; input on what to do when their chosen paths have left them empty and unfulfilled; and advice on what they can do about it.

When I look back, I realize that others seeking my advice started early in my life. When I was a young boy in grade school, my classmates often looked to me to put things in order or reach a conclusion on a particular issue. This included something as simple as selecting a playground activity during recess to more important advice on what might be the best process to use or what path should be followed.

Two events still vividly stand out in my mind even though they happened so long ago. The first event came in 1957 when I was a fourth grader. During a recess period, none of us could decide what to do as the period was ending. A classmate yelled, "Robert, you decide." Spontaneously I said, "We have time to let the boys race the girls from one end of the field to the next and back before lining up to return to class." Without questioning the choice, that was what we did, and we had fun in the process. If I remember correctly, the girls won. While this seems simple, so were the games we played back then. As we lined up to return to class, a good feeling surged through me when I thought about the outcome of that simple act. Every student on the field participated in the race I suggested.

The second event occurred two years later when I was not giving advice but was instead following advice being given to me from within—the whispering of an inner voice I will talk about later in the book, especially in chapter 15.

In the early years, I resisted when my advice was sought. However, as I grew and began to take note of one of the great tragedies in life—hopes and dreams deferred and in other cases left totally unfulfilled—I became comfortable with my God-given talent as an enabler, a helper. I do not possess magical powers of any kind, and it humbles me when I am told, "Through your ideas, encouragement, and ability to empower, you are touching lives that have not been born yet." Such comments have caused me to realize how life is to be lived when you follow what seems to be your destiny. As a willing vessel of all that has been poured into me and as a good steward of all that has been made available through me, I am fully persuaded that with such investment in me, my time, talent, and treasure are not my own but rather are to be used to help others fulfill their own destinies in life, looking for nothing in return.

The words printed on these pages are from a grateful heart and are born out of my own experiences and struggles. It is useful for readers to know that my own life has had its twists and turns, its setbacks and disappointments, its highs and lows. It was during one of the darkest times in my life that I found myself caught in a telephone booth, not knowing which way to turn. But oh how I remember when a still small voice said to me on two separate occasions, "I will never leave you nor forsake you."

When such a powerful event happens to you, your simple and quick response should be, "Yes, Lord, I am listening. What would You have me to do?" Such a response will most assuredly set you on a path toward fulfilling the destiny that has been set for your life, and you will most likely fulfill it in stages and step-by-step. As you read the pages of this book, you will come to know me more deeply and personally. It is my deepest hope that this book will put you on a path toward a brighter and more rewarding future as you embrace your own God-ordained destiny, day by day, moment by moment, and step-by-step.

Chapter 1

THE BEGINNING

I WAS BORN A DREAMER. DREAMING often got me in trouble in my early years. When I was supposed to be working, I was somewhere dreaming. My grandmother Sally said I did more daydreaming than completing my chores. I once asked her what daydreaming was. She said it was sleeping when you are lazy and do not want to do something and that I was a professional at it. One day I looked up the word and found that it had a different meaning than the one my grandmother had given me—a pleasant visionary, usually a wishful creation of the imagination. I liked this definition better than my grandmother's because my mind sometimes captured visions of a different place in a different time that led me to think it was all my imagination. But what was I really dreaming, or daydreaming, about? Who knows? My grandmother said it was just my way of avoiding work. I hung around her a lot, so she may have had it right.

For example, one hot summer day my grandmother called for me to meet her on her front porch. We had a telephone but lived so close to her house that calling me in a loud voice from her back porch was the easiest way for her to summon me. So she let out a loud call: "Come up here, Robert Louis." I never liked being called by both my first and middle names, but that was the way adults and some others summoned me without fail. So I got use to it.

We lived in a neighborhood in the house built by my father on land my grandmother had divided between each of her seven children. The neighborhood was so close together that my grandmother had to specify to whom she was calling because if she didn't, anybody could come from at least six different houses, which were all within a short walk to her front porch.

"Why?" I shouted back. As the word left my mouth, I realized it did not sound good at all.

"Do not ask me any questions. Just get up here and bring with you that BB gun you got."

Realizing the error of questioning an adult, and my grandmother at that, I now had a good reason to be concerned. Having heard of her legendary skills with a shotgun, not only was I worried of her reaction to me questioning her, but I also immediately wondered why my grandmother wanted me to bring my BB rifle.

With rifle in hand, I slowly walked the twenty-five or so yards from where we lived to my grandmother's front porch. Knowing I was a good shot with my BB rifle, my grandmother pointed out the spot in the top left corner of the roof of her house, where a blackbird was about to build its nest. She told me to sit in the brown chair on the left side of her front porch and when the blackbird returned, to shoot him.

I sat down in the brown chair. My grandmother told me I needed to sit still and concentrate to see the blackbird when it returned to the corner of the roof. With my rifle in hand, I sat in the soft brown chair on the left side of the front porch as instructed and waited for the blackbird to come. I waited and waited, but no blackbird came anywhere in sight. I sat so long with my eyes fixed on the left top corner of the roof that I grew tired and restless. To not lose my focus, I stood up and walked from the left side of the front porch where I was sitting to the right side, careful to keep my gaze in the flight path the blackbird would have to take to land at the top left corner of the roof.

After walking from the left side of the front porch to the right side for about ten minutes, I finally sat back down in the brown chair and continued to look for the blackbird. With no bird after nearly thirty-two minutes of waiting and watching, I could no longer keep my eyes gazing up at the corner of the roof, so I drifted off into a trance, looking straight ahead. I had not been in this state a full minute before my grandmother came out the front door. She just stood and looked at me; I was now in such a deep trance that I was not even aware of her presence.

While standing there staring at me as I stared straight ahead, motionless, my grandmother looked up and saw the blackbird with twigs in its beak land in a tall oak tree some ten yards to the left of the rooftop of her house. After staying perched on a limb high in the oak tree for about a minute and a half, the blackbird swooped down and went into the left corner of her roof. Without saying one word to me, who was in one of my famous trances (daydreaming, my grandmother called it) and was oblivious to what was going on around me, my grandmother turned and walked back into the house, mumbling, "He is just like I said—lazy."

I grew up in a small town named Garner, North Carolina. The town is some three miles southeast of the state capital of Raleigh. The town has a rich history. Established in 1847, it was said that Garner—like most towns that had settlers before the Revolutionary War and before the Civil War—got its start with the coming of the railroad. Despite my scant knowledge of how Garner was established, even as a youngster, to me Garner possessed all of the ingredients and elements suited for dreaming. The town was small. Some people sat on their front porches and chatted among family and friends. Others walking along the unpaved roads would come up and join in the conversations. This was especially true on Sunday afternoon following church service and Sunday dinner. And for me, this routine lent itself to wondering and having visions about distant places and imagining what it would be like somewhere other than Garner.

I grew up in a close-knit community. The community was named after my grandmother—the Sally Whitaker Subdivision. My grandmother was born in Garner in 1899. She was an amazing lady who lived to be one month short of ninety-three years old and passed away in November 1992.

Her husband—my grandfather Willie, who she married in 1924—passed away in 1941 after a lengthy illness, but not before making sure family land matters were in order.

My grandmother was left a forty-two-year-old widow with seven small children to raise—including Verna, my mother— ranging in age from seventeen years old to four years old. Back then the neighborhood came to the aid of a widow with that many small children. My grandmother appreciated the support from the community, church family, and others, but as a person with a year of college training, she set out to develop a strategy for how the family would survive and even thrive following the death of her husband. Her success in moving the family forward originated from land ownership matters that had been taken care of prior to the death of her husband. As I grew into adulthood, I came to understand the tremendous importance of the planning my grandparents must have done early in their marriage.

Her Christian faith was exceptionally strong and served as the taproot in her life and also in the lives of her seven children. She understood the connection between secular education and the strong Christian training she had received from her own parents and grandparents. A good example of her strategy was the process she used in the education of her seven children. My grandmother believed men were ordained by God to be leaders in every aspect of their lives, especially leading their households, which would result in the total family being anchored by a solid spiritual foundation.

This philosophy resulted in three of her four sons being college educated, while education for her three girls ended with a high school diploma. Her oldest son was not able to go to college because he was needed to help work the fields after the death of his father and was soon after called off to the war. Like his three sisters, his education ended when he graduated from high school. Christian development linked to a formal education was the bedrock of the Whitaker family, and I was keenly aware of the importance of this two-way connection.

My great-grandfather William was a land baron. His land was situated in St. Mary's Township. It is reported that he once owned nearly half of

Garner. The origin of his land is not known for sure, but some family members said that land ownership by Blacks was passed down from their slave ancestors and it was possible his land was acquired this way. This landownership and transfer process helped his family maintain a high degree of stability, and the land transfer continues today to the current generations.

With education as the second tenet of the family legacy and segregation being so entrenched at the time, my great-grandfather donated some of his land to start the first school for Blacks in the town of Garner. The school was located on the current site of the Rand Street Christian Church. Great-Grandpa William was one of the first Black persons on Garner's original board of education, which oversaw the administration of the town's first Black school. In 1917, the school moved from the Rand Street United Church of Christ (new name today) to a larger building on Old Garner Road. The school was renamed the Garner Colored School in the 1940s.

Not having known my great-grandfather, I always imagined that his land ownership probably produced a measure of respect between him and white citizens. Although there was separation among the races in every way imaginable—from housing to eating facilities, schools and churches, drinking fountains, and more—there appeared to be a type of unique relationship among the Blacks and whites of Garner, and I was proud of the role my great-grandfather played in bringing about some measure of respect for him and others in the community who followed him.

As I pondered the relationship between my great-grandfather and the town elders and the subsequent seemingly benign relationship among the whites and Blacks of Garner, I wondered if my sense of well-being was because of my ignorance of the real state of affairs that simmered just beneath the surface in the town during the period I came along. After all, Blacks made only a fraction of what whites earned for the same jobs or because they could not or did not aspire to work beyond these jobs because of race and class. My sense of well-being also may have resulted from the fact that Blacks "stayed in their place," which seemed permanently fixed upon the landscape of my life. Still further, it could be associated with the proximity of Garner to the capital city of Raleigh, which provided a fair amount of comingling and respect, at least in the service workplace, so I became increasingly aware as I matured into young adulthood and continued to process what was going on around me.

I had a sense of fairly peaceful race relations in Garner during my youth. I remember that although life was hard at times, its simplicity brought with it a kind of peace, a sense that life for me would be filled with great possibilities. This peace manifested itself in a sweet pleasantness that came from the routines of my life that were as constant as the seasons that came and went and the stars that shined brilliantly in the expansive night sky.

And as a testament to how a town and its surrounding areas can expand their horizons to improve the lives of all residents, the tobacco fields where I worked as a young boy now hold upscale high-end housing developments. Blacks who were once subject to discriminatory housing practices now live harmoniously—it appears—with whites and many other races and cultures on the same land once owned and farmed by their white ancestors. Also, today, at the entrance to Garner, small shopping malls and business centers line the highways where cotton gins and farming trade stations once stood. Unless you miss the "Welcome to Garner" sign, one cannot tell where Garner ends and the capital city of Raleigh begins.

As the years progressed, my grandmother passed her land down to her children. Today, some family members live on the same plot of land. Family members who established their lives in other parts of the country have sold their land to other members of the family. My uncle Thurman sold his land to me, and I have now transferred it to my son. This landownership and transfer process has helped my family maintain the land of their ancestors down through the years. Growing up in this kind of environment—on the land that bears my grandmother's name—provided a high degree of constancy in my life and suggested to me that things were good. So to me, Garner possessed all the ingredients and elements suited for dreaming—or daydreaming, as my grandmother would say—about future possibilities.

As I grew older, I continued to dream. My dreams became one with my soul, embedded deeply into my being. They matured as I matured and somehow seemed to be symbolic of the changes that so completely enveloped my community in the 1950s and early '60s. This maturity allowed me to notice what seemed to be the limitless commitment of Garner's Black educators, ministers, civic leaders, and families to make a better life for the Black community despite what seemed to be insurmountable challenges at the time. I came to understand that through all these challenges, my community somehow managed to survive and even thrive, at least from my vantage point.

In examining how the community met these challenges, I found a common thread of strong faith and hard work that wove itself relentlessly throughout the lives of those in the community, which became a tightly woven quilt of love for and pride in family and community that made Garner such a special place, a place that continues to bring great joy to me even now, more than sixty years later. I would later learn that this beginning help set the foundation for me.

By the way, I did not kill the blackbird that day but did shoot and kill it days later before nesting season started.

Chapter 2

MY MOTHER AND FATHER

I N 1927, MY GRANDMOTHER BORE her third child and named her Verna Mae; she in time would become my mother. The name Verna is of Latin origin, but the actual genesis of my mother's name among her family ancestry is unknown. The meaning of Verna is "spring green." In the mid-1800s, Verna and its different forms were very popular first names for girls. The popularity of Verna as a first name for baby girls declined in the 1960s. Looking back at the ancestral line, with no other known family member named Verna, it is probably accurate that the name was given because of its popularity at the time and her personality.

Verna Mae, as she was called, was a brilliant child, and she lived her entire life in Garner. Early in life, Verna Mae demonstrated a love for life. Well into her adult years, she often revisited her childhood days affectionately, speaking of her love of reading, singing, and dreaming. I was keenly aware of my mother's desire for something greater out of life. Her love of reading and singing was balanced with a passion for education. My mother was a star student throughout her twelve-year high school education. She often said to me and my six siblings that if she had come along in more modern times, her education definitely would not have stopped with a high school diploma but would have ended with a college degree and possibly even more advanced training.

An avid reader from her youth, my mother extended her training and enlarged her world through books. As a child, she traveled to distant lands by dreaming of the many places she discovered through the various books she read. Her love of reading continued into her adult life. Her favorite reading material was the hardbound version of *Reader's Digest*, which became required reading for me and my siblings. It was in *Reader's Digest* where my mother found what would become her favorite poem, "The House by the Side of the Road" by Sam Walter Foss.

Because of the time in which she lived and the family's condition (her father passing away when she was fourteen years of age, leaving her mother to raise seven children by herself), my mother was never afforded the opportunity to soar as her spirit often yearned. As a result, her real world was often lived through books as she dreamed and imagined how things could be for her if the situation and circumstances had been different. After being baptized at the age of twelve and becoming a member of Wake Baptist Grove Church, my mother turned to the Bible to find answers to the many questions she had about her circumstances and station in life.

From reading the Bible, my mother found that God honors faithfulness and that one should trust in Him completely. She also discovered that with God, nothing was impossible, but without Him, nothing was possible. One biblical message she believed with all her heart was that for a person to become all that God had destined for him or her, that person has to leave the place of his or her birth and establish him- or herself elsewhere. One of her favorite sayings to her children was, "If Jesus was not accepted in Nazareth, the place of His birth, followers of Jesus will not be fully accepted in their hometowns, and that includes you if you are a follower of Him." I heard this saying often and really took it to heart as I examined and processed the lives of others who had been fixed in Garner since birth.

My mother truly believed that individuals could make greater contributions in an environment different from their birthplace. She often said that in a person's birthplace, he or she never overcomes being seen as "the carpenter's son." She would say to me that here in Garner, people would always see me as Louis Shepard's boy, and that designation would carry with it many things that had nothing to do with me and never take into consideration my own personhood of who God made me to be and the steps I was taking to get to what He had already destined me to become. She would say the distance from one's birthplace could be as simple as changing church locations and relocating to a new neighborhood across town.

Following graduation from Garner Colored High School in 1944, Verna Mae was hired in the laundry department of the Raleigh Linen Company. While there, she met Louis Lee Shepard, a fellow employee who had recently moved to the area from Snow Hill, North Carolina. This young man had come to Raleigh to live with his aunt in search of a life different from the one he and his family were living in Snow Hill as sharecroppers on a white owner's land. After a short courtship, in 1946 Louis and Verna Mae were united in marriage.

With the birth of me, my two brothers, and my four sisters in the late 1940s through the mid-1960s, it was as if my mother's spirit leaped with excitement as she realized her dreams might be possible through her children. This excitement stemmed from three sources: (1) her belief that God honors one's faith, (2) her love of family, and (3) the encouragement she received from reading about distant places. The combination of these three elements led her to know with certainty that the segregated world, especially the Jim Crow South she had known all her life, was not going to last forever.

My mother took her faith in God, her love of family, and her vision of the world to come and gave her total self to her seven children. She made sure her family lived like some of the white families whose houses she worked in. For example, the breakfast table was always set beautifully. Six-ounce juice glasses with designs of oranges and strawberries on the outside and a matching pitcher sat in the center of the table. We ate breakfast and dinner on beautifully designed high-quality chinaware. Her belief was that just because society felt Blacks were second-class didn't mean we had to live that way. Another of her favorite sayings was, "Reach for the stars."

My mother saw to it that we all joined Wake Baptist Grove Church at an early age. While growing up in Garner, the Shepard children could always be found attending Sunday school and Sunday worship service, participating in church plays, singing in junior and young adult choirs, and attending Vacation Bible School (VBS) in the summer. There was no compromising in the area of church attendance and participation. She believed that Christian training coupled with a solid secular education allowed us to become aware of God's plan for our lives and that this combination would guide us in fulfilling our destinies in life.

My mother made sure her children were exposed to music, art, and reading. And during our early years, she made provisions so all of us had the opportunity to become accomplished musicians, singers, and piano players in church and in school. The exposure was for all the children, no exception. That is why as a third grader I was taking piano lessons in 1956 along with my oldest sister, Ruth Ann. As the only boy taking piano lessons in Garner, it became difficult to continue because the other boys made fun of me. I pleaded with my mother to please remove this burden from me. This was one of the few times my mother reversed a decision she believed was for her child's benefit, even if the child could not understand it at the time, as I would later come to realize.

When I stopped taking lessons, my mother started my sister Maxine, the next child under me, on the piano. Ruth and Maxine became excellent piano players. If I had continued, my mother's first three children would all be talented piano players today instead of just two, an unheard of thing for Black children in the small town of Garner. Today when I am in the presence of outstanding male piano players, I regret that my mother did not stick to her wisdom and force me to stay with the piano, recognizing that the mocking would one day pass away.

Ruth Ann has passed away, but at her homegoing service her contributions to society through her playing of the piano were acknowledged from across the United States. In time, I came to understand that my statement of being the only boy in Garner taking piano lessons was not correct. It would have been correct if I had said I did not want to be the only Black boy taking piano lessons in Garner at the time. There were probably some white boys my age taking piano lessons.

My mother believed that possession of a good name and work ethic from an early age were important. Therefore, she required that chores around the house be performed by the boys and the girls. We were required to complete our tasks on time and correctly. If not, there were consequences metered out in return. Time out? There was no such thing. Never heard of it.

Work ethic and being responsible was not gender based. My mother stressed with deep passion that "Negro boys must be taught how to be leaders so they will be equipped to carry out God's divine plan that men lead." Therefore, my brothers and I could not, as she put it, "run wild all over Garner" while young girls were home being trained to become respectable young ladies and future wives. She would say to us, "Where will the respectable young ladies find good young men to marry if boys are not trained on how to be respectful and responsible?"

About the good name, my mother would say one's name was important because it would be what the outside world would use to characterize who that person was. She constantly stressed that an individual's name, good or bad, went ahead of him or her and that the first impression on paper was all others would ever see and know about you. My mother stressed many things, but the foundational teachings she emphasized all the time can be summed up this way:

» Keep God at the center of your life.

» Get a good education (which meant a college degree).

» Leave your hometown if you want to grow and expand your horizon, because if Jesus had to leave His own hometown for people to take Him seriously, you will certainly have to leave yours, no matter what.

» Learn to do an honest day's work if you want to succeed.

» Be willing to lend a helping hand to others.

» Never think you are better than anyone else.

» First, give God 10 percent. Afterward, save something for yourself for a rainy day.

» Always maintain integrity and good moral character. Your name is all you got, and it always arrives ahead of you.

» Wherever you go in this world, do not ever stray from your spiritual upbringing.

I pushed back on some of my mother's concepts early in life, but we all took her teachings to heart. As time passed, I came to appreciate my mother's teachings as values that were richer than anything life had to offer, including fine gold. Today, I cannot deny that all seven of her children became well-educated adults who were and are productive citizens. Maxine, Janice, Roger, and I received college degrees, with Maxine receiving an advanced degree at the master's level and me continuing on to earn a PhD. My fifth sibling, Celestine, continued her education after high school by graduating from a two-year business college. After completion of her sophomore year in college, our oldest sister, Ruth Ann, decided not to continue and instead married and started her family with her husband, making a career in the military.

My sisters Maxine and Janice are educators, and Celestine works in the administrative division of the city of Raleigh. My brother Roger, who wrote the foreword to this book, is a high-level executive in corporate America. Our younger brother, Anthony, said the family was too "educationally oriented" and decided to take a different path than college after high school, but today he is still a productive individual working for the city of Raleigh. More is worth noting about my baby brother, Anthony.

Following his taking of the Scholastic Aptitude Test (SAT), he received a personal letter from the dean of engineering at Boston University, who had identified him from a minority student database, offering him the opportunity to continue his postsecondary education in Boston. I really wanted him to enter the engineering program at Boston, but Anthony wanted to do something different. With his brilliant mind, he recognized that the world he would operate in was quite different than the one his older siblings had grown up in.

Technology was on the horizon. He talked often about the capability of the computer and his belief that it would be at the center of his world in the future, more so than it had been in the past. Anthony became fascinated with the inner workings of the computer. To gain a better understanding of this technology, he enrolled in a computer and electronic design training program at the local community college. After two years of training, he entered Saint Augustine's College but quickly realized that path was not for him. He continued to self-teach himself things he found interesting.

My brother's talent can best be described by recounting a real-life situation. When big screen televisions arrived, I purchased a sixty-inch home theater system equipped with surround sound. Company experts delivered and set up the system, including the five-speaker surround sound unit.

Initially, the system appeared to work okay. I say okay because for what the system cost it didn't deliver the high-quality sound in my basement that had been advertised in the showroom. Even accounting for environmental differences between the showroom and my basement, the system never produced the quality of sound it should have. There were many hours of discussion with company experts about the lack of system sound quality. The company sent who they referred to as their top surround sound system expert to fix the problem. After a year of going back and forth with the company expert on what now had become a time-consuming issue, I decided to call Anthony and ask if he could come to my home in Maryland and take a look at my surround sound system. I explained that since experts had installed and checked out the system, it could be my ears and nothing was wrong with it, but I believed the sound clarity was missing. In his customary laid-back voice, he consented to come and check it out.

Anthony had not seen or heard my new system and upon entering the basement and listening to it he immediately said, "Big bro, your sound system is definitely not as clear as it should be." I asked if the basement environment was causing the difference between the sound I heard in the theater room at the store and what I was now hearing. He said, "No. There is a malfunction in the system." I asked how he knew the system was not functioning properly without hooking up any test gauges. He said he had listened to so many speaker systems that he could tell from listening when wires were crossed.

After joining us for dinner, Anthony returned to the basement and began to strip my system down, pulling wires from everywhere. My sixty-inch surround sound television system sat in the middle of my basement floor with exposed wires running in every direction possible. When my wife got up enough nerve to venture into the basement, she quickly retreated back upstairs to ask me, "What in the world is Anthony doing?" I mumbled, "He's fixing the problem the company experts could not resolve."

I must admit that with it being late Saturday night and Anthony having to leave sometime on Sunday to return to Garner, I was just as concerned as my wife by what I saw in the basement, but I dared not say anything. I did get the courage to return to the basement to ask Anthony if he thought he would be finished by Sunday. He gave his patented response: "Don't worry, big bro. It will be just fine." With that, I said goodnight to Anthony, went upstairs, and climbed in bed. I wanted to remain with Anthony, but looking at the wires and speakers lying around disconnected in the middle of the floor was more than my mind could take.

When I arose the next morning, Anthony was in bed fast asleep. I nervously went to the basement to check the status of the project. To my complete surprise, everything was put back together with wires clamped and running more neatly than the way Anthony had found them. I sat down and marveled at the exterior of the job but had some doubt that the system would work any better than before. I nervously turned on the system and heard a melodious sound like I had never heard before in my basement and certainly on par with the sound I had heard in the theater showroom. All I could do was sit, listen, and think about the talent of my young brother upstairs in bed. I continued to listen until Anthony came into the basement.

With his patented smile and customary coolness, he asked, "How does it sound, big bro?" Shaking my head in disbelief, I told Anthony that the system was delivering the sound it had been designed to produce.

While I was thinking it could not get any better, Anthony reached over and handed me his hand-written instructions describing what he had done. As I sat looking at the notes of my brother's neatly written, clear, and concise description of his work, Anthony said I could use the information to retrace what he had done if the system malfunctioned again. He said the root source of the sound system problem was a crossing of wires and the need to add a booster to the system to increase power outage to the kind of surround sound speakers I was using.

After a hardy breakfast and chatting with family, my baby brother was in his truck and on the way back home to Garner. There is much more I could share about Anthony, especially his talent as a self-taught artist. With eighteen years difference in age between my baby brother and me, building and maintaining a strong bond was important to me, and it was my responsibility to make it happen. And happen it did.

At age forty, my mother was diagnosed with rheumatoid arthritis. From then on her health was up and down. My mother continued to run the house until the arthritis condition confined her to a wheelchair in 1997 at the age of seventy. On the decision of my father—who was an avid cook—Mom remained in the home and was cared for by the entire family, headed by her husband, my father. My mother passed away on April 20, 2001, at the age of seventy-four. Her legacy is firmly etched in my mind and in the lives of her children and grandchildren.

With my mother gone, the question in my mind was, *Will the family continue the tight bond that was our strength?* With the glue and central figure no longer around, the impact of Mother's influence was even more pronounced. My siblings and I have always been close. As each new child was born, he or she immediately bonded with the older siblings, or more appropriately, the older siblings quickly bonded with the new addition to the family. The bond was genuine and lasting.

I knew my mother had done her job well in holding the family together and was confident the siblings would remain close and not scatter and lose connection with each other. Less clear in my mind was the direction my father would take since he had not always been around during our formative years. With Mom, we were accustomed to family dinners and other family gatherings from which my father had been absent, mainly because he worked all the time.

My mind was put at ease as I got to know my father much better following the death of my mother. My mother's presence kept the home focused on spiritual, cultural, and educational happenings. My father was not into such goings-on in any significant way. He worked, and he fished. These two pursuits were my father's life. I did not know the joy fishing brought him until I began to spend more time with him after Mom was no longer around. Fishing moved my father into a trance. Pure ecstasy!

My father was born in 1929 in Snow Hill, North Carolina, a small rural town in the eastern part of the state. Snow Hill could be compared in many ways to Garner. The town was small, quiet, and not much activity outside of work and going to church. One major difference is that the town was more rural than Garner and comprised mainly of farmers.

My father was a short stocky man of few words. His education was limited to the third grade because of his family circumstances. His parents, Mandy and Will, my grandmother and grandfather, lived and worked the fields as sharecroppers all their lives. Visiting with Grandma Mandy and Grandpa Will introduced me to an entirely different world than the one I was accustomed to growing up in the Sally Whitaker subdivision. Corn and soybeans surrounded the house my father grew up in as a boy. In the summer when these two crops were at their peak, the acreage was so vast that it appeared the two extremes of the crops and skyline actually met way off in the distance from where I stood.

My father shared that when he was a young boy, he and his siblings worked in the fields from sunup to sundown. He spoke with resentment of not being able to attend school because of fieldwork for the white landowner. The other aspect of my father's home in Snow Hill that I did not enjoy when visiting was how pitch-black the surroundings were once the sun went down. Growing up in Garner near a major city like Raleigh with its many street lights, I did not feel comfortable in a place where the starry night sky was all the light I could see. As I grew into my teen years and into sports and other activities, my visits to my father's boyhood home diminished. The one thing I cherished was the fact that I knew my grandparents on my father's side and always felt their love when I visited with them.

My father was a hard worker who probably didn't make an annual salary that exceeded $6,000 by the time he and Mom had their first four children. That was the life of a Black man in America in the South during the 1940s and '50s. Dad was in the prime of his life when the segregated South was at its worst. With his meager earnings, Dad did the best he could to provide for the family, and he did a good job. Like any head of household who had little in terms of financial resources, the pressure got to Dad at times, and in those moments his relationship with Mom and the rest of the family left much to be desired. The cupboard was never bare, but when it was almost bare you could hear that something was bothering Dad. At those times, Dad would drift away from the family. As I grew, I understood and appreciated to a greater extent what I think Dad was feeling.

By 1964, before Blacks were allowed to vote in America, there were seven children, a wife, and only God knows who else internal and external to the family looking to Dad for financial and probably other forms of support. During this time, adults did not talk about other goings-on of this nature within earshot of children. However, it was common knowledge within the Black community that when one moved on to a better life, he or she was expected to reach back to provide assistance to other less fortunate family members. And moving from Snow Hill to an area close to the capital city marked "moving to a better life" for my father.

So Dad worked. In the late 1940s and early '50s, he worked as a painter and repairman from seven in the morning to five in the evening at the Capital Club Building in Raleigh. After leaving his first job, he would drop by the house to hurriedly scoff down a bite to eat and then briskly walk to the Wrought Iron Works Company in Garner. Work at the Wrought Iron Works Company was hard and long. Most times the rest of the family and I would not see Dad until the next morning when the cycle started all over again. Dad always worked second and third jobs in the evening that took him way into the night.

Dad came home one afternoon and announced that he would no longer be working in the evenings at the Wrought Iron Works Company. He announced that a lawyer working in the Capital Club Building had asked if he would be interested in doing janitorial work at the North Carolina Bar Association (NCBA) building after he got off of his regular day job. Dad was only sharing the information with us. He had already accepted the job and was not seeking family approval. From that point, it was not often that Dad came home after his regular job to eat; instead he would go straight from his job at the Capital Club Building to the NCBA building and do janitorial work until 9:00 or 10:00 p.m. and sometimes even later depending on the level of cleaning required.

One of the NCBA executives liked how Dad worked cleaning the building and asked if he was interested in doing light yard work consisting of grass cutting, gutter cleaning, snow and leaf removal, and trimming hedges. Without hesitation, Dad accepted this now third job. He could be found in the yard of the NCBA executive every Saturday from about 8:00 a.m. to 3:00 p.m. I joined my father and learned a lot about what it took to produce a beautiful yard. Dad kept the NCBA executive's yard beautiful, and I mean *beautiful*. He would stand back and look over his work with much pride and then look at me and say, "Now this is what a yard in a neighborhood like this should look like."

I also found myself looking at the magnificent homes in this northern section of Raleigh, with the tree-lined streets that were well lit and absent of holes in the pavement, contributing to the beauty of the entire neighborhood. Dad often observed me from a distance leaning on my rake, broom, shovel, or other work tool, dreaming rather than working. He knew me well enough to know that I was asking myself if it was possible for me to one day have a yard like the one we were working in and even having the audacity to dream of living in a neighborhood of such magnificence. From listening in on conversations I had with Mom, my father understood that I was a dreamer, but to him my dreams were simply fantasies.

It was clear to me at times from Dad's response that to him my dreams could not be real because of the predicament in which he found himself. Sometimes Dad wanted to discourage me from such thoughts in fear that I would end up with nothing but broken dreams and heartache. In my father's mind, such change seemed so far out of reach that he could not construct a mental profile of what such life would look like. He only worked in such places. This was a place in which he could never live. The world my father knew consisted of hard work from sunup to sundown, leaving no time for dreaming—or daydreaming, as my grandmother called it. So in my father's mind, it was simply time to go back to work.

Even as quietly as it was kept, Dad had dreams because Mom moved him in that direction. For instance, Mom always dreamed of having her own home. Since marriage she and Dad had rented someone else's house. My mother dreamed of better, especially since she had grown up in a house owned by her mother and father. My mother wanted a family home built on the land her mother had given to her children. Without the financial resources between her and my father, at times imagining home ownership seemed too far out of reach for even Mom. So the thought of home ownership drifted in and out of her mind, sometimes leaving quicker than it entered. So it seemed to me that nothing changed in the Shepard-rented household, and Dad continued to work from sunup to sundown, which appeared to be his lot in life. I was right. There were two things in my early life that seemed to remain constant, and they were my mother's hope and my father's work.

My mother believed it was possible and could never let go of her dream of someday living in her own home. She thought of the beautiful furniture neatly placed in the white folks' homes she had worked in. Windows open, ushering in fresh outdoor air. As someone who loved to prepare family meals, she thought about moving around in her own kitchen. She spoke often of having a large picture window in the living room of her home. She would tell us that we would have our own rooms in our own home. She said the house would have inside running water and an inside bathroom, because where they rented none of these luxuries existed.

As the family gathered as often as possible on Saturday night for our standard Saturday meal of pork and beans and hot dogs, my mother continually repeated her home ownership dream during the prayer before the meal as the family ate and watched Lawrence Welk on our first television. By the way, Lawrence Welk was my mother's favorite television program. In addition to beautiful listening music, she thought the music was educational because of the explanation Mr. Welk gave about its source. If there was talking when Mr. Welk was explaining the music, Mom would snap, "Be quiet. You might learn something." I hated *The Lawrence Welk Show* and did not like the music his band played or the songs his guests sung as he counted off, "A one and a two." It made me laugh on the inside, but I would not dare show any such outward disrespect in Mom's presence.

I said my father was a quiet man. Mom did most of the talking, leading, and just about everything else when we were small children, and my father, well, he was there but more in the background. Mom pushed the spiritual and educational side of life. Dad was simply there, not offering much by way of sharing and going the extra mile on behalf of other family members. That was left to Mom. Sometimes I wondered if the role my mother took on—the role of being responsible for so much of the family matters—contributed to the onset of her illness at such a young age, with her failure of health at forty and death at seventy-four. I think about it a lot and advise young married women not to take on the role of the man, no matter what. After Mom passed away, my father participated in more family events than in the earlier years when Mom was alive and the children were young.

The more I hung out with my dad, the more I got to know him. It was not until many years later that I learned that as a young man, my father desired to make his wife's dream of homeownership come true. The many evenings after the sun had gone down and no one was around, my father would go to the vacant lot given to Mom by my grandmother and try to imagine what it would take to build a family home himself. It was revealed years later, the many nights he sat and pondered the idea of building a home. Long periods of time passed between my father's wondering if what he had in his mind about building a home was possible in light of his working from sunup to sundown. The two extremes sometimes left him in a quandary. I was told that a year after I was born in 1947, my father found the will to take the first steps toward building a family home.

He started by bringing home a square-end shovel that had been given to him and digging by hand what would become the foundation of the Shepard home. Shortly after receiving the shovel, my father was given a few cinderblocks taken from a building that had been demolished near the Capital Club Building where he worked. With no transportation, my father asked a friend if he would go with him to pick up the unwanted blocks and if he could use his truck to transport the load from Raleigh to Garner. Without hesitation, the friend said yes. People helped each other a lot during that time.

The first corner of the house that would eventually be erected went up at the hands of my father. With the corner standing and not falling, my father's confidence in being able to build a complete house increased. The electrical wiring that would be required was the one thing that caused early doubt in his mind, causing him to question if he had the ability to build a complete house. Recognizing this fear, he returned to the building next to the Capital Club Building that was being demolished to see if he could find someone to answer some questions he had about electrical wiring. An electrician working at the site explained what was involved in wiring a new house but quickly added that a person needed to be a licensed electrician to handle the job. My father explained that he understood that a properly trained electrician would be needed. No more free cinderblocks were available, and neither was there any more free advice on how to handle the electrical work needed in building a house.

Many years passed before any new building materials came to the site where the lone cinderblock corner stood. The unfinished corner stood so long without anything happening that some people in the neighborhood would ask, "What is Louis Shepard trying to do? Does he ever plan to add anything to that corner?" The talk grew louder and insensitive as the years passed with the corner just standing in what seemed to be a desolate field.

In the spring of 1952, my mother's dream of home ownership returned but this time with a flame that could not be extinguished. It was an early Saturday morning, and I was in the front, sweeping the yard of the rented house where we lived with a handmade straw broom, when I heard a bone-chilling scream coming from the kitchen. I froze in fear and could not move. Realizing the scream was getting louder, I dropped the broom and dashed into the house to find my mother jumping and screaming uncontrollably. She was so out of control that it frightened me before I saw her and even more when I did. Her movement was the same movement of people I had seen in church when, as they say, the Spirit had overtaken them, leading them to scream, jump, and shout for joy.

My mother was carrying on like the church ladies, but the difference was she was in her kitchen and not in church. My mom was looking out the window through the dense patch of trees, whose leaves were just beginning to show that spring was near, and saw a delivery truck bringing a load of cinderblocks to what would become her homesite. My mom was so ecstatic that morning that it spilled over into me. I joined her in the jumping and shouting because the delivery was made possible through sacrifices by all of us. Dad could now continue the building process and add to the longstanding lone corner.

I was always big and tall for my age, so at six years old, I helped my father build the family house that still stands today at 221 1/2 Haywood Street. Here was a man with only a third-grade education, and he built the house I grew up in. My father built it from ground up, dug the foundation with a square-end shovel, laid every cinderblock, and hammered every nail on the inside. From the time the first corner was erected in 1948, the house took ten years to complete. It was one of the great unacknowledged accomplishments witnessed in the Black, and probably the white, community of Garner at that time. The greatest and most heroic thing my father did was tackle and complete all of the electrical wiring in the house once he built it.

As a young boy, I watched my father take a light bulb, two pieces of electrical wire, and a battery and work with this three-piece system until he taught himself how electricity flowed. Once he got the bulb to come on, he looked at me, smiled, and said, "I know how to make it work now." Once he figured out how to make the light come on, his next challenge was trying to understand how to place a switch in the system to turn the lights on and off. I remember the joy that came over my father's face and flowed over into me when he figured out how to place a switch in the system and still have the lights turn on and off. All of this was done after a sunup-to-sundown day at work and second and sometimes third jobs on the side. Although building the house resulted in him working into the wee hours of the morning, Dad never complained because this was for him and his family.

Armed with the information received years earlier from the man when the building in Raleigh was demolished and his self-taught education in electricity, my father wired our entire house. The job was completed, and it was then time for the Garner electrical inspectors to inspect the wiring work. I was by my father's side when the electrical inspector showed up. To appreciate the magnitude of the situation, a frame of reference is essential: (1) the year was 1957; (2) it was a small Southern town in North Carolina; (3) a Black man with a third grade education and no electrical license or union pedigree stood in the presence of a white licensed electrician, claiming he wired a 1,200 square foot house; and (4) a young Black son stood by his father's side, taking it all in.

With no acknowledgement of my father and a statement of "I am here to inspect the electrical wiring done on this house," the white inspector walked past both my father and me and went into every room, nook, and cranny of the house. He climbed into the attic space and stayed up there for quite a while. My father looked at me and said he hoped the man had not gotten electrocuted up there. The inspector finally came down out of the attic and moved to the outside. He proceeded to squeeze through the door to the crawl space that led under the house.

Then the inspector came out from under the house, went to the junction box outside, and opened it to inspect all the wires my father had connected inside the box. When the junction box door was closed, there was only a single wire extending from the top of the junction box that was now ready for connection to the electric pole that stood some seventeen to twenty yards away. During this time, the inspector did not acknowledge my father as he moved from one area of the house to another.

The man apparently completed his inspection of the wiring job and finally spoke to my father about the wiring of the house. He asked, "Who did the electrical work on this house?"

Father answered, "I did."

He asked, "Do you have an electrician's license?"

Father answered, "No."

"Then who did this work?" the inspector asked.

Father repeated the same answer: "I did."

The inspector asked, "Who helped you with the electrical work?"

Father answered, "A man some years ago showed me how to do it, and I did it."

Without any more questions, the man said he had to leave but would return in a little while. He got in his truck and left. My father was confused by the man's actions. My father said the man was strange. The man returned in about thirty minutes. This time he had a second white man with him. The second man did not inspect the house but came to my father and asked the same type of questions the first one had asked. My father's answers were the same.

The second man identified himself as the supervisor of electrical inspectors for the town of Garner. He told my father that they could not find anything wrong with his wiring but could not understand how he did the wiring work without a license. He then said that they were going to pass his electrical job but warned my father not to do any electrical work again in Garner without a license. My father acknowledged that he would not ever do it again. The man said someone from his department would come out to connect the wire from the junction box to the electrical pole within the next few days and that it would not be necessary for anyone to be at the house. My father thanked the men, and they left.

After the two men were out of sight and out of hearing distance, Dad said to me, "They do not know it, but I never intend to build another house in Garner or anywhere else and won't need to do it again." I laughed but recognized that I was in the presence of a really smart, industrious man. The family moved into the new house built by my father in the summer of 1958.

It seemed to me that my father was old when he was building the family home, but he was only thirty years old when we moved in. He continued to work two and three jobs after building our house. In 1966, at the age of thirty-eight, my father got the job of his life. That year he became a campus police officer at Saint Augustine's College in Raleigh. He rose to the rank of sergeant and remained at Saint Augustine's for thirty years before retiring in 1996.

My father passed away December 21, 2010. I learned much from my father—how to build things from scratch, how to do "modest" electrical work around my home, and how to make major and minor repairs to my cars. When working on small electrical jobs, I think of the courage it took my father to teach himself to deal with something as delicate, complicated, and dangerous as electricity without formal classes or training.

I remember the feeling that came over me in 1984 when I rebuilt the Rochester Carburetor on my 1975 Corvette Stingray. After the job was done and I cranked the car, a big smile spread across my face because I had performed the job the same way I had seen my father do his car work when I was a young boy in Garner. I was excited when I was able to show my father my successful major auto mechanic job. I learned as a young boy that my father was a brilliant Black man who had not been afforded opportunities to show forth his brilliance through what was considered normal, traditional, and acceptable channels. To me, my father was the most brilliant man I ever met.

At his homegoing service, we spoke with pride about how our father had moved in to fill the gap left by our mother when it came to family gatherings and demonstrating genuine love for each other. In his later years, my father cherished the time he spent with his children, grandchildren, and great-grandchildren. My father was a true jack-of-all-trades who mastered them all: painter, carpenter, locksmith, electrician, plumber, auto mechanic, kitchen chef, master griller, and landscaper, just to name a few. My father applied all his talents to the benefit of family, friends, and the entire community. No hour was too late. No job was too large or small. And no job was too hard, because he was truly a "friend to man." As time passed, it became clear that Mom had married a man who showed all the characteristics of being a "friend to man" as described in the words of her favorite poem, "House by the Side of the Road."

Chapter 3

ENTERING A CHANGING WORLD

I WAS BORN ON CHRISTMAS DAY in 1947, a time when hospitals in the South did not admit Blacks. Most Black children were born in the home, aided by a midwife. However, this was not the case for my siblings and me. My older siblings and I were born in Saint Agnes Hospital for Negroes on the campus of Saint Augustine's College in Raleigh, North Carolina. Saint Agnes opened in 1909. It was the only major hospital for Blacks in between Washington, DC, and New Orleans, Louisiana. Saint Agnes closed in 1961 when the large and more technologically sophisticated Wake Medical Center opened. My baby brother, Anthony, the seventh child and last addition to the family, was born at Wake Medical Center in 1964.

The year I was born, the nation and the world were undergoing rapid changes in many areas. The year saw some significant inventions that would impact how people lived for many years to come. The transistor and the mobile phone were both invented during this time. Other noteworthy events that year included the breaking of the sound barrier, the first instant camera, the turning on of the first digital computer, the invention of the Frisbee, Goodrich inventing the tubeless tire, and Jackie Robinson taking the field for the Brooklyn Dodgers, becoming the first Black to play in Major League Baseball.

When I was born in 1947 there was much unrest in the world. For example, the Cold War started the same year and lasted close to four decades as the world's two superpowers at that time, the United States and Russia, worried about world dominance and the politics and influence of each other on the other countries in the world. Also that year India and Israel gained their independence and Palestine was divided into two separate states, one Arab and one Jewish.

This unrest among nations gave rise to buildup of military forces in the United States and elsewhere. For example, in 1947 military-related institutions were set up, including the Central Intelligence Agency (CIA) and Atomic Energy Commission (AEC); the AK-47 weapon went into production in the Soviet Union; and the US Air Force, the US Navy, and the War Department were brought together to form the massive US Department of Defense (DOD). The US Department of Army was also created during this time with the passing of the National Security Act (NSA). These were some of the world-changing events that took place the year I was born.

With the world events on center stage, there was no distraction from my birthday. My day was very special on many levels, the most notable of which is that it was the day that many people across the entire world celebrated the birth of Jesus Christ, and not just in 1947; the day was celebrated before 1947 and will continue to be celebrated every year into the future. It was also special because the area had not seen a measurable amount of snow on Christmas day for years. But on my birthday the Garner area received 0.4 inches of snow.

There was an event under discussion the year I was born that turned out to be a major incident in my neighborhood but would not play itself out until years later. The discussions eventually led to the Federal Aid Highway Act of 1956 for the construction of thousands of miles of the Interstate Highway System over a ten-year period making it the largest public works project in American history at that time.

These advances in the nation's highway system became extremely important to the neighborhood where I was born. US Highway 70 would be one of the first major highways affected by these new laws. As can be derived from its number, US Highway 70 is a major east–west highway of the southern and southwestern United States. It runs for 2,385 miles from eastern North Carolina and ends in east-central Arizona. It formerly ran from coast to coast, with the current eastern terminus near the Atlantic Ocean in North Carolina and a former Western terminus through Los Angeles to the Pacific Ocean in California. Before the completion of the interstate system, US Highway 70 was sometimes referred to as the "Broadway of America" because of its status as one of the main east–west thoroughfares in the nation.

The highway passed right through Garner, about 135 yards from my grandmother's front property line. With my grandmother's property line so close to the US 70 highway, it was thought that by 1950 her land and the land of others in the community would suffer at the expense of progress.

The stretch of the highway running through Garner had two lanes. In 1956, because of an increase in road traffic, local and state officials responsible for highway expansion chose to widen US 70 from two lanes to four lanes. This decision set off a firestorm of disputes and disagreements in Garner because the proposed expansion route would have the greatest negative impact in the Black community.

The highway was already close to the property of Black homeowners, and widening the highway on either the right or left side would negatively impact property owned by Blacks. The widening process would require the enforcing of easement laws, which gave the Federal Highway Authority the right to take property in the name of progress in places where property posed an obstruction to the highway widening process.

The clash about the widening of US 70 reached a high-water mark when it was deemed necessary to remove and destroy the home of Mr. Sherwood Haywood. Under no circumstances was Mr. Haywood going to allow, as he put it, "thievery of his land" to take place. Mr. Haywood was a well-respected citizen of Garner. He owned a grocery store and served much of the Black population in town. There was another store owned by a Black man named Mr. Jeffries. His store did not stock as large of selection and variety of food, medicine, auto products, and many other needed items as Mr. Haywood's store. The building Mr. Jeffries operated his store out of doubled as the home for him and his wife. Mr. Jeffries lived directly across US 70 from Mr. Haywood. If the policy had been to widen US 70 on Mr. Jeffries's side, his home would have been taken and destroyed, not Mr. Haywood's home.

Mr. Haywood's home was on the east side of US 70, not more than fifteen yards from the center of the highway. His property was in the direct path of the widening project, but he vowed he would not move and the federal government had no right to force him to find somewhere else to live. The matter went to court, where it was settled that the federal government had the right to the land based on state, county, and local laws and regulations governing highway expansion. The final outcome was that Mr. Haywood gave in and followed through on finding land to build a new home.

The federal government provided some funds for Mr. Haywood to build a new home, but it was never known if the federal resources covered the entire cost. His new home reflected the progress that had been made in that it was equipped with more modern appliances and had a more appealing outside appearance. Mr. Haywood secured land far away, where there would be no remote possibility or threat of road expansion, new business development, or any other form of progress advancement in the area during his lifetime.

The expansion experience left a lasting negative mark on Mr. Haywood. He vowed that an event like that would never happen to him again, and it did not. I was aware of the impact the widening of US 70 had because every time I visited Mr. Haywood's store he would mention how wrong it was for the federal government to take his land and for the town of Garner to stand by and let it happen.

I always believed that what Mr. Haywood experienced with the highway project changed him in many ways. While I was on a routine trip to his store to pick up some food items for my mom, Mr. Haywood expressed his desire to build a gym in Garner where big strong boys like me could train to become boxers. Mr. Haywood told me that as the biggest

Negro boy in Garner for my age, I should consider becoming a boxer and that if I did he would back me financially.

I was so impressed with the idea that I ran home to tell my mom of Mr. Haywood's offer. I did not think through all the consequences of what it meant to be a boxer. I never considered that this offer might have resulted from Mr. Haywood's personal experience, which had led him to become dissatisfied and bitter with his beloved hometown of Garner. I did not think of such things. My initial thought was that becoming a boxer would put me on the newly expanded US Highway 70 and take me to places I was now only able to dream about.

It did not take long for me to snap back to reality when I shared the information with my mother. Verna Shepard would have no part of such nonsense and did not want such wild thoughts put into her son's mind. She was furious that Mr. Haywood would equate size with a violent sport like boxing and not balance his comment by saying how I could use my mind, since it was known in Garner that I was an outstanding student. Mom quickly shifted the discussion about boxing to the food items I held in my hands. She asked me why I had returned from the store with her food in my hands and not in a grocery bag.

Without giving me time to respond, my mother said, "Do not ever walk through Garner with my grocery exposed in your hands for every Tom, Dick, and Harry to know what I am preparing for dinner." My mom's next visit to the store allowed her the opportunity to convey to Mr. Haywood the importance of encouraging me and every child who came into his store to become all they could be by using their minds to the fullest.

The boxing offer was gone from the recesses of my mind, but I never forgot what happened to Mr. Haywood's home. I thought about it often when I sat on my grandmother's front porch, watching the big eighteen-wheeler trucks traveling up and down the now widened US 70 highway that had changed Mr. Haywood's life forever.

As time passed, I thought less of the widening of US 70 and more about the possible end routes for the many trucks and cars traveling on it. With the highway running from the Atlantic Ocean to California, I longed to one day travel along US 70 to see what the other parts of the country were like. While I did not know the final destination of the many trucks and cars traveling the route, I felt certain they were all going somewhere far different than where I sat tucked away in the small town of Garner.

Those days sitting on my grandmother's front porch I dreamed of only one thing—having a career as a long-distance truck driver so I could drive on US Highway 70. I dreamed that one day I would drive one of those big trucks to its final destination, where fascinating things would be happening—a place that was simultaneously imaginary and real but most of all far beyond my grandmother's porch. It was not a matter of being unhappy with where I was during that time in my life. After all, life was good. Eating good country meals, working small jobs to earn money, playing sports during and after school until dusk, learning my part in school and church plays, doing my schoolwork, playing in the school band—all of this excited me. Yet I just wanted to see what was happening wherever those big trucks were going and going in such a *hurry*.

I was not sure when I came to understand fully the unwavering commitment my mother and father had in creating and maintaining a strong sense of family. I was certain that the foundation established by my parents was the source of my siblings' and my belief that we could achieve whatever we desired in life. But I also was aware that success was tightly bound to the siblings holding on to each other no matter what, holding on to our morals, remembering that work was our friend, staying true to our faith in God, and reaching back to help others. I remember my mother as one who was in constant search of becoming better today than she was yesterday and how she transferred this belief to her children. If there was a school field trip but no money for us to participate, Mom would borrow the money from wherever she could find it so we would not, as she put it, "miss out on such a golden opportunity." She stressed going on field trips because she believed reading really came alive once you could see it. She constantly said to us, "Reach for the stars, and while you may not make it, you will end up higher than where you started from."

Although I was aware that I could not eat in the same restaurants as whites or go to the same schools, theatres, or restrooms, I was perfectly content in a world filled with happiness, the warmth of a family's love, and great possibilities. I would later come to understand and appreciate the struggle when my father talked about the reason he had worked multiple jobs for most of his life and why my mom took the extra care she did so that we would grow into thriving, successful adults.

My mother was a focused individual with wisdom I needed to heed if I was to navigate the changing world into which I had been born. The combination of my mother's insight of a new world coming—and in many ways already evolving on the horizon—and my father's steady work ethic, coupled with their collective sacrifices, provided a solid foundation upon which me and my siblings stood.

Chapter 4

BETWEEN THE ROWS

My PARENTS WERE NOT RICH. This meant I had to work for my spending money. I was expected to also share some of my money to help with family financial needs. Dad served as a good role model, so it was easy for me to embrace work. I was good at finding work in rare establishments in the Jim Crow South and was disciplined at striking a balance between working and studying.

My first paid work outside the home was in the summer of 1957 at the age of ten in tobacco fields. My job was to unload and spread the tobacco brought in from the fields out on a long table. The empty sled cart would then return to the field where it would again be filled and brought to the barn for me to empty. This cycle was repeated for a twelve-hour workday that included a one-hour lunch break and two fifteen-minute breaks, one in the morning and a second in the evening. Since my work in the tobacco field was limited to Monday through Friday during the summer months of June, July, and August, this provided an opportunity for me to find other jobs.

In the spring of 1960, when I was in the seventh grade, I was hired as a shoeshine boy in the town's white barbershop. I worked after school from five in the evening to nine at night. My mother allowed me to work the job provided I kept up my grades and handled my other chores around the house first. The routine went like this: I arrived home from school around 3:30 p.m., completed chores and homework, and then walked to the barbershop some two miles away, arriving by 5:00 p.m. This worked fine for me. I got my chores done around the house, and my schoolwork did not suffer at all.

Being paid once a week by the barber and receiving tips every night, coupled with being a conservative spender as taught by my mother, allowed me to open my first Passbook Savings Account at First Citizens Bank in Garner. Life was good. In fact, life was very good. My account began to grow, but I was not acting on what Mom had said sometime earlier: "The first 10 percent of what you earn must go to the Lord in the form of tithes." This sounded fine when there was nothing to give but became hard to do once money was in my hand. I put an offering in the plate when I attended Sunday school, but during those times I did not check to make sure the amount I gave was at least 10 percent of the total amount I had earned.

After three weeks on the job, I arrived at the barbershop just like before, but this time I was called over to the barbershop owner before starting my setup routine. The owner informed me that his customers were complaining about getting home and finding black shoe polish on their white socks. The barber made it clear that the problem needed to be corrected right away. I promised to be more careful and pay closer attention when applying polish to the customer's shoes to avoid the black polish ending up on their white socks.

Less than a week after the first meeting, the owner summoned me again, quoting the same message about the customer's complaints. This time there was no discussion or warning. I was fired and told not to start work that day and not to return to the barbershop as a shoeshine boy. Devastated at being fired, I walked down the short hallway to the small room where I took off my apron, returned the shoe polishes to their bins, and sat for a moment to collect my thoughts about what had just happened to me. I had never been fired or removed from anything.

In that moment sitting in the small room I replayed the words the boss had said to me: "Do not start to work today, and do not return to my barbershop as the shoeshine boy." The echo of the words from the lips of my boss was impactful on me. The event caused me to reflect on many things about life in general, the town I was growing up in, and more specifically my personal life in a deeper and more profound way. *Enough of this*, I thought. It was time to go.

I rose from the stool I was sitting on and packed my belongings. I left the small room and closed the door as I took one last look over my right shoulder. While seated in the small room, I had first thought to simply exit the barbershop without saying a word. But I knew this was not the right thing to do. My mother taught me to always at least try to take the high road in life and never let circumstances and situations make me bitter.

Reflecting on my training and upbringing, I walked over to where the barbershop owner was standing and thanked him for the weeks I had served as shoeshine boy. I extended my right hand while asking the owner if there was anything else I could do around the barbershop other than shining shoes. He said no. But it was clear from the expression on his face that he was surprised by my response to his firing me. All the same, the owner grasped my extended right hand with his right hand as I looked him in the eye and then turned and took the few steps that led out the front door.

As I exited through the front door of the barbershop, I smiled inside about how this final exchange had made me feel. I was not bitter about the firing experience. The anxiety I had anticipated never surfaced. Some nervousness did rise as I walked home. I did not want to mention to my family, especially my mom, that I had been fired. I began to wonder if I would have been fired if I had done what Mom had said years earlier about the first 10 percent of what a person makes being given back to the Lord in the form of tithes. I thought some more about this matter. My mom expressed her delight in how I had handled the situation and encouraged me to not be defeated by the incident but to use it as a learning experience.

After the firing, I did not look for any work after school for the remainder of that school year. I immediately started searching for work once school closed, but initially I had no success. After a few rejections, I again was successful in finding rare work in the Jim Crow South in the summer of 1960. This time it was yard work and other chores around the home of the librarian for the town of Garner. This job was on Saturday mornings from 7:00 a.m. until noon.

The librarian's home was located on the side of the railroad track where the white folks lived, away from the side I lived on. I was already quite skilled at designing and beautifying yards based on working with my father years earlier in the yards of the beautiful homes of the white folks who lived in North Raleigh. Now I would get to put my own touch on the librarian's yard, and I was anxious to see what my father would say about my own yard work.

I was disappointed on Saturday mornings when rain kept me from my yard work. It had to be raining hard for me not to be crawling in that yard. Light showers and drizzles could not keep me away. I brought along a dry shirt to change into since Mom always stressed not letting wet clothes dry on me in the tobacco fields. I knew the same thing applied to working in the librarian's yard. I could easily find an isolated place off in the distance to quickly change out of a wet shirt.

Like I had seen my father do, I stood back and looked at the beauty of the cut, the trim of the hedges, the weeding of the flowerbeds, and I dreamed of the time I would stand back and admire my own yard. I would quickly snap back to the job at hand, for there was not much time for standing idle and dreaming; there was work to be done. I arrived at my yard job on time every Saturday. In fact, I always arrived at least ten minutes early and sat in a chair that had been placed out back for me as I waited for the 7:00 a.m. start time.

I always liked to be on time for everything I was involved in, and while sitting and waiting to start, I began to think about all the times I was late because of others. I wondered why my family always found ourselves standing in the vestibule of the church, waiting to enter because we did not arrive before the 11:00 a.m. worship service started. While standing in the vestibule, I would think that it was just as easy for us to have arrived at 10:55 a.m. as it was to arrive late at 11:05 a.m. Arriving five minutes after the start time caused us not to be able to enter the church until after completion of the call to worship, the invocation, and choral response. Depending on the length of the three activities, it could be anywhere from 11:13 a.m. to 11:17 a.m. or longer before we and all the other laggards were allowed to enter the service.

To make matters worse, late arrivers had to search for somewhere to sit. Of course, worshippers who understood the importance of being on time had taken the good seats and now looked at me and the other stragglers with contempt, as if to say, "These folks can never be on time." Eyes would now be firmly fixed on me and the other latecomers who had little regard for printed times as they roamed around the church, looking for somewhere to sit.

I had great difficulty understanding why a pattern of five minutes after the start time could not be reversed to five minutes before things got started. It was because of these kinds of experiences early in life that I made up my mind that when I became responsible for myself I would not be known as one who could not be on time.

With my yard work, my Passbook Savings Account continued to grow, but not my thoughts about the 10 percent that should be given back to the Lord in the form of tithes. One Saturday morning I stood to stretch from the backbreaking work I was doing in the yard when I noticed my mother and father pulling into the driveway of the librarian's home. This had never happened before, so I was sure something terrible had happened. My heart felt as though it had stopped beating. My mother stepped from the passenger side of the car with a somewhat sorrowful look on her face. I could not move or find words to say. So I just stood and looked into my mother's eyes.

My mother came closer and then spoke: "Robert, I hate to come to you with this, but we do not have the money to pay this month's mortgage on the house and need your help to make the mortgage payment." By this time, tears trickled from her eyes.

I quickly rushed to my mother and said, "Mom, not a problem, and do not worry. It will be all right. How much do you all need?"

She said sixty-seven dollars. I said I had more than that in my savings account, so it wouldn't be a problem. My mother constantly reassured me that the money would be returned to me as quickly as possible. I repeated to my mother that it would be all right and for her not to worry.

I knocked on the back door, and when the librarian answered, I told her I had to go with my parents to transact some business but would return to my job shortly. Knowing my trustworthiness and timeliness, the librarian said it would be fine. I jumped in the backseat of the old family car with my parents and drove off. We arrived at the bank before it closed at noon on Saturday. I went into the bank and withdrew sixty-seven dollars and gave it to my mother. We drove back, and they dropped me off to continue where I left off.

As I continued my work, I felt good about being able to help my parents. After the car pulled away, I propped myself up on the rake and thought about the impact of what I had just done. I wondered if God viewed my action in the same way my mother talked about the 10 percent tithe going to the church. Again, I did not take the time to calculate if the $67 was 10 percent of the total in my saving account. I was certain that in this transaction I had given enough to be in good standing with God for many years to come because at the time of withdrawing the sixty-seven dollars, I had nowhere near $670 in my Passbook Savings Account.

The one thing I often reflected on about that episode was how my father didn't say a word during the whole transaction, from the time they arrived in the driveway of the librarian and my mother stepped from the front seat of the car to the time I exited the backseat of the car after handing the money over. His actions really made me mad. But as I grew older and reflected on this event some years later, I came to the conclusion that my father had been so hurt by the need to come to his young son that he had been unable to find words adequate enough to express his true feelings at the moment.

The matter was never mentioned again, and—you guessed it—the funds were never returned to me. Not a problem. It was much more than 10 percent. I continued doing yard work for the librarian for another summer. With the success of my finding work in all-white establishments like the barbershop and the yard at the home of the white librarian, as well as thinking about my father's jobs at the Capital Club building, which was filled with white lawyers, his work at the NCBA building, and both of us working in the yard at the home of the white lawyer in North Raleigh, I concluded that work in the Jim Crow South was important in moving forward, regardless of one's race.

To test my theory about work even in the segregated South, one evening when walking home from my yard job at the home of the librarian, I stopped off at the drugstore and asked the pharmacist who owned the store, if he needed some work done. Before stopping, I was aware that Blacks could only purchase items and get prescriptions filled at the drugstore. No sitting at the counter. If Blacks purchased items like milkshakes, ice cream cones, popsicles, or sodas, they would give their orders and then step back from the counter, where whites were sitting on stools, and wait to be handed their items. I knew all of this before stopping in to ask about work. I also knew I had a theory about work being an equalizer that needed to be tested.

Looking somewhat dazed and shocked by the exchange, Mr. Lloyd— the owner—said, "Yes, I do need a good worker."

My immediate response was I was a good worker and needed a job. Mr. Lloyd described what he needed done. He said he needed a worker who could arrive at the drugstore just before the 9:00 p.m. closing time to scrub the floor and take out the trash. He said the job would require about an hour and no more than an hour and a half.

Already knowing that I had learned to balance my study with work, I asked, "When do you want me to start?"

Mr. Lloyd said Monday night.

I said, "I will be here." I left the drugstore more confident than before that work was everybody's friend but equally sure that everybody was not aware of this fact.

All the while I worked on Saturday mornings in the yard of the librarian and weeknights in the drugstore owned by Mr. Lloyd, I continued to work Monday through Friday during the summer days in the tobacco fields from sunup to sunset.

During the time I worked in the fields, tobacco was an important crop in North Carolina and was one of the main pillars of the state's agricultural industry. Tobacco was planted about two feet apart, which provided enough space between the plants for an average size human body to move between the rows. There was one sled row for every four normal tobacco rows. The sled row was an open space of two to two and a half yards, just wide enough for a mule-pulled sled to pass through the field to collect the armpit loads of tobacco leaves from the primers (the men who removed the tobacco from the stalk).

This planting sequence was repeated until the acreage was totally filled with tobacco plants in accordance with federal guidelines set by the Agriculture Stabilization and Conservation Service (ASCS). These plants could grow to six feet in height. Farmers who could afford it usually used four to six primers. A total of four primers would all prime on just one side of the sled row. Six primers would allow four on one side of the sled row and two on the other side. Obviously, six primers removed tobacco from the field quicker than four.

Farmers with more financial means than others replaced mules with tractors. The planting sequence did not change to accommodate tractors. It was still the four normal tobacco rows to every sled row or tractor row. When harvesting in the tobacco fields became mechanized from mule to tractor, the manual backbreaking labor I experienced did not change. The federal quota on acreage of tobacco allowed for farmers, as set by the ASCS, did not make any difference on the farm I worked on; the tobacco always joined the skyline out in the far distance. The tobacco rows to sled or tractor row opening ratio was critical because a farmer's income was based on how many pounds of tobacco were sold, and the poundage was a direct result of the number of acres grown.

When I first started working in tobacco, my job was removing the tobacco from the sled and placing it on a long table-looking platform. Taking tobacco from the sled and placing it on the long bench was just one phase of the hard work I experienced in the tobacco fields in North Carolina. Since I lived in an area that no longer grew and harvested tobacco, I had to be picked up and transported some twenty miles from home to the tobacco field worksite.

While I entered the tobacco process in June as a field-worker at harvesttime, I was in school when tobacco farmers began preparing their seedbeds in January for the next planting season, followed by planting and transplanting from the seedbeds to the fields in April. As the plant grew in the field, it produced a flower in the topmost part of the plant. If the flower was not removed, the plant would grow seven to eight feet tall.

I often entered the tobacco fieldwork right after school closed in mid-May. My job would be to snap the flowers from the top of the plants. This process was called topping the "suckers" from the plant. This was not backbreaking work because I could stand while breaking the suckers from the tobacco plant. Removing the suckers became a strain on my back when they emerged in the lower part of the plant. I did not have problems with this part of working in tobacco but had great difficulty with the pest called the tobacco worm. My stomach turned every time I came upon one of the big, green, nasty looking irritants. Each plant had to be inspected from top to bottom for the worms, which had to be removed by hand. If the unsightly nuisances were not removed, they would eat the leaves of the green plants until the plant was totally destroyed. So this part of the work had to be done.

The backbreaking work in tobacco that often started in mid-June was hard for me. After spending one summer at the barn, unloading the sled coming in from the field, I moved to the field to become a primer (i.e., a field-worker) that removed the tobacco from the plant stalk and placed it in the sled. In the tobacco world, this was viewed as a step up, a promotion—if there was anything of the sort in this kind of work. It just meant that I would move from being paid fifty cents an hour to seventy-five cents. In 1960, this was a raise in pay and a big one at that. I would no longer start working at the barn area around the women but instead would go straight to the field where the men removed the leaves from the tobacco plants.

The farmer's truck came to pick me up in the morning around 5:00 a.m.; it was still dark. Now that I was working in the field, every morning before I climbed onto the back of the pickup truck, my mom would always check to see if I had a change of clothing. She knew that by 7:00 a.m., I would be completely wet from the early morning dew coming off the plants. One of her greatest fears about me working in tobacco fields was that one day I would develop health problems, especially arthritis in my bones stemming from wearing wet clothes that were allowed to dry on me.

She would send a change of clothes and say, "Robert, please do not forget to change out of those wet clothes around 8:00 a.m. after the dew has evaporated from the tobacco plant. Please, son, do not keep those wet clothes on and let them dry on you."

I assured my mother that I would make the change of clothes. After the farmer's pickup truck pulled away from the house with my two sisters and me, I would wave bye to Mom as the truck faded into the darkness. My sisters rode in the cab part of the truck, with one seated next to the passenger door and the other seated in the middle next to the white driver. The other male workers and I all rode in the truck bed, wrapped in blankets and other coverings to keep the cool morning wind from us.

When the truck arrived at the work site around 5:30 a.m., the other field-workers and I were dropped at the edge of the first row of what seemed to be a field of tobacco the size of the whole town of Garner. The pickup truck continued with my sisters still seated in the cab for another half mile or so until it arrived at the first barn, which would soon be filled with tobacco from the field. The farmer dropped off my sisters at the barn, where later they would be steady at work preparing the tobacco arriving from the field for packing in the curing barn. He then parked the pickup truck and walked the short distance from the barn to the stable where the mules were housed. Workers were seated in front of the barn.

After he had a brief discussion with the workers, one of the workers would enter the stable and soon exit with what sometimes looked like an old worn-out mule that was about ready to collapse and die. The worker would hitch the old mule to a long, narrow sled made of boards mounted on wooden planks shaped like snow skis for easy gliding over the sandy white soil in the field. The sides of the sled were filled in with coarse cloth from old fertilizer or feed bags. This material was tacked to the wooden frame. Using cloth material as part of the construction rather than wood kept down the weight of the sled, resulting in less work for the old mule. I was fascinated by the resourcefulness of tobacco farmers.

I often thought about how with limited resources farmers made good use of the things around them. They threw away very little. Items were reused until they had no further use, and the creativity of the farmer stuck with me. I wondered if many of the ingenious products could be patented, but I knew the farmer I worked for never had time to explore such thoughts.

The workers would then guide the old sled-pulling mule from the stable and barn area to the field where the other workers and I had been dropped off. Tobacco harvesting began once the mule arrived with the sled in the field around 6:00 a.m.

The job I did for many years—called working in tobacco—began at this point. The process for removing tobacco leaves from the stalk was called priming, and the other field-workers and I were called primers. The priming process began by first removing the leaves from the bottom of the plant. These leaves were called sand lugs. I was always frightened of removing the sand lugs because at times we would come upon snakes using the big leaves to shade themselves from the hot sun.

I remember one occasion when a worker shot past me, running perpendicular to the rows because he had come upon a snake that did not want to move. The fast-running worker who was trying to escape the snake's path crushed tobacco plants to the ground as he shot through the field. When asked if he knew that he should have run parallel to the rows to keep from damaging the tobacco, the worker said he did not have time to stop and think about running parallel to the rows or anything else. His first instinct was to get as far from the snake as possible, and since the snake was gliding along the ground parallel to the tobacco rows, it seemed important that he run in a perpendicular direction away from the snake. This scene was repeated many times in tobacco fields when sand lugs were removed from the base of the plants.

We broke the large dew-saturated leaves from the base of the plant and swung them swiftly under our armpits. A right-handed worker primed with his right hand and swung the leaves under his left armpit. The reverse would be true for a left-handed primer. Primers continued this process down the row, stripping the leaves from their stalks and tucking the leaves under their armpits in a steady cadence, moving from one stalk to the next without straightening their backs until an "armpit load" was obtained. An armpit load was defined by how many leaves could be carried before a worker had to straighten up, walk to the sled, and unload the leaves, careful not to damage them. All primers were instructed not to fill their arms to the point of crushing the leaves. Damaged leaves were lost money because tobacco buyers were interested only in leaves that were intact.

Armpit load size differed based on the arm length of the worker. I was tall and had long arms, so I could gather a large armpit load before having to straighten up and walk over to unload the leaves into the sled. I always noticed when I straightened up that I was much farther down the row than the other shorter-armed primers. But long arms were a disadvantage because the mule-pulled sled was always farther back down the row where the shorter-armed workers were. This meant that once I straightened up with my arm loaded with leaves, I had to walk back down the row to unload into the sled.

Moving farther down the row in front of the other field hands also meant that I got to the end of my row first before the other workers arrived. To keep from being idle, I usually started priming on their rows back in the direction of the other primers. In other words, if a count had been made of the number of leaves unloaded in the sled per day per worker, I would have ranked at the top.

I never forgot to change out of my wet clothing around 8:00 a.m. My mother's clothes-changing system worked well, and I never complained of aches like some of the workers did as they stood wet from head to toe with the clothes drying on them around midday. The clothes-changing system did not work at all on days when afternoon or morning showers rolled in after the change. The only option when this happened was to allow my changed clothes, which were now wet, to dry on me like the other workers. My mother never asked me if wet clothes had dried on me because I usually arrived home wearing dry clothes with the wet ones I had changed in the morning in a bag.

The process of priming the plant to remove additional leaves continued about every two to three weeks after removal of the sand lugs, depending on the weather conditions and the ripening of the leaves. The priming process continued until the last leaves were removed from the top, leaving the plant stalk stripped and bare in the field. Removing these last leaves was easier than when taking off the lower ones. I did not have to bend to remove the last leaves. I could now stand straight up.

At the close of the field-harvesting season, I knew another summer of backbreaking fieldwork was over. In an average summer from mid-June to late August, I returned to the fields to prime tobacco anywhere from thirty to fifty times depending on the acreage of tobacco planted by the owner. The farmer knew tobacco leaves were ready to be removed from the plant stalk when the leaves were light yellow in color.

If long hard rains came and saturated the field to the point that the mule could not pull the sled through the rows, priming would not take place. I also did not have to worry about going to the fields to prime if the rain was continuous enough to turn the light yellow tobacco leaves green again. These conditions slowed the harvesting process, which sometimes kept some students from reporting to school on opening day. Those students were required to delay attending school for a week and sometimes longer in order to get the tobacco from the field to the barn and onto the market. This never impacted me; Verna Shepard did not allow any of us to miss school because of working in the fields. I was always present the first day of the new school year and did not work in the tobacco fields beyond August.

Mules pulled the sled full of tobacco leaves out of the fields and up to the barns, which could be quite a distance depending on the size of the acreage of tobacco. Once at the barn, mules with attached sleds stood like statues. The mule remained in this frozen state until every leaf was taken from the sled by the workers doing the unloading job I used to do. The tobacco leaves were placed onto a long bench scaffold, where my sisters and other young workers, usually females, stood ready to start their part of the job. After the last leaf was removed from the sled and placed on the bench, the mule would take off as if he knew the sled was empty and return back to the field.

Depending on the number of mules the farmer owned, there was never a time at the barn when a sled full of tobacco was not being unloaded. The timing sequence was such that as an empty sled left the barn on the way back to the field, a full sled was at the barn being emptied. It seemed the farmer had to perform several calculations to keep tobacco field-workers from sitting idle, waiting for empty sleds to return from the barn to the field. If the distance from the barn to the field was very long, there were times the primers had to wait for the mule to return.

Sled drivers walked alongside the mules as opposed to riding the mules. Going back and forth in the hot sun made it important to conserve the mules' energy and to give them lots of water. The driver of the sled was also responsible for maintaining a steady supply of water for the primers.

The precision of the tobacco harvesting process fascinated me because there was usually never a chance to rest for any length of time. Tobacco harvesting work required speed, and speed is what I had. The process could slow down slightly with increased distance between the field we were working in and the barn because as the day progressed the mules walked more slowly on their return from the barn to the field. I was glad when this happened because it gave me and the other field-workers time to catch our breath. It was clear to me that if the farmer could have afforded at least three mules in the rotation at certain times rather than his customary two mules, there would have been no downtime for me and the others. When the owner replaced his two mules with two tractors the next year, the speed increased significantly, leaving little room for standing idle.

The process that started at 6:00 a.m. would continue until sundown. During the time I worked in tobacco, North Carolina observed Eastern Standard Time (EST) year-round. There was no Eastern Daylight Time (EDT) back then. So my work in the field stopped at 5:00 p.m. each day to reserve time for me and at least three other primers to move to the barn, where we would become hangers. It was important that the start and stopping times in the field were well timed because under EST, it was totally dark in August by 7:30 p.m. The tobacco had to be in the barn before dark.

At the end of the day, the other primers and I all climbed on the sled taking the last load of tobacco from the field to the barn. Sitting on the edge of the sled, looking off into the distance, I wondered when the time would come when I would not have to work all day in the heat of the day. The backbreaking fieldwork caused me to do a lot of dreaming. As the old mule slowly pulled the sled along the dirt path toward the barn, I looked at the tobacco field. Everywhere I looked tobacco was touching the evening skyline, just as it had in the twilight of the early morning. I looked around the edge of the sled at my fellow field-workers, wondering where they all would one day end up.

My eyes would soon give way to shutting, but the sled would hit a bump, causing my eyes to fly open. Once again I would gaze off into the distance, dreaming of a better time that was to surely come. My mind would go to my grandmother's front porch. Again I closed my eyes in exhaustion, and in my mind I would see the big eighteen-wheeler trucks traveling up and down US Highway 70 with their final destinations just about to come into view. Then the sled would hit another bump, jarring my eyes back open. Even though while in my trance the last scene was interrupted, I was sure that all of the trucks and cars traveling west on US Highway 70 ended up in California, a place I dreamed of going to see one day.

Seeing the field-workers on the sled was the only indication to the workers at the barn that their day was about to end. The remaining task for the day was to move the strung tobacco on the sticks from the outside racks to the curing barn. The process is referred to as hanging. Once the sled arrived at the barn, the field-workers were responsible for hanging the tobacco. I did not like hanging and certainly not hanging from the bottom position. Therefore, if I was asked to be one of the hangers, I would quickly move to the top tier of the curing barn. Hanging tobacco was hard. And after ten hours of hot, backbreaking fieldwork, I found hanging tobacco in curing barns to be worse than being in the field.

Two hangers climbed into the barn, one scaling to the top and the second remaining on a lower tier. The hangers balanced themselves by placing one foot on a tier on the left and the other on a tier on the right. The heat inside the barn was almost unbearable. The farm owner recognized the possible danger of workers inside the barn suffering from heat exhaustion. So before barns were filled, there was a change of hangers.

The first hangers would fill half the barn, and fresh hangers would be brought in to hang the remaining half. The top of the barns reached in excess of 100 degrees.

Even with the extreme temperatures in the top of the barn, I still preferred being the hanger in the top tier. My sisters were concerned for my health and asked why I always chose the top. I explained to them that when I stood on the bottom tier, sweat would fall on me from the body of the top hanger. I preferred the excessive heat in the top to body sweat from the hanger in the top and sand from the tobacco leaves dropping onto my face and eyes.

A production type assembly line was needed to get the hanging job done. The line included all workers, including my sisters. If the barn was close to the outside racks holding the strung tobacco sticks, a long line was not needed. This would free my sisters from having to take part in the tobacco hanging process. Strung tobacco sticks were heavy. I told my sisters before leaving home in the morning for the tobacco field to keep back and not rush to get in the strung tobacco stick handling line. They followed my instructions and were not usually involved in the hanging process.

In addition to being hard, hanging presented a hazard for me and the other hangers. I always checked the soles of my shoes to make sure they were dry enough to stand on and grip the narrow planks that were about forty-eight inches wide. Slipping and falling was a concern for hangers.

The heavy sticks were passed along the line through the entry door to the barn. The last person in the handing line was inside the barn, standing directly under the hanger on the lowest tier. The stick strung with heavy tobacco was lifted up to the hanger, who was bent over and reaching down between his legs to grab hold of the stick. The hanger on the lower tier received the stick, straightened up, and passed it to the hanger directly above him in the top of the barn. The hanger in the top of the barn bent down, reached between his legs, and took the stick from the lower hanger. The top hanger straightened up and placed the stick on the top rack in the barn.

I checked the hung sticks several times before reaching for the next stick to make sure the stick was secure on the tier and would not fall on my head, causing me to fall and bringing possible great physical harm to me and the hanger below me. This process continued until all strung tobacco on the racks outside the barn was put on the tiers inside the barn, keeping enough space between the hung sticks so heat could move through the sticks to dry out the tobacco leaves. After the last stick of strung tobacco was hung, I climbed down from the tiers and exited through the door of the barn. The day's work was finally over.

Reaching this point meant workers would now be paid for their day's labor. I always found the pay process interesting. All Black workers would stand somewhere around but never too close to the farm owner, who would be holding a large bundle of money in his hand. I would take a seat on the scaffold that had been full of tobacco leaves a couple of hours earlier. Once a worker's name was called, he or she would briskly go over to the farm owner to receive his or her pay. This was the sight every day of the tobacco-harvesting season. I knew very little about federal and state income taxes, but I knew nothing was taken from my pay.

Payout went something like this for each worker. In my case, the farm owner would say, "Robert, for the ten hours you worked today at seventy-five cents an hour, here's seven dollars and fifty cents." I would take my money and return to my seat on the scaffold. I thought of how hard it would be for me to find work that netted that kind of money and paid out at the end of the day's work without anything taken out. Seated on the scaffold, I looked at the seven dollars and fifty cents in my hand and thought about the hard labor I had put in both in the field and in the hot barn. I looked at my pay again and said in my mind, *My time between these rows in the heat of the day will not last forever because a better time is coming for Robert Louis Shepard.*

As I climbed my bone-tired body into the back of the tobacco farmer's pickup truck to return home to Garner, I would snap back to reality and think of how my seven dollars and fifty cents was going to help my Passbook Savings Account continue to grow after I had given a portion of it to my mother to help with housing expenses. I would practically fall down onto the metal floor in the bed of the farmer's pickup truck as my sisters climbed into the cab beside the white farm owner.

Before the truck moved, I often looked around and wondered if the truck had ever traveled west on US Highway 70. If it had gone west, I wondered where it had gone. I never asked the tobacco farmer about the travels of his pickup or about the farmer's own travels away from his fields of tobacco that touched the skyline. As the pickup truck pulled away from the barn and onto the road leading back to Garner, my mind first replayed the day's work between the rows with the hot sun beating down on my back. My attention then turned to my mom's teachings about tithing and how the 10 percent of what you earned should first go to God. These thoughts flooded my mind as I rode home, nodding in the twilight of the evening while seated on the metal floor in the bed of the tobacco farmer's pickup truck.

Chapter 5

CHOOSING NOT TO STAY

I N SEPTEMBER 1953, I ENTERED first grade as a five-year-old. Back then, North Carolina laws governing the age of a child entering first grade stated that the child's sixth birthday must fall on or before October 1. If the child's sixth birthday did not fall within this timeframe, the child could not start school until the following year. With my sixth birthday not coming until December 1953, by law I should not have entered first grade until September 1954.

People often asked my mom how I was able to enter first grade at age five when my sixth birthday was not until December, well past the October 1 cutoff date. Her patented response was, "Robert was such a big boy and growing so fast that I had to think of a creative way to get him in school and not wait until the following year. He would be so much larger than the other children." Then she would simply say that she wrote on the entry form that I was six when I was really five. When I graduated, I found that the creative method my mom used to get me into the first grade at age five was not telling the truth. Therefore, I graduated from high school in the class of 1965. If North Carolina laws had been followed, I would have graduated in the class of 1966.

In the early years of my education I did not work as hard as required to excel to the fullest of my potential. A lackadaisical attitude toward school continued for the first six years of my education. Then a life-changing event happened to me in the fall of 1959 at the start of my seventh grade year. I started the year in an overcrowded classroom taught by the school's band teacher. It was well known that the band teacher was not rigorous and was labeled as "easy." In fact, everyone was aware that he preferred to conduct band activities rather than teach academic classes.

None of the students in the overcrowded class were aware that a new teacher had been hired during the summer to relieve the crowded conditions. On the first day of class, the principal simply walked into the room with the new teacher and introduced her as Mrs. Crenshaw. His statement went like this: "Today, this overcrowded class is going to be split into two classes. Half of you will remain with Mr. Judkins, and the other half will go to Mrs. Crenshaw's class. Let me see a show of hands of those who will volunteer to go to Mrs. Crenshaw's class. If there are not enough volunteers to split the class, I will randomly pick students to go."

I was sitting with my longtime friends. The back of the room where we were seated was known as the location where nonserious students sat. The group I was sitting with stopped talking, and we all turned around and looked toward the front of the room, where the principal was standing. None of us knew anything about this new teacher Mrs. Crenshaw. Never heard her name. Never seen the lady before. My buddies and I did know that remaining in Mr. Judkins's class meant a pretty relaxed seventh grade year, academically speaking.

Without warning and without a word being spoken, my hand went up as one of the students volunteering to go to the new teacher's class. My hand simply rose high into the air. It went up and stayed up as if I had no control over what I was doing. I heard the guys around me saying, "Shep, man, what are you doing? Do you know this new teacher? Why do you have your hand up? You do not even know this new teacher, but in Judkins's class, you know we won't have much to do this year. Man, put your hand down."

As they encouraged me to lower my hand, my hand remained high in the air. I remember it like it was yesterday. With my hand high in the air, my mind thought of how this meant leaving the guys I had been in class with since first grade and choosing not to remain with them but instead to go to the classroom of a new teacher I did not know. As I sat with my hand still raised high in the air, signifying my choice not to stay with my buddies but to go to the new teacher's class, I felt sadness. With my hand still high in the air, I began to think about how long I had shared space in the same classroom with all of the guys seated around me, and now I was about to leave them behind. I began to question myself, *Why is my hand up in the air?* I really began to question whether this was the right thing to do. I was so torn about choosing to leave my buddies that for a brief moment I thought about putting my hand down.

While I was in the midst of being preoccupied in a dreamlike state about the uncertainty of leaving and of the commitment I was making, the principal suddenly called for me to get in line with the other students who had volunteered to go to Mrs. Crenshaw's class. I settled in my mind that this would not be a lasting separation. I was not sure how often I would see my old buddies during the day since recess was no longer a part of the school day for seventh graders.

As I lowered my hand and rose from my chair while still mulling over my decision, I briefly stood beside my desk and looked around at my buddies as they sat with confused expressions on their faces. Then I walked over to the left side of the room where the line had formed and got in line with the students leaving to go to the new class. As I walked toward the line, I glanced over my right shoulder to again look at my buddies, who all were staring at me in disbelief.

Before the students in the line left Mr. Judkins's classroom to report to Mrs. Crenshaw's classroom, the principal once again explained what was happening. He said his intent was to improve the education delivery and academic performance of all students from first grade to the twelfth. With both Mr. Judkins and Mrs. Crenshaw listening, the principal said that reducing class size was the first way he was starting the improvement process.

He continued his speech by saying, "Starting this year, teacher performance also will be evaluated more closely than it has been in the past."

Standing in line, I thought about the discussion my buddies and I had just minutes earlier about not having much to do academically if we remained in Mr. Judkins's seventh grade class. Based on the principal's comments about evaluating teacher performance, I concluded that my buddies were going to be required to work harder than they thought in Mr. Judkins's class because Mr. Judkins would need to work harder than he had in the past to demonstrate his competence as a teacher. While standing in line, waiting to go to my new class, I was satisfied that my choosing to move on was the right decision.

It was in Mrs. Crenshaw's seventh grade class that my life forever changed academically. I developed a love for education like I had never known before, and this love never left me.

Mrs. Crenshaw laid down the law once we arrived at her classroom. She said, "If any of you are not ready to work hard and do not want to learn, then get up right now and report to the principal's office." She said this with a force that let me know she meant every word.

During my first six years of school, I had never heard any teacher say that. I was shocked. I began to think that the buddies I left behind had been right and that I had made a big mistake by choosing to leave Mr. Judkins's class. After listening to Mrs. Crenshaw, I could hear my buddies' chant in my mind, *Put your hand down and stay with us.*

From the seventh grade on, I excelled for the remainder of my schooling. I was elected president of my eighth, ninth, tenth, and eleventh grade classes. My senior year, I campaigned for another classmate from Mrs. Crenshaw's seventh grade class to be president of the senior class. I was busy on other things while excelling in my academics. There was simply a lot going on in my life after the transformation I experienced during my seventh grade year. My seventh grade year was exceptional academically speaking. I continued to raise my hand to participate in class activities. I really surprised myself when I raised my hand to volunteer to go to the blackboard to work out difficult mathematics problems. I had never done that before entering Mrs. Crenshaw's class. I wondered how much of this change was because of my maturing or the love for learning I was introduced to in Mrs. Crenshaw's class. I could not tell which it was. Whatever the reason, I continued to excel academically.

I was excelling in another area as well—basketball. I was steadily growing taller. Grownups in the neighborhood said, "Robert is growing like a weed." That was an expression often used when children shot up in height. And boy did I shoot up. By the end of the seventh grade I was already approaching six feet in height. I was growing so fast that some of my classmates made me feel uncomfortable by always asking when was I going to stop growing. My weight was proportionate to my height, so I looked physically fit. Outside the classroom I put my height and weight to good use by becoming a very good basketball player.

During the summer after my seventh grade school year, before my tobacco fieldwork started, I could be found playing basketball at various outdoor courts in Garner. With a good grip on my academics and a solid handle on balancing work and school, I was not aware of it, but I was acquiring the ability to become a multidimensional individual—an academic scholar, a scholar-athlete, and a reliable worker. The combination of my talents began to be noticed by persons inside and outside the school. Without knowing it and without trying, I had found the ingredient to what would become the way I would forever move through life.

I transitioned through all my activities with relative ease. I demonstrated an ability to juggle many balls without a single one falling to the ground. In other words, I was good at taking on tasks and completing them all and doing so at a very high level. As my school activities flourished, so did my basketball skills. I was getting better and better as a basketball player. With many positive comments about my skills coming from star basketball players on the varsity team, I began to focus on strengthening my body and improving my skills even more.

I concluded that I could use my combination of gifts to be successful now and probably in the future. For the remainder of my junior high and high school education, I continued to demonstrate leadership skills beyond my years. I was now fully convinced that my choosing not to stay with my buddies in Mr. Judkins's seventh grade class was the right decision and one I would never regret.

Chapter 6

FALL ON THE KNEES

I ENTERED THE NINTH GRADE IN the fall of 1961 at age thirteen. I stood six feet tall and weighed 185 pounds. Already established as an academic scholar, the label scholar-athlete was soon to be added as another descriptor. The high school coaches were knowledgeable of the skills I had displayed on the various basketball courts in Garner. I was so good at the game that the coach of the junior varsity could hardly wait for my arrival in the ninth grade. The coach was sure that adding me to his junior varsity squad meant his team would be set to win the conference championship.

I came with a reputation of a skilled basketball player before I even tried out for the team. Even the upper classmen were excited about my arrival in high school. With very little effort I earned a place on the team. Not only did I become a member of the junior varsity team, but I was also immediately selected as one of the five players to start. In other words, I was selected as what is called in the basketball world "one of the starting five."

I continued to perform outstandingly well academically. I took great satisfaction in being able to be a star member of the basketball team while grasping the highest-level academic courses in mathematics, biology, history, English, and geography. I even enjoyed my civics and physical education classes. Life was good for me.

The junior varsity season opened like a breeze. The team easily won the first three games with me leading the way in scoring. In addition to my scoring ability, I was a tenacious defender. The varsity team did not fare as well in its season openers. Out of the first three games played, the team lost two. One of the senior varsity players hinted that he was going to recommend that I take his uniform when he graduated. The varsity coach had more current ideas about me, not after the senior graduated but while I was still a freshman.

The varsity coach raised the idea of me moving up to his varsity squad as a freshman. The junior varsity coach was not going to hear of such an outlandish thing. He had been coaching for many years and was a good judge of talent and was confident his junior varsity team would win a championship this year with me on the squad. Besides, no player before had ever been elevated to varsity their freshman year, and the junior varsity coach did not want it to start now. He had a plan for how his team could take the championship. All summer the junior varsity coach had looked forward to the start of the 1961–62 basketball season. Now the talk of losing his star player was too much.

The idea of me moving up to the varsity squad set off a firestorm in the school's athletic department. There were accusations on both sides. The junior varsity coach maintained that the varsity coach was attempting to sabotage his effort to win a championship. The matter was raised to the principal, who was a basketball fanatic. Both coaches were aware of his love for the game and his desire to see a championship at any level come to the school. After discussing the matter with both coaches, the principal concluded that moving me up to the varsity presented the best chances for the school to have a winning season.

I moved up to the varsity and became a starter after sitting on the bench for several games. This event was impactful in a personal way. My uncle Thurman Whitaker, my mother's brother, was the junior varsity coach in search of a championship that year. The event became a point of discussion on my grandmother's front porch every Sunday for a while.

Uncle Thurman knew it was not my fault that I had been moved from his team to the varsity team. I was relieved to know that my uncle never held it against me, but my grandmother would not have allowed dissension even if Uncle Thurman had wanted to make a big issue out of the situation.

I summed up the entire transaction this way: my basketball abilities were very good, and it was more prestigious to the school if the varsity won a championship than the junior varsity. A senior and a couple of sophomore players on the 1962 varsity team were very good, and the addition of at least one more skilled player to the varsity team provided a chance at the team winning during tournament play. Increasing the chances for the varsity team to at least end with a winning season had required that I be added to the starting lineup. To me it was a simple matter of connecting all the dots. While there seemed to be a little friction about my being moved from junior varsity to varsity, everyone in the school and the athletic department's fan base knew the purpose of the transaction.

I joined the varsity team and had a very good year. The team contended for the conference title game but lost. After developing a name for myself during my first year in the conference, I returned in 1962–63 for my sophomore year as a solid starter on a strong team with three other starters from last year's squad. The first half of the basketball season was exceptional. Going into the Christmas break, the team had lost only a couple of its conference games and none outside the conference.

I continued to perform in my classes and on the basketball court. Everything was continuing to go well for me, and with my fifteenth birthday coming up, I planned to pressure my dad during the break about getting some driving practice. Well, I turned fifteen, but there was no driving practice. My dad said that with only one car and him needing it to get back and forth to work, there was no way the car could be left at home for me to experiment with learning how to drive. I thought I knew how to drive. I had learned by driving tractors in the tobacco field. But Daddy said driving a car was different than driving a tractor and that I would need to take lessons, especially if I ever expected to drive his car. For now, this brought an end to the discussion of me driving.

At the beginning of the spring semester of my sophomore year, I returned from Christmas break and picked up where I left off in the classroom and on the basketball court. My play on the court continued to be noticed. It had gotten to the point that my name preceded me when I arrived at the gym of opposing teams. With my gym bag swung over my shoulder and my team jacket on, I could hear whispers coming from the stands as spectators attempted to point out which one Shepard was. I could not help but feel exuberant when I heard my name coming from the lips of so many people I had never seen and did not know.

As I strolled with my teammates to the visitor's locker room, I remember my mom's words: "Never let anything go to your head." She taught my siblings and me a simple lesson about such by saying, "The same people who are praising you today will be the ones cursing you tomorrow. Therefore, let your actions speak for you and move on." In my head I was saying, *I understand Mom's point, but hearing my name like this excites me.* I wasn't prideful or arrogant about my status in the classroom and on the court but more in a state of disbelief that this had happened to me so soon.

In March 1963, the team had a very tough home game against one of our top rivals. The two teams always packed the gym, and that night was no different. Midway into the game I drove hard to the right of the basket and leaped high into the air in anticipation of making a routine finger roll layup into the right portion of the rim. A player from the opposing team could not get to the spot to challenge my shot in a legitimate defensive manner. So in a desperate move to stop the play, he ducked underneath me, cutting my legs from their vertical position and turning my body horizontal in midair. While in midair, I recognized the potential seriousness of what was happening and thought quickly of the position that would result in the safest impact to reduce the impending damage of this fall. I quickly fixed in my mind that I had to avoid my head and neck hitting the floor first at all cost. I shifted in air so that my hands and legs would take the bulk of the impact.

When I crashed to the gym floor, silence came over the gym as if time stood still. Both knees hit the floor as I landed, with my left knee receiving the main force of the impact. The left knee hit the floor so hard that some in attendance described the sound as similar to the firing of a gun. Upon impact, I let out a scream that was probably heard throughout the whole town of Garner. The entire crowd in the gym stood on their feet, most with their hands over their mouths or clasped to the sides of their heads, watching in disbelief and praying the fall was not as serious as it appeared from the bleachers.

In what many later described as a single leap, my mother was at the center of the court on her knees, bent over me as I rolled from side to side screaming in pain and agony. Coaches and trainers from both benches ran to see how bad the situation was. The referees tried to control the pandemonium and bring calm to the crowd. From the bench and from the stands, the fall had not looked good, and the attention turned from a basketball game to the seriousness of the fall. Had Shepard sustained an injury that would be permanent? Would this scholar-athlete be able to walk on his own? Not to mention would he ever again be able to play the game of basketball?

Another critical issue was that my team was in the hunt for the conference championship. They needed to win the next game over a powerhouse high school named Cooper High School from Clayton, North Carolina. Cooper High had won the first game we played against them, and Garner Consolidated High needed to win the second contest to stay even with Cooper. The coaches and fans all worried about my health condition at the moment, but the importance of the game in two weeks with a team we had to beat was fresh on everybody's mind. If I was not able to play, there was no way the team had a chance at winning the conference title.

Sweat poured from all over my body as I continued to lie on the floor in pain. My mother said to the referee that she had seen the whole play as it unfolded and that she believed the foul was flagrant and malicious with the intent to injure. The foul against me was building up in my mom. She had expressed to me on several occasions that it appeared to her that opposing players were playing very rough against me in my second year on the varsity team and that the referees needed to put a stop to this roughness before somebody got badly hurt.

She told the referees and whoever else could hear her that she attended all games and had noticed that since I had become a good player, the opposing team members went out of their way to play rough against me, and the referees were not calling fouls. To say the least, my mom was not happy with what she saw happen to me, and the player on the opposing team felt the sting of her stare. He quickly moved away from the scene and faded into the background among his fellow teammates.

After some ten minutes, I was able to sit up on the floor. When I did, the entire gym, both friend and foe, erupted with loud applause and screams as they continued to stand on their feet, observing every movement. I remained sitting in a 90-degree position for about two or three minutes. I then motioned to the trainer that I wanted to try to stand. With my teammates watching closely, I was helped to my feet. With one arm wrapped around the shoulder of the trainer and the other arm wrapped around the shoulder of a teammate, I was able to put light weight on my right leg but none on the left. As the crowd screamed and clapped, I was carried to the bench with my right leg touching the floor ever so slightly but without putting any weight on the left leg that had crashed hardest into the floor. After the coach and trainer questioned me about how I felt, I was immediately taken to the locker room, where icepacks were administered to my knees. Needless to say, I did not return to the game.

As I sat alone in the locker room with the trainer keeping close watch on me, I played back in my mind as best I could what had happened and was thankful I was holding icepacks to my knees and not strapped motionless in the back of an ambulance, being taking to a hospital for severe neck and head injuries. After replaying the events in my mind, I did not believe I used all my intelligence on the court to avoid the severity of the incident. From my peripheral vision while in the air, I had been able to see the moves being made by the defender and could have twisted my body in midair to avoid much of the contact. But by choosing at the start of the play to make the basket at any cost, I disregarded the quick movement I could have made that would have prevented much of the impact of the fall.

I made the decision that night in the locker room to remember in the future that it was more important to pull back than continue to pursue an athletic outcome, or any other outcome, when my well-being was threatened or jeopardized.

Two weeks after the fall, I was on the court, but my knees were still in a fair amount of pain. My team played their hearts out but to no avail. Cooper High beat us for the second time in regular season and ended up the Wake-Johnson Athletic Association (WJCAA) Conference champs. This 1963 team possessed outstanding talent in every position on the basketball court. I had extreme difficulty that night with my rival Charles Heath, a smooth forward who played the game as fundamentally sound as any player I had come up against. The other Cooper High player that actually drove the stake into my team's heart was a six-foot-nine center named Willie "Babe" Watson.

That night belonged to Cooper High School, and I wasn't sure if things would have been different if my knees had been in excellent condition. It was simply a matter of our team being good but Cooper High being better. Both teams would now have to wait for their placement in the regional tournaments leading to the 1963 North Carolina State Title.

At the conclusion of the game and before drawings were held for the pairing for the district tournament, my mother took me to the family doctor. The Black doctor examined my knees and sent me home with prescribed medication and instructions to continue applying icepacks to both knees coupled with soaking them in warm water. But the medication, icepacks, and soaking treatments did not work. The pain remained, becoming more intense when I sat for long periods of time with both feet placed flat on the floor. The left leg continued to be more painful but not enough to keep me off the basketball court for practice and scrimmage games.

The pairing for the district tournament was set. I was absolutely stunned when I saw the printout and both Cooper High and my team were paired in the same region. After thinking about it, I began to feel good about the pairing. Since Cooper had emerged on top in the two regular season contests and the conference finals, I figured that if the two teams met a fourth time in the regional playoffs, it would be highly unlikely that they would win a fourth time against a team I was on. Teams I played on had never lost four times in one season to the same team. This fact raised my confidence about our team going to the regional finals and winning the most important game of the season between the two teams.

Apparently I looked too far down the road before stepping onto the court. I was right about one thing: Cooper High would not beat my team for a fourth time in a single season. My team lost in the quarterfinals and did not make it to the final regional game, but Cooper High did. Cooper High proved to be the best. Not only did they win the regional title game, but they also went on to win it all. In 1963, Cooper High School won the North Carolina High School Athletic Conference 2A Division State Men's Basketball Championship. In addition to the feeling of accomplishment on the part of the town of Clayton and Cooper High School, winning the state championship was a proud moment for the entire all-Black WJCAA.

Not satisfied that my knee was still hurting a month after the fall, my mother said a specialist should take a look at it. My mother was always concerned about the future. She told me that proper care now for my condition would avoid possibly having to deal with more severe and intense pain and problems with my knees as I got older. In my mind, this was not something that was necessary. My mother was as concerned about what might happen in years to come as she was about anything of importance, especially relating to health. I understood her concern over the long term and followed through by having an examination of my knees done at the Raleigh Orthopedic Clinic, the top clinic of its type in the area.

As we were driving to the clinic, I looked out the window and recognized that it was in the section of Raleigh where I had worked with my father in the lawyer's yard. All the yards were beautiful. They looked manicured, with care taken to edge and remove the grass from around the driveways and sidewalks. I could not help wondering if it would ever be possible for me to live in a place that looked so stunning. My mind drifted back to my aching knees.

Once we arrived at the clinic, it was clear that we were in a place not found on the side of the tracks where we lived. Everything on the outside and the inside of the clinic was as beautiful as the yards that had turned my head from side to side on the drive. Floors were clean and shiny. Beautiful pictures hung neatly on the walls, arranged as if they were meant to tell a story. I wondered about the story behind the pictures on the walls. Before I took a seat in the waiting room, my eyes became fixated on the picture on the wall near the receptionist desk. It showed a waterfall cascading down a mountainside. The picture was incredibly beautiful. I wondered if this was a place the receptionist had visited, or maybe the doctor himself knew the place well. My mother motioned for me to take a seat beside her. She had no idea what was going through my head about what I saw and what I wished for in my own life.

We were the only Blacks in the waiting room. The white patients did not want me to see them staring at us, but our presence was unavoidable. I was not sure if their stares were because of my height and overall physical stature or simply my color. I did not have time to process why they were staring. I believed that trying to process such unknowns in the minds of others required too much of my energy, and in the end expending such energy yielded nothing for me personally. I had other plans for how I would use my energy in a more productive manner and trying to process how others perceived me was not part of my plan. As we left the waiting room to head for the doctor's office, I passed a large window on the left side of the office that offered another view of the striking yards surrounding the clinic, which was nestled in the center, giving it the appearance of a park.

Unlike the family doctor's examination of my knees, this doctor asked many questions. I had never experienced a doctor's appointment in which I was asked so many questions that did not seem to have anything to do with the purpose of my appointment. The doctor asked if I did things other than play basketball. He asked how I was doing in my schoolwork. The next question was what subjects I was taking. At one point when the doctor turned away from me to retrieve something from his desk, I looked at my mother as if to ask, "What is going on here?" My mother quickly nodded in an affirmative manner as if to say to me, "Just listen and answer the questions."

This was the first time I had been examined by a white doctor for something like my injury. In the Jim Crow South, my uneasy feeling was quite normal. A white doctor had administered flu shots to me, but that was done to all of the students at school. I did not know why the doctor was asking me so many questions, and I did not have the time or energy to try to figure out the motive behind all the personal questions. But briefly thinking about it, I figured my mother was right: just answer.

I was astonished at how comfortable I felt in the presence of the white doctor. Deep down I liked him because of the personal interest he took in me, since it was not related to my performance as a basketball player. As I sat in the chair, waiting for the doctor to continue his questioning, I wondered if the questions were because of curiosity about Blacks on his part or if he saw something special in me that I did not see in myself. I quickly snapped back to the situation in the room because trying to figure out why others, white or Black, did what they did began to require too much energy.

When the doctor turned back around to face me, he had a writing tablet in his hand and began to scribble on the tablet as I responded to his questions. After about thirty minutes of probing—this was my recollection of the investigative questions—the doctor asked my mother to return to the waiting room, where he would come to get her when he completed the examination and was ready to explain his findings. I waited for my mother to tell the doctor that she needed to be in the room during the examination. To my surprise, my mom agreed without comment and left the room. The doctor then motioned for me to go to the examining room next door, remove my trousers, and lie down on the examining table. The doctor entered the room, and this time there was no talking.

The doctor hunched over me while I was lying faceup on the table. He began to twist my leg, first to the left and then to the right. The doctor asked if I experienced any pain from the twisting. I answered not on the right leg but that there was pain when he twisted the left. The doctor then continued with twisting, bending, rotating, and stretching my legs where they joined at the knees, steadily asking if any of the actions produced pain in any area of my joints? I felt pain only in the bending of the left leg toward my back beyond the 90 degree point and circular rotation. The X-rays did not show any fractures or other internal damage in and around my knees.

The doctor wrote his report and gave me a regiment of leg exercises to do while lying on the floor. He recommended that I not return to the basketball court for at least two weeks and requested that I make a follow-up visit if the pain continued beyond thirty days. The fall on my knee and subsequent examination performed April 12, 1963, by the doctor at the Raleigh Orthopedic Clinic would turn out to be one of the most significant events in my life.

Chapter 7

DISCONTENT WITH SAMENESS

I RETURNED TO MY NORMAL ACTIVITIES at home and at school. I continued to outperform all my classmates except Patricia Turner. Patricia was simply brilliant. She was tops in all subjects. I did not want to get comfortable coming in second to her all the time but was never able to beat her in anything. I refused to get comfortable with being second to her. I never said in my mind that being second to Patricia was fine. I made every attempt to come out on top at all times.

I believed that if I strived just to be second to her I would slip further behind because of not working to my full potential. But I realized the only thing I could win out over Patricia would be a basketball shoot around called horse. On one occasion in the ninth grade, I tried hard to get matched up with her in a physical education class on basketball. If the teacher had matched us up, I would have challenged her in a game of horse. I rationalized that Patricia was so smart in the classroom that she might lose some confidence once I emerged victorious for a change. On second thought, Patricia was so smart that it was possible she might beat me on my own turf. Nothing doing. The game was never played.

As my sophomore school year came to a close, I continued performing at a high level in my classes. At the year-end awards assembly, I received the award of athlete with the highest scholastic average. All athletes from all sports were in competition for the award. When my name was called, I was shocked. How could a sophomore win such a distinguished award over the junior and senior basketball, baseball, and track athletes? The football program did not start at our school until I was a senior. I was shocked to the point that it took several seconds for me to rise from my seat and walk toward the stage. My classmate seated to my left asked if I had heard my name called. I was so speechless that I simply looked over and smiled at my classmate and then got up and walked toward the stage to receive my award. As I walked straight ahead, I thought about what was happening to me and decided it was too much to comprehend at the moment. After receiving my award, I turned to walk back to my seat but was called back to the stage. The award this time was for high-level performance in all of my academic classes. I took the award with my left hand, shook the principal's outstretched hand with my right hand, and this time returned to my seat without being summoned to the stage again.

The basketball season was over, and we had the entire summer to think about our performance on the court versus that of Cooper High. I started preparing for my junior year. Over the summer of 1963 I continued to work in tobacco and pick up odd jobs around Garner. My focus was on finding work that increased the strength of my body. With no facilities open to Blacks for bodybuilding, working in the tobacco field served as an adequate alternative for me. I did not sit on my grandmother's front porch as much as I used to wondering where the traffic was going up and down US Highway 70.

The different trips I had taken during basketball season my freshman and sophomore years opened my eyes to new possibilities. Distances as short as twenty-five miles away from my hometown of Garner were filled with elements of something new. I met new people and saw new places. In the tobacco field, I spent less time dreaming about one day driving an eighteen-wheeler truck to some destination along US Highway 70 and devoted more of my time to thinking about how to use my junior year to improve in every area of my life. I thought of ways to improve in my studies. I really desired to come out ahead of Patricia Turner and convinced myself that studying more next year might get me to the top.

I thought often about how the precise execution on the court of every aspect of the game by all the Cooper High players had led them to the state basketball championship. With the loss of two key players from the 1963 squad, I wondered if our team was capable of winning the championship my junior year as Cooper High had done my sophomore year. What would my junior year team need to do to win a state basketball championship? During the summer before the start of my junior year, I thought about my future on and off the basketball court.

I was ready for the start of my junior year. School opened September 3, 1963, a day after Labor Day. I was excited because on December 25 I would turn the magical age of sixteen years old. I had completed driver education training, so that meant I would be able to get my driver's license when the motor vehicle office opened after Christmas day. With only one car and my father needing it every day to go back and forth to work, I was not sure I would have anything to drive once I got my license, but I wanted it anyway. My classes were more challenging and demanding, especially chemistry and trigonometry classes, but I wasted no time. I picked up where I had left off my sophomore year.

Early in the start of the year, Patricia Turner and I, along with two of our other classmates, were called to the counselor's office. When we arrived, a representative from Saint Augustine's College in Raleigh, North Carolina, greeted us. The representative began to explain a new two-year academic program the college had developed for high-achieving students in their junior year of high school. The twelve-week program was a one-night-a-week class at the college in which students would be introduced to college-level academic subjects. The class met for an hour and a half every Wednesday night starting the second week of September through the second week of December. The subjects offered were English and writing, mathematics, science, and world history.

The four students selected from Garner Consolidated High School would join students with similar academic backgrounds from four other schools in Wake County. The students were responsible for their own transportation to the class. Depending on their performance, some college credit could be given for participation in the program. The students were instructed to take the information home and share it with their parents. I had to talk to my parents because I was only fifteen years old, and if I participated someone would have to drive me to and from the class. Participation in the advanced studies was based on a student's overall grade point average (GPA), daily classroom performance, and extracurricular activities. I became one of sixteen students in the advanced program.

The work was challenging, but I enjoyed participating in the class, especially the attention I was getting from a female student from another high school in the area. She seemed to "give me the eye" every time the class met. I needed to remain focused and worked hard not to make eye contact with the class beauty too often. We all made it through the class and were introduced to lots of new and advanced academic material.

The closeout activity for the class was an invitation to be the guest group on the WRAL-5 television station's highly watched *Teenage Frolic* dance program. The program could be seen every Saturday morning from 10:00 a.m. to noon. To my surprise, the girl who always gave me the eye in class asked if I was going. I said I was not sure considering I did not have a driver's license and was always obligated to work somewhere on Saturday. She responded that she hoped I would be there. I started working immediately on getting a ride to the station for the special activity.

I explained to my mother that everyone in the class was going to be at the television program and that I did not want to be the only one not present. She wanted me to participate but also did not have any idea how I would get there and back. The school counselor announced that if we could meet her at school at nine o'clock, she would transport the four of us to the station and bring us back to school for someone to pick us up. My mom agreed. My father dropped me off at the school on his way to one of his Saturday side jobs. I told him I could walk the three miles back home from the school if there was no one to pick me up at noon. There were probably some other students not going to the dance, but I had to be there to see the pretty girl this one last time before the class ended. I had a good time at the dance with the young girl and some of the other students. I experienced a great feeling of excitement and delight when I returned to school and my classmates and teachers said they saw me on television.

I excelled in every phase of high school. Well, not exactly. French gave me trouble. Chemistry was the class I really loved. To me, it was the best. Mr. E.L. Sanders was my chemistry teacher, and what a fine teacher he was. Early in the class Mr. Sanders noticed how easily chemistry came to me. Because of my high achievements in chemistry class, Mr. Sanders gave me responsibility for making stock solutions for all laboratory experiments. I enjoyed the new assignment. I excelled in demonstrating my understanding of chemistry concepts.

At the start of a class one day, Mr. Sanders announced that all of his chemistry students must enter a project in the local science fair competition. He then passed out a sheet with a list of ideas for science projects. The list had projects involving making soap, acid-base reactions, analyzing water and unknown samples, temperature and heat measurements, and various other topics. I looked at the list, and nothing on it appealed to me. With only the short period of time I had been in the chemistry class, nothing on the page captured my imagination. I surveyed the list several times hoping to find a topic that appealed to me.

I put the list of topics to the side and cracked open my chemistry textbox. I flipped through the pages, hoping to see something of interest that was not on the list. Seeing nothing, I shut the book and went back to the list. Since I did not have to select a topic right away, I decided to return later to the job of finding a topic for a science fair project. The list of projects and topics in the textbook sounded the same, and I wanted something different. The more I shuttled back and forth from the list to the textbook, the more discontented I became with the fact that I had seen some of these same projects at previous science fair competitions. I did not quite know what I wanted to do, but I knew I wasn't happy with what was before me.

Basketball practice was beginning in two weeks, and I was in shape and ready to get back into the conference competition. I was voted co-captain along with a senior standout. The team seemed very competitive, but my concern was the lack of depth. Skilled talent had been added to the team to replace the outstanding players who graduated from last year's squad. The problem was that the new talent needed to gain some experience.

In the opening game, I knew we were in for another tough season. We won the game and went on to have a winning season, including winning the WJCAA Conference Title and making it to the district tournament. The team once again lost in the district; however, this time in the second round. I was named to the all-WJCAA Conference team and the all-District 9 Tournament team. Just like the year before, at the school's year-end awards assembly, I received the award for the athlete with the highest scholastic average. Unlike my sophomore year when I had sat motionless for a moment, asking myself, *How could a sophomore win such a distinguished award over the junior and senior players,* I expected to win the recognition my junior year. At this point in my academic success, the players, students, and teachers inside and outside my school knew I was a scholar first and then an athlete.

One day when my father stopped to get gas for the car, as I looked out from the front passenger seat, I became transfixed by the octane ratings written on the gas pump. One sign read regular grade, the second said midgrade, and the third said high test. Sitting there looking at the ball in a glass bubble on the gas tank go round and round as my father pumped the regular grade (cheapest gasoline) into his car, I wondered if there was a difference in the three grades of gasoline written on the tank. I decided the best way to find the answer was to ask Mr. Sanders.

I was in the stock room the next day when Mr. Sanders checked in on me. I asked him what the different octane ratings on gas tanks meant when it came to miles per gallon and price per gallon. I also wanted to know if the different octanes required different engine temperatures to burn in a car's engine. Mr. Sanders said research was required to answer such questions. I asked if I could run some experiments in the laboratory to search for answers to my questions and use it as my science fair project.

Mr. Sanders was somewhat stunned at my idea for a possible project. He asked where I had gotten it. I explained my fixation at a gas station when my dad was fueling his car. Mr. Sanders liked the idea. We then bounced around some ideas of a title and finally settled on "Are There Differences in Gasoline Based on Octane Rating and Pricing?" We exited the stockroom, agreeing the project was an excellent idea. Mr. Sanders turned to me and said that I should continue planning for the project but that he would have to check the idea out further.

In this project, I wanted to look for answers to three questions I had about gasoline:

1. Did octane ratings (i.e., regular verses midgrade verses high test) burn differently in a car's engine?

2. Was there a difference in quality based on pricing (i.e., was lower-priced gasoline of lesser quality than the high-priced gasoline)?

3. Was there a difference in brands of gasoline (i.e., Esso, Humble, Gulf—these were the names of the gas stations back then)?

Seeking answers to the questions required that I run some distillation experiments in the laboratory. Mr. Sanders knew that working with gasoline could be a potential problem. His major fear was the danger of an explosion in the laboratory. He told me he needed to check with the principal and probably others outside the school to see if gasoline could be brought into a North Carolina high school chemistry laboratory. He said there was a high degree of danger in what I wanted to do but that he would check into what would be required to run such an experiment, if it could be run at all.

The decision to perform an experiment of this nature had to be decided by officials at the central office level. Mr. Sanders assisted me with the research and write-up of the experiment and presented it to the central office officials with me present in the room. I listened intently as Mr. Sanders explained my write-up and especially paid attention to his response to the most critical question of how the experiment was going to be conducted to avoid a disaster like an explosion. He responded that care would be taken to avoid distilling the solutions to dryness and the use of a sand bath would help control the temperature profile over which the experiments would be carried out to make sure there would be no explosions in the laboratory. The officials asked a few more questions and then asked both of us to leave the room.

We were escorted to a small sitting area, where we sat for about thirty minutes. We were summoned back and told that the committee had ruled that the experiments could be done in the school laboratory under some strict conditions. First, the experiments would have to be conducted under supervision of a qualified professional. Second, the experiment would have to be performed after normal school hours with no one in the building but the student conducting the experiment and the professional supervising the work. The state officials ruled that Mr. Sanders was qualified to serve as the supervisor of the experiment. Mr. Sanders thanked the committee, and he and I left the room.

For a brief moment, we stood in the hallway, looking at each other with a twinkle in our eyes. When the silence was broken, Mr. Sanders said, "They are really going to let us conduct these experiments." He asked me if I was ready to get in the laboratory and go to work. I responded, "Yes sir, I am."

As we walked down the long corridor toward the door leading to outside of the building, I asked myself if what I had just witnessed was the process used to move forward with new ideas. All I could think about were the steps that had been taken to bring something different from an idea to reality—from looking at a gas tank as gasoline was being pumped into a car, writing up the idea, presenting the idea to a committee for evaluation, the idea being blessed, and now the idea becoming a reality. I would use the process I was part of that day back in October 1963 with the state officials on many occasions over the next half century of my life as I continued to move from sameness to creating things that were different.

I thought I was ready to get in the laboratory and start my science fair project. Mr. Sanders had other ideas. He sat me down and explained how much library research was going to be required before I would be ready to perform any lab work. My research started with the Encyclopedia Britannica. The first important information I focused on was that the gasoline engine was powered by burning a mixture of gasoline vapor and air by an electrical arc generated by the spark plug.

I further researched the term *mixture of gasoline*. I was aware from earlier studies that the term *mixture* meant more than one substance. This was my clue of what should come next in my research. Gaining a detailed understanding of a mixture was necessary. From a chemistry standpoint, I found that a mixture was the composition of two or more substances that were not chemically combined with each other and were capable of being separated by physical means.

This research led me to a term called *distillation*. Mr. Sanders had introduced the term to the class, but I had no practical experience with it. I found that distillation was a common method used to separate mixtures into their individual components. I continued my research and found that in distillation, applying heat demonstrated that some molecules of a liquid possessed enough kinetic energy to escape into the vapor phase (evaporation) and some of the molecules in the vapor phase returned to the liquid (condensation). This set up equilibrium with molecules going back and forth between liquid and vapor. I gained confidence as I continued to research the topic.

At first glance, the distillation apparatus was intimidating. Upon examining the schematic drawing of a distillation apparatus more closely, I began to appreciate more fully Mr. Sanders's point of the potential danger of experiments involving gasoline. Applying heat to a container filled with a highly volatile substance like gasoline caused me to rethink my idea. I asked Mr. Sanders for a meeting to share what I had learned, particularly my concerns about performing such dangerous experiments involving heat, and get his directions on how to proceed.

Mr. Sanders was pleased to hear me express my anxieties about the project. While he knew I was capable of doing it, he also knew that my worries would keep me from becoming overconfident in a laboratory setting that could turn dangerous very quickly. Mr. Sanders was thrilled with my research findings so far. He instructed me to go back to the library, but this time I was to research the following topic: what is a sand bath, and when is it used?

I wondered if a sand bath was real or just something Mr. Sanders made up. I would soon find out. To my surprise, a sand bath was a common piece of laboratory equipment made from a container filled with heated sand—hence, the term *sand bath*. It was used to provide even heating for another container, usually during a chemical reaction. The key was that the sand was heated rather than applying an open flame directly to the bottom of a flask containing gasoline or other volatile substances. The literature pointed out that the most difficult part of distillations was setting up the small apparatus properly and attaining a slow steady heating rate. Maintaining a slow and steady heating rate would be critical to the experiment I would run. I was ready to go to the laboratory and practice what I had learned from the literature before performing the distillation on my gasoline samples.

The science fair competition was not until March 1964. Since it was exactly five months away, I was comfortable that I had plenty of time to do the project and do it with excellence. I shared my research on sand baths with Mr. Sanders. He agreed that I had done good research. He commended me on my research notebook. He cautioned me not to lay aside my work just because the competition was five months away and said that he and I would go into the laboratory at least twice a week and start working on my experiments. I did as Mr. Sanders suggested. I continued to work hard in all my classes at school, in my night class at Saint Augustine's College, and on the basketball court. But I was soon to witness, along with the rest of the world, a life-changing event.

The day was Friday, November 22, 1963, a day the nation and the entire world would never forget. I was sitting in my chemistry class when I heard someone screaming. There was no mistake about it: the scream signified death. It was like nothing I had ever heard before. But the sound echoed death. As the voice got louder and came closer to the classroom, I could make out what the screamer was saying: "President Kennedy has just been shot, and the report is he is dead. President Kennedy has been assassinated."

A deafening silence came over the class. Mr. Sanders told the class to stay seated. He left the room and caught up with the screaming person running in the hallway. The hysterical screamer was Mrs. Toole, the school's guidance counselor. She and the librarian, Mrs. Hunter, were the first persons at the school to receive the dreadful news because they had been standing in the library, discussing the day's activities when the special report had come on the television.

My classmates and I sat speechless in disbelief of the early news from Mrs. Toole. When Mr. Sanders returned to the class showing emotions on his face none of us had seen before, he announced, "The president just died. He was assassinated, and a person named Lee Harvey Oswald is in custody and is believed to be the one who shot President Kennedy."

Mr. Sanders was a teacher who did not show his emotions, and he did not have disruptions in his classroom. But on this day, he could control neither the shouts of disbelief nor the sadness that showed on his face and the faces of students in his class and classrooms throughout the entire school.

There was no containing of anything or anybody on this day. Pandemonium and disbelief broke out in the classrooms and spilled over into the hallways. Students and teachers burst from their classrooms, running through the hallways toward the library to catch a glimpse of the special report coming over the television. The small black-and-white television in the library was the only television in the whole school. Every student, every teacher, the janitor, and all the cooks from the cafeteria stood in the hallway and library around the small television, trying to see what was going on. With the volume on the television turned to its maximum, you could hear a pin drop as every single person assembled listened while reporter Walter Cronkite delivered a minute-by-minute update on what was unfolding in Dallas, Texas, where President John Fitzgerald Kennedy, the thirty-fifth president of the United States, had been assassinated at 1:00 p.m. Central Standard Time (CST).

Because the chemistry classroom was not too far from the library, I was able to be inside the library, stuck over in a corner, listening and watching every detail of the events as they developed right before my eyes. I kept telling myself that what I was hearing and watching could not be true, that it was all just a dream and everything would return to normal in a few minutes. President Kennedy had talked about a lot of things, but his focus on Black people and their dreams for the civil rights movement was what was on the minds of everybody in school that day. *Grief* was the only word to describe the mood in the school and on the television. The grief captured in the Garner Consolidated High School library that day had captured the entire nation. Even Walter Cronkite, a man known for his stoic, unemotional delivery of the CBS Nightly News, choked up.

There was a cloud that would not go away. Nothing was normal that day, and as I stood in the corner of the library, I wondered what would follow in the days, weeks, and years to come as a result of the assassination of President Kennedy since he had been such a strong supporter of the civil rights movement. School closed early, and buses began transporting students home. If there was anything that could be pointed to as good in my mind about Friday, November 22, 1963, it was that I was glad no one had to report to school for the next two days because it was the weekend.

When my siblings and I arrived home early that Friday, we found Mom sitting in front of our small floor model black-and-white television. She turned and looked up at us. On this occasion, we were not surprised to see her crying. We had witnessed the same scene among the teachers and students before leaving school.

"What will happen now?" she asked.

I could not answer her since I did not know what was going to happen. On this day, I felt that nobody knew what was going to happen. We all stood speechless as Mom turned back to stare at the television. I usually had words of comfort, but that day there were none. I wanted to say to Mom that everything would be okay, but I could not help but wonder if things would ever return to what could be considered okay. So we all remained silent as we looked up and saw our dad coming through the front door with the look of sadness on his face.

Events continued to unfold during the three days following President Kennedy's assassination, suggesting to me that things were getting worse. My family remained crowded in the living room on Saturday, November 23, to watch television and find out about all that would take place as events kept emerging. We watched President Kennedy's body being transported from the hospital to the airport to be flown from Dallas, Texas, back to Washington, DC. We watched as the flag-draped coffin was moved from an ambulance and carried into the East Room of the White House, where the president's body would lay in state.

On Sunday, November 24, we remained transfixed to the television set, watching everything unfold. If the pain of the tragedy we were watching wasn't enough, more pain was soon to be inflicted on the entire nation.

In the midst of watching the caisson moving in what seemed to be slow motion, every person in proximity to a television witnessed the networks switch from Washington, DC, back to Dallas for the surreal on-camera murder of Lee Harvey Oswald, the accused assassin, being shot and killed by a man named Jack Ruby. A frightening silence fell upon the living room where we were gathered. Transfixed on the television, I was convinced that the pandemonium and dark and gloomy events happening right before my eyes were a sign that the world was nearing its end. Up to this point, I loved watching old Westerns on television. But on this day I was not thinking about Westerns. I was convinced that even if my beloved Westerns returned to the television, it would never be as before. In fact I didn't believe anything would ever return to normal again.

There was no break from the television coverage of the assassination. The only break from television was to turn off the set and find something else to do. My mother did just that on one occasion. With Thanksgiving only three days away and it being such a big family event in my household, I thought maybe the holiday would help change things somewhat. I was glad for the early break from school that lasted a full week leading up to the Thanksgiving holiday. Not having to return to school until Monday, December 2, was good. I did not know if other schools in the area were doing the same, but I felt it was needed for the students, teachers, workers, and everybody in the community. Soon it would be Christmas. The festive activities would surely help bring some calm to the great loss I was feeling, especially since Christmas day was my birthday.

As the days and weeks passed following the assassination, the veil of unspeakable pain seemed to linger on. Before the Christmas break, the English teacher mentioned the upcoming writing contest for high school students. This was an annual announcement I never really concerned myself with. But this year was different. Individuals, the nation, and the world were still trying to deal with the assassination of President Kennedy.

I met with my English teacher, Mrs. Newkirk, to express an idea I had for entering the writing contest. With the assassination so fresh in everybody's mind, I thought about a unique topic for the writing contest project. I wanted to conduct door-to-door interviews of persons in the Garner area where I lived to get their reactions on the day the president was assassinated. Mrs. Newkirk said the topics to write about had not been sent to the school yet and that maybe I should wait and look at that before deciding. I said I would look at the list when it came but would go ahead and conduct my interviews during the Christmas holidays.

I informed Mrs. Newkirk that I had the title of my report. She asked what it was. I said, "The Day America Slept." She asked where my title had come from. I had come up with it during the assassination coverage when a commentator mentioned President Kennedy's senior year thesis titled "Why England Slept," which was a reference to Sir Winston Churchill's 1938 book titled *While England Slept*. Mrs. Newkirk was not sold on the idea at all. Again, she suggested I give it more thought during the Christmas holiday and take a look at the contest list of topics when I returned after the break.

At the moment, I was recalling the list of suggested topics for science fair projects. Something in me seemed to always push back when presented with ideas developed by others without an opportunity to explore and create for myself. It was not that I resisted authority. It dealt more with a belief that I had the ability to create and a high degree of discontent with repetitiveness and sameness developed by others.

Christmas day was on a Wednesday in 1963. I was excited for the day because this was the day I would turn sixteen years old. I did not say anything on my birthday about getting my driver's license, but the first thing from my lips the next day was asking when could I go and try for my license. I quickly learned that I would not get the opportunity to take the test until the school year was out. With that settled, I focused on the door-to-door interviews I would do to get the reactions of persons in the community the day President Kennedy was assassinated.

I visited about ten homes within walking distance of my house. I lived in an all-Black community; therefore, all the homes were occupied by Blacks. I contacted someone in the homes ahead of time to explain my desire to conduct an interview for a school project. All persons contacted were anxious to share their thoughts and feelings on the events of November 22, 1963. The responses were very similar, so it was easy for me to condense the information into a manageable grouping. All persons interviewed said about the same thing: I was in disbelief. It was incomprehensible. I am still confused by it all. It changed my world, changed the whole world. What will happen to all the good things he was trying to do, especially all the civil rights changes he wanted to make? It seems like it happened yesterday even though it happened over a month ago.

One person explained in detail how he had been returning from his lunch break when his fellow workers asked him if he had heard that President Kennedy was dead. At that point, he had not heard because he always brought his lunch from home and went off into a corner of the building that was totally isolated. He said that he first took it as a joke but quickly saw in the eyes of his coworkers that it was no joke. As he looked closer around his workspace, he saw fellow workers crying and holding on to each other as if in need of support to stand up. They all looked as if they were in need of support from somewhere and somebody. As he got more and more information, he was sure the world had moved toward its end. How could things go on without President Kennedy, the man asked me.

As I conducted this interview with the gentleman, he was seated at his kitchen table looking out the window. He said what came to him that day was the same sinking feeling he had experienced on Christmas day, December 25, 1958, when he received the call from his brother saying their sister had just passed away. He experienced the same feeling on the day President Kennedy was assassinated. I collected his information and the stories of others and used the balance of my holiday time to write up my report.

The Christmas holiday was soon over. To me, it appeared that the nation would recover from one of the greatest losses of modern times, so I remained committed to continue even in the midst of a life-changing, world-shifting month of events. When school opened after the Christmas break, I was eager to get my paper to Mrs. Newkirk to hear her thoughts about what I had done.

Overall, she thought the paper was good but that it needed some minor touch-ups. She noted that the content reflected substance and careful thought. She was quite surprised at how clear and concise I was in my delivery of the material. With the report showing only two references (i.e., door-to-door interviews and a single newspaper article), Mrs. Newkirk indicated that the report was the most creative work she had seen in her years of teaching. There was no doubt in her mind that my report would stand apart from other entries. There were a few problems involving the mechanics of grammar, spelling, and punctuation that required some minor corrections.

Mrs. Newkirk also thought the report was too long at seven and a half pages. She wanted it cut down to five pages or less. The final report was five pages. In looking it over, she pointed out areas that could be cut without negatively impacting the report's content and creativity. Mrs. Newkirk expressed to me again her concern about the limited number of references: (1) a report from the *Raleigh News & Observer* and (2) personal communication with persons in the local community. Though she was uncomfortable with a report containing only two references, she entered it into the contest anyway.

In early February, Mrs. Newkirk was notified that my report had been judged as one of the contest winners and that the North Carolina State Chapter of the Phi Beta Kappa Honor Society would be honoring me in two weeks at a formal dinner at the YMCA on Hillsborough Street in Raleigh. My award was a copy of John Kennedy's book titled *Profiles in Courage*. I remember the announcer saying to the audience, "The report of our next winner had only two references—an article from the newspaper and personal interviews from persons in his neighborhood. Even with limited references, the judges found this report to be creative, interesting, and quite unique."

I looked at Mrs. Newkirk. She knew exactly what my look was about: her comment, "The reference section of your report will be uniquely different and not what the judges are accustomed to seeing." The statement about the report being unique stuck with me because prior to entering the competition, Mrs. Newkirk had been concerned that the work was so novel that it might fail, not because it wasn't good but because it did not look like the other work entered into the competition, especially with such limited references. Nevertheless, she had sent it, and it won.

A noteworthy aspect of this event is that coming from the background that I did, the Black high school I attended was very excited about my honor. The whole school was interested in the overall growth and development of its students. Knowing I had never attended a formal dinner of this nature, the home economics class (known as HomeMake) stepped in. The teachers knew that no student from their school had ever participated in such a grand affair and were going to make sure I would be at ease in the elegant environment for that evening. The HomeMake class changed their normal classroom activities to develop a special class for me.

Two weeks before the event, I was brought into the class, and the students turned into waiters and waitresses to step me through what the setup of the formal function would be like. I was introduced to how forks, knives, spoons, cups, salad plates, wine glasses, and water glasses were arranged on the table for a formal affair. They walked me through a demonstration to show me the order and how to use each item around my plate. They showed me books filled with pictures of table layouts for formal dinner gatherings.

While viewing a picture in one of the books given to me, I said in my mind that the dinner I was attending would not have that many utensils around the plates. I thought that out of the excitement of the moment, the HomeMake class was exaggerating what I could expect. One picture in the book showed a dinner plate with two forks on the left side of the plate, a knife and spoon on the right side of the plate, and another fork and spoon at the top of the plate. Then there were two glasses shown at the upper right corner of the plate. One of the glasses had a napkin tucked down inside it. Still in the right corner was a cup sitting on a small plate. Positioned in the upper left corner was another small plate with a small knife on it.

I smiled inside because I was certain that such a setup would not be at the dinner I was attending. I even voiced this sentiment to the teacher. She explained that the textbook showed pictures of many different types of elaborate formal table settings and that one of them would be the setup at the event I was attending. The teacher and her students took me through the exercise several times so I would not mess up or become confused at the dinner function.

The twice-weekly training under the supervision of the HomeMake class was soon over. No more pictures to look at. No more opportunities to get it right. No more training sessions. I knew my accomplishment and award were not only for me and that my presence at the event would represent the entire Black school I attended and the community in which I lived. The event was next week, and I was ready to face whatever the evening would bring. The challenge and experience of exploring new and unchartered territories always energized me going into any situation, and the final weekend leading up to the event was no different.

The etiquette training was needed because the table setup at the elaborate function held at the YMCA on Hillsborough Street in Raleigh included lots of utensils around my plate, just like some of the pictures in the books. The other key point to note here was that while my parents could not attend the grand occasion with me, my English teacher and coach, Mrs. Newkirk, and I were the only Blacks at the dinner among the "sea of strange white faces."

It was evident four months after the assassination of President Kennedy that the sting of the catastrophe would be with the world forever. So I and everyone else had to move forward.

The time came for the science fair competition. I had returned to the chemistry laboratory along with my teacher Mr. Sanders every Tuesday and Thursday evening for the months of January and February 1964. Mr. Sanders reminded me to use boiling chips; concentrate on maintaining good regulated heating of the sand bath; and keep my eyes on the distillation to guard against the flask drying out.

I completed my distillation experiment on the various grades, brands, and prices of gasoline the first week of March. The most difficult parts of the distillation were setting up the small apparatus properly and attaining a slow steady heating rate to the sand bath. I kept close watch on the temperature of the sand bath and a steady visual inspection of the flask. Maintaining proper heat control was essential. Mr. Sanders made sure I got this message by periodically dropping by the laboratory commenting, "To avoid the possibility of a dangerous explosion around here, do not let the flask dry out. If it appears to you that the flask is going to run dry, stop the experiment." I kept my eyes on the thermometer and hand on the temperature control knob to guard against letting the temperature rise too rapidly.

As heat was added, the gasoline mixture moved from a liquid to a vapor. The vapor rose in the distillation column and condensed back to a liquid that was captured in a beaker. I noticed at one point in the experiment that there was no liquid captured but the temperature rose. Then additional drops were collected. This pattern continued for three different temperature ranges. I recorded in my notebook the three liquid fractions I collected as each liquid distilled off at a different temperature and volume of drops.

At an all-Black high school, there was no analytical equipment available to analyze the different fractions to see what these various components were. I used the information to plot a graph of the temperature versus volume. The graph showed a step function of the fractions as the temperature rose. The temperature plateaued for about one to two minutes, rose rapidly, and then plateaued again as a second liquid fraction distilled off. This happened three times. I explained what was happening as the first liquid distilled off after it reached its boiling point. A transition took place between the first liquid fraction and the start of the second. When the second liquid distilled off, it had reached its boiling point, and so on for the third liquid fraction.

Mr. Sanders said the gasoline mixtures were comprised of at least three components but that more work was needed to find out what the components were. He said that a simple experiment not involving any additional laboratory work would be for me to find out what compounds had the boiling points of the liquids collected. According to Mr. Sanders, that would be proof of the identification of the compounds.

The findings provided answers to some of my questions. For example,

I determined that gasoline is a mixture of components that have different boiling points. This suggested that to vaporize all of the components requires the car engine to operate at the highest temperature required by the last liquid fraction collected from the distillation experiment. To be specific, the three fractions reached their collective boiling points over the range of 62°C to about 110°C. The first component dripped into the tube at 62°C. I removed that tube and placed a new tube under the distilling collection head. The second component started to drip into another new tube at 89°C. The temperature rose to about 110°C, which was when the first drip of the final component dropped into another new tube. This pattern was the same for all brands and grades, all of which produced the same distillation results.

I explained in my report that with these results I believed all gasoline grades and brands were made up of mixtures that burned at different temperatures. I then hypothesized that to burn all components of gasoline required a car's engine to operate at an internal temperature of at least 110°C. When it came to questions about pricing, I was not able to say if there was a difference in gasoline quality based on cost.

I entered my project into the science fair competition, and it won a first-place ribbon and a gold star at the competition. I remember receiving many questions from both the judges and the regular viewers that streamed through looking at the various projects. With Mr. Sanders's guidance and mentoring on this project, I had gained many new insights about science, especially on how to conduct laboratory experiments and the importance of literature review and recording results in a laboratory notebook. In carrying out the project, I also realized that additional research was needed to increase my understanding of how gasoline burned in the engine of a car. I had come across terms like *compression ratio, combustion-chamber influences, knock, expansion stroke,* and many more. I planned to continue my work on gasoline after the science fair competition was over. Mr. Sanders encouraged me to definitely continue my research on gasoline but to do so under the careful supervision and guidance of a qualified scientist. Many years and lots of training later I again would be engaged in laboratory experiments on gasoline products.

After a productive junior year, I entered my final year of high school poised and confident about my future. I had built up an impressive record of academic achievements and extracurricular involvement and demonstrated leadership potential, and I wanted to continue my education at one of the major institutions in the Raleigh-Durham-Chapel Hill area. To my disappointment, North Carolina State University, the University of North Carolina at Chapel Hill, and Duke University all said my scholastic aptitude test (SAT) score was not sufficient for me to attend their institutions.

When I opened and read the three rejection letters, I was extremely disappointed and felt the reviewing committees had made a mistake by not giving me the opportunity to continue my education at one of their flagship institutions. I used none of my energy trying to figure out what was in the minds of the committees that had caused them to reject me. The letters said I was not qualified to attend their institutions, and that was that. I did not lose any sleep over the rejections and was satisfied with my future possibilities and quickly returned my thoughts to completing my senior year on a high note academically and on the basketball court.

My senior basketball season reminded me of my sophomore year. The team was solid. We started the season winning, and I was voted team captain. In addition to our team and a once again strong Cooper High School, there was another powerhouse who had always been on the fringes but never at front and center of the league. Their start of the season suggested this was going to be their year. The team was Berry O'Kelly from Method, North Carolina, a town some thirty miles from Garner. Led by all-American Lawrence Dunn, Berry O'Kelly won the North Carolina 2A Division State Title in 1965.

The same scenario that happened in 1963 when I was a sophomore repeated itself in 1965 when I was a senior. Berry O'Kelly, with its star Lawrence Dunn and his fellow teammates, were too much for Garner in the two regular season games, the conference title game, and the district tournament. After knocking my team out of the district playoffs, Berry O'Kelly represented the WJCAA Conference and then went on to win the North Carolina State Title.

Teams from the WJCAA Conference had won the North Carolina

State Title twice in the past three years. I understood the game of basketball and knew that any of the teams I was a part of during my high school years had all the elements to win a state title. That was why even after I graduated I was in the stands, cheering the team on when my school, Garner Consolidated, was playing for the North Carolina State Title game in 1967. That year, the team I had once led played their hearts out but was no match against Central High of Whiteville, North Carolina, with its star Reggie Royals. The all-Black WJCAA Conference that included Garner Consolidated High School had powerful basketball teams and players. In a five-year period, three high schools from the conference competed for the North Carolina State Basketball title. The three schools were Cooper, Berry O'Kelly, and Garner. Two won the title game. When Garner was the runner-up in the title game in 1967 season that marked the last year for the segregated conference before desegregation started in 1968.

There was no doubt in my mind that if media coverage had been available to my community during the days of segregation like it is for the desegregated schools today, more athletes from my era would have won scholarships to play for major schools and possibly would have continued on to the National Basketball Association (NBA) like David West, who also played his high school basketball career at Garner Magnet High School.

My twelve years of schooling in the Garner Public School System ended in 1965. Some things had followed me throughout my journey, and my senior yearbook confirmed it. I had been the tallest student in the lower grades and was still the tallest in my senior class. Other superlatives included the student best exemplifying the cardinal principles of leadership, scholarship, character, and service. My classmates voted me most athletic, most scholarly, most promising future, and most of the best. I really loved the fact that my classmates also voted me, along with one of the other males in the class, as most handsome.

Added to my sophomore and junior years, I received the award as the athlete with the highest scholastic average. Three years in a row I had won this prestigious honor. I also valued that the valedictorian and me (the salutatorian) had both come from Mrs. Crenshaw's seventh grade class. This was a pivotal moment in my life.

It was now May 30, 1965. At seventeen years old, I stood six feet three and weighed 192 pounds. I was seated on the stage in the gymnasium with the valedictorian, dressed in my graduation cap and gown and honor ribbons, waiting to introduce the speaker for the baccalaureate address that was soon to be delivered by Dr. Paul Johnson, Pastor of Martin Street Baptist Church in Raleigh. As I sat there, my mind drifted back to days long past. As I listened to the principal's welcome and comments from the mistress of ceremony, Mrs. Toole, as she described my class through the years and drew audience laughter at times and applause at others, I used the time to reflect.

I reflected back on the day in the fall of 1959 when I raised my hand and could not put it down and declared to my friends that despite all of the unknowns, I could not stay and had to go to the new class. I thought about Mrs. Crenshaw's first words that had made me wonder if I had made the right decision: "If you do not want to learn, get up now and go to the principal's office to be reassigned to another class."

My mind went back to the day President Kennedy was assassinated. My thoughts then drifted to the discussions I had engaged in with people in my community during my door-to-door interviews for my writing project. I thought about my mother and what all her lessons and training had meant to my being seated on the stage and what I had gained from my father as well. I thought of all the days of working in the tobacco fields and with my father in the yards of the fine houses in North Raleigh. I reflected on my science fair project, and when I put it all together, my graduating class motto came to mind: no gain without pain. I thought of how true that statement was. And I would never forget how Mr. Sanders returned in the evenings to work with me on the science fair project after a full day of teaching. The thoughts began to overwhelm me, and tears began to fill my eyes.

It was time for me to introduce the speaker, and I needed to compose myself. I did. As I rose to introduce the speaker, I realized that it was God moving in my life and that all of this was part of my destiny being fulfilled. I could not help but mention some of the things that had gone through my mind while seated on stage and reflecting on my life before introducing the speaker. As I spoke, I often looked at my mother in the process. The audience applauded enthusiastically after I made my comments. I then introduced the speaker.

Dr. Johnson congratulated me on such fine comments and on my introduction and then delivered his powerful baccalaureate speech titled, "Go Out into the Deep." When Dr. Johnson stood and gave the title of his speech, I was sitting there asking myself, *How can I go out into the deep if I can't even swim? Where is this going?* But what Dr. Johnson said that day was that as high school graduates we had started our journey to take our places in the larger world. He used a body of water as the metaphor to make his point. He stressed the need for us to move from the comfortable but shallow shoreline of the river out into the parts that are more hidden and concealed.

He said that to go further required us to continue holding on to God's unchanging hand and taking our education beyond our high school diplomas. As I listened, I first thought, *That sounds exactly like what Mom tells us all the time.* By the time Dr. Johnson finished his message, I concluded that while I did not know where going into the deep would take me, I committed to myself that day to try to do two things moving forward: (1) to hold on to God and (2) continue my education.

Commencement was next Friday, June 4, 1965. The valedictorian introduced the speaker, Dr. Prezell Robinson, president of Saint Augustine's College, who led the institution I would attend after receiving my high school diploma following the commencement address. After his message, Dr. Robinson made a major announcement that involved me. When he announced that I had been awarded a four-year academic scholarship to attend Saint Augustine's College starting in the fall of 1965, I looked out and saw my mother slowly rise to her feet and stand with her arms folded. As she applauded with others in the audience, I could see her mouth moving, and while I could not read lips, I had no doubt that she was saying, "Praise God for what You have done today," or something like that.

I had not been aware that such an announcement was going to be made. Both my counselor and principal were aware of the scholarship but had wanted it to be a surprise to the audience and obviously to my family and me. We were all surprised as the president motioned for me to come to the podium. As I stood and made my way over to where Dr. Robinson was standing, I knew for sure now that I was taking a giant step toward fulfilling my destiny.

Chapter 8

SAINT AUGUSTINE'S COLLEGE AND "THE PEARL"

I HAD LOTS OF INTERACTIONS WITH Saint Augustine's College (STA) before entering as a freshman. Three of my mother's brothers were graduates of the institution. My Uncle Thurman had been the first to graduate from STA in 1957. He had been immediately hired as a teacher and coach at Garner Consolidated, the high school he had graduated from in 1953. He was my junior varsity coach for the short time before I was moved up to varsity when I was in ninth grade. He also taught me biology and driver education when I was in tenth grade. Uncle Thurman loved STA and was a very active and supportive alumnus. A servant at heart, Uncle Thurman worked to create new ways to give back to the institution that had meant so much to him in his educational development and professional life.

Unlike Uncle Thurman, who had been able to go straight from high school to college, when his older brother Howard graduated from high school in 1948, he could not afford college, so he enlisted in the army. After a four-year tour of duty that included involvement in the Korean War in 1950, Uncle Howard returned home to Garner. He was fascinated and proud that his younger brother Thurman had saved enough of his own money to enroll in STA immediately after he graduated. Uncle Howard definitely wanted to continue his education. He remembered that before being discharged from service a friend had mentioned how he planned to get a house on a fairly new military benefit called the GI Bill. He also remembered that others talked about returning to school to complete their education using money from that same GI Bill.

Uncle Howard continued to probe and ask questions. His diligence to find out more about the GI Bill was fruitful. He found the bill was a law that had provided a range of benefits for veterans returning home from World War II. The benefits included low-cost mortgages; loans to start businesses or farms; cash payments of tuition and living expenses to attend college, high school, or vocational education; and one year of unemployment compensation. The Bill was available to him and every veteran who had been on active duty during the war years for at least ninety days and had not been dishonorably discharged. He continued his research, especially when he found that millions of veterans had used the GI Bill education benefits to attend colleges or universities and that millions more used it for other kinds of training programs.

Recognizing that these benefits were opened to him, Uncle Howard moved quickly to complete the necessary paperwork to enroll in STA with his younger brother Thurman. They both marched in the graduating class of 1957, with one brother paying for his education out of his own resources and the other aided by the federal program called the GI Bill. At the time of graduation, Uncle Thurman was twenty-one years old, and Uncle Howard was twenty-six. Uncle Howard always said he was blessed to have benefited from the GI Bill program just before the program was terminated in 1956.

Upon graduation from STA, Uncle Howard was not blessed with a job like his younger brother Thurman. Not willing to return to the job he had worked before going to college and listening to his coworkers tease him about being educated but having to work at their level, Uncle Howard decided to enlist in the US Navy in the spring of 1958. He served a four-year tour of duty in the navy and was discharged in 1962.

In 1960, Uncle Howard was seated in the stadium in the Olympic Village in Rome, where he witnessed a young fighter named Cassius Clay (now known as Muhammad Ali) win a gold medal in boxing. He wrote home that he had seen a young fighter from Louisville, Kentucky, in the Olympics and that all in the United States should pay attention to him because he was going to be known by the entire world very soon. He said for my grandmother to share the letter with his brothers because what he had witnessed in Rome was going to shock the world, and he wanted them to know about it first. Uncle Howard said that nobody had ever seen a boxer with the footwork and hand speed of Cassius Clay. He said the young boxer was popular inside and outside the Olympic Village, with people following him and listening to every word he spoke, and that he was only eighteen years old.

Uncle Bobby was the baby in the Whitaker family. He graduated from Garner Colored School a couple of years after his brother Thurman. Like his two older brothers, he continued his education at STA, graduating in the spring of 1959. Loving the fact that his brother Howard was seeing the larger world by being in the military, he decided to enlist in the army. With the Korean conflict over and no knowledge about what was going on in Vietnam in 1960, the world seemed to be at peace from his standpoint, so he decided to enlist in the army with the intent of traveling the world.

He also heard stories from his brother Claywood, who was the oldest of the four boys, about his travels and his involvement in World War II. Unlike his younger brothers, Uncle Claywood hadn't been able to go to college because most of his time had been spent working. With their father passing away when all seven children were young, Uncle Claywood had to work to help the family financially. I was aware of this and admired him for what he sacrificed so his mother and siblings could have a better life.

Uncle Bobby had been encouraged even more about his decision when young President John F. Kennedy took office in January 1961. His feelings changed somewhat when three months later the Bay of Pigs invasion took center stage in news around the world. He began to second-guess his decision to join the army, but the incident never escalated to a United States war effort.

The traveling he hoped to experience did not happen either. The fascinating thing about Uncle Bobby's army experience was that when he first enlisted, he scored so high on his entrance examination that he was offered the opportunity to attend flight training school to become a military pilot. He was excited about the opportunity to not only travel the world but to be able to see the world from high in the clouds.

Unfortunately, my grandmother pleaded with her baby son not to take the offer. While he knew that not being able to pursue the opportunity to become a pilot would bring about a low point in his life at the time, he turned down the offer. All of my grandmother's children tried to honor her wishes, including Uncle Bobby. When I learned years later of the opportunity for him to pursue becoming a pilot, I thought he should have been allowed to follow that path. Uncle Bobby shared with me that he never had any regrets about it because other aspects of his life had turned out so well and were so fulfilling.

After two years, Uncle Bobby was discharged from the army while he was stationed in Philadelphia, Pennsylvania. Unlike his older military brother, he took a path similar to his brother Howard and never returned to Garner. Uncle Howard made his home and raised his family in Roanoke, Virginia. Uncle Howard entered the Roanoke Public School System as a teacher and retired as an elementary school principal.

Uncle Bobby remained in Philadelphia, where he met and fell in love with his future wife. The area became his home and the place where he raised a family. Uncle Bobby took the path of his older STA-educated brothers and also became an educator. After working in the Philadelphia Public School System as a high school teacher for a short period, he returned to school and obtained a master's degree. After receiving his advanced degree, he was asked to apply for a faculty position at Cheney University of Pennsylvania, a public coeducational institution that is part of the nation's historically black colleges and universities (HBCUs). I was aware of the history of the institution where he worked that is now a part of the Pennsylvania State System of Higher Education. I was thrilled when my uncle became a professor at the oldest HBCU in America. He retired from the institution in 2005 after thirty years of service. Today he still lives in Pennsylvania.

With such a rich and long-time family association with STA, I was excited and eager to continue my education at the institution. My older sibling, Ruth Ann, had already continued the family tradition by enrolling in STA in 1962 but fell in love her sophomore year, married, and started her family. I was determined to graduate from STA, as were two of my younger sisters, Maxine and Janice. Our brother Roger said enough of STA and looked to continue his education elsewhere.

Much had changed in the Jim Crow South by the time Roger came along. So much so that upon graduation from Garner Senior High School in 1973, Roger was recruited by and attended Wake Forest University. He accepted Wake Forest's prestigious George Forster Hankins Scholarship and rejected a chance to attend the University of North Carolina as a Morehead Scholarship finalist. My mother knew the world was changing and witnessed it in our family with the University of North Carolina's rejection of me in 1965 and my brother's rejection of UNC in 1973.

Having been part of STA's advanced program for high school students and coming into contact with the basketball team on a regular basis, I was already familiar with the institution of higher learning when I entered in the fall of 1965. I had made up my mind about what I would do at STA before arriving on campus. I was going to major in chemistry, minor in mathematics, and try out for the basketball team. I was comfortable with my decision in all areas. My academic scholarship award from STA legitimized me for the classroom. My performance as a star basketball player who was well known in and around the region of North Carolina in which STA was located validated me for the basketball court. I was ready for my college years.

Founded in 1867 by prominent Episcopal clergy and laymen in the historic Oakwood District of Raleigh on a striking 110-acre campus, the private college had evolved into an institution of higher learning noted for both academic excellence and a wide range of city, state, national, and international services. At the time of my entry, the institution was ninety-eight years old, with a student enrollment of 1300 and 102 faculty, 71 percent of whom had doctoral degrees. I was impressed that the student body consisted of students from thirty-seven states, the District of Columbia, the US Virgin Islands, Jamaica, and thirty other foreign countries. The campus was absolutely stunning. With its twenty-three buildings built with gothic stones that included some faculty housing, I viewed the campus as a pretty place to continue my education.

I was awestruck at the background of the faculty in the Division of Natural Science and Mathematics. Being in the presence of a collection of that many persons with PhDs was an enthralling experience for me. The professor chairing the Chemistry Department would become my dominant professor. He was a Black scientist with an undergraduate degree in chemistry from Wiley College, a small HBCU in Texas, and his PhD from the University of Texas. He captivated me with his knowledge of organic chemistry, especially the synthesis part of the course.

The professor who taught me physical chemistry was a German scientist who received his PhD from the University of Vienna in Austria. The professor down the hall chairing the Biology Department was a Black scientist with a PhD from Saint Bonaventure University. The professor who taught me most of my mathematics classes was a female mathematician with an MS from North Carolina State University and who had a joint appointment at STA and North Carolina State University. She was close to completing the requirements for her PhD when I graduated from STA. My physics professor was from India and had obtained his PhD from Calcutta University. I found the science faculty to be top-notch in their areas of specialty. Taking courses from such a collection of science and mathematics talent demonstrated to me how well my high school had prepared me. I was delighted when I saw my high school chemistry teacher, Mr. Sanders, on campus in the afternoons, teaching the electricity and magnetism course to college students. I quickly pointed out to anyone in hearing distance, "Mr. Sanders was my high school chemistry teacher."

The academic scholarship I received allowed me to stay on campus even though the institution was less than five miles from my hometown of Garner. The short distance from the campus to my home did not bother me. I was not caught up on distance from one location to the next but rather the new experiences I expected to gain by interacting with people from other parts of the country and the world. I was aware of the growth from intermingling with others based on my participating in the Wednesday night class at STA while in high school. I viewed my staying on campus as providing me the opportunity to grow even more by associating with a much larger mix of individuals. And grow I did.

After settling into my academic studies and receiving several of my tests results, I was confident that my study skills in college would continue to produce good results as they had done in high school. My most time-consuming course was chemistry. Because I loved the subject, it never seemed to require as much time as it actually did. In my general chemistry class, lectures were Monday, Wednesday, and Friday. The laboratory component of the course on Tuesday and Thursday, with its requirements for recording meticulous notes of findings, made this the course to which I knew I would need to apply myself at my very best.

I had also learned from high school that chemistry had lots of connecting material and subunits and that the subject matter moved along at a somewhat fast pace. At such speed, I was aware that it was important that I start well in chemistry. I knew from my days working and talking with Mr. Sanders in my high school chemistry class that once you got behind in a chemistry course, it was very hard to catch up. Not that it couldn't be done, but I knew that to do so would require taking valuable time away from some other important areas of my time chart. While my mathematics course was time-consuming, none of my other courses were as demanding as my chemistry course. This was because none of the other courses had a laboratory component. So I got a quick start on time management relative to my studies. Physics had what I considered a modest laboratory component, but that would come later. I did have one complaint though.

On Tuesday and Thursday, the days I had chemistry lab, I always missed my dinner meal. My experimental procedures required close attention in apparatus setup and getting the experiment started and started correctly. Then came the disassembly of the experimental setup and cleanup. I had learned from my high school chemistry class the importance of taking my time to try to do my laboratory work correct the first time. Experience had shown me that speed had no place in a chemistry laboratory. I was aware how quickly a two-hour laboratory exercise could expand into a much longer drawn-out event. I knew to pay close attention to the tiniest of details. I was familiar with the sting of experimental failure and the need to do it all over again.

From my high school teaching, I was keen on the importance of conducting laboratory work in a manner that increased safety and reduced—or even eliminated—the destruction of expensive equipment and chemicals. On the occasions when my laboratory work went well, additional time and care was needed to perform cleanup and proper disposal of used materials and solutions. The point is, it never failed that on these days, the combination of conducting the experiment with care, the length of time required for the experiment to run its course, restarts when everything failed, proper recording of results, and laboratory cleanup and closeout required so much time that the dining hall was closed when I left the laboratory. It did not take long for me to avoid hunger pains at around 6:30 p.m. on Tuesday and Thursday by building up a personal food supply in my locker in my dormitory room to cover the times when I missed out on dining hall meals.

I was off to a good start with my academics. I was only a few months from having played several WJCAA Conference games, the conference tournament, and the District 9 Tournament in the Raleigh area, where I was recognized as an all-conference, all-District 9, and all-star basketball player. My basketball skills were well-known in the STA basketball community. When I asked to meet with the coach, I was not surprised when he said, "I know who you are. I am very aware of your accomplishments on and off the court in high school. Welcome to STA."

He also knew that I was enrolled at STA on an academic scholarship. I was somewhat surprised by the coach's detailed response before I could state why I had come to visit with him. I began to wonder why, since he knew so much about me on the basketball court, he had not offered me an athletic scholarship. At that moment I wanted to ask the question, but I did not.

I quickly figured that it was probably a matter of economy of scale; by not giving me a scholarship, the Athletic Department had more resources to use on other potential players. Why should his department offer me an athletic scholarship if I already had financial resources coming from some other department within the institution? I was presumptuous in my standing there thinking the coach ever had in his mind to offer me a basketball scholarship. Nevertheless, that was my quick summation of the situation. The coach invited me to sit down. I explained my intention to try out for the team even without a scholarship. The coach said he was hoping I would make a run for it. He said basketball practice would start in mid-October and for me to be there and ready to go.

I was there to try out for STA's basketball team, a member institution in the Central Intercollegiate Athletic Association, better known as the CIAA. I knew I had played some strong competition in the WJCAA Conference. I figured the skills I would bring to the STA team would add to the team's overall competitiveness in the CIAA. Anyway, I had followed the CIAA since my freshman year in high school. An annual treat for my high school basketball team was to be given time from school to travel to the CIAA tournament. Back then the tournament was always played in the Greensboro Coliseum. The level of competition was like none other.

Back in those days, in my mind the Atlantic Coast Conference (ACC) basketball players could not compare to the high-flying, acrobatic moves demonstrated by the Black players in the CIAA. Because the ACC had no Black athletes until 1966 when Charlie Scott put on the sky blue and white uniform of the University of North Carolina's Tar Heels, I did not find watching ACC basketball that interesting. I began watching more ACC basketball because the play of Charlie Scott paralleled that of other great Black players in the CIAA during those days.

These CIAA players included Freddie "Curly" Neal of Johnson C. Smith University (Harlem Globetrotters), Hugh Evans of North Carolina A&T State University (St. Louis Hawks and NBA official), Mike Gale of Elisabeth City State University (Chicago Bulls), Bobby Dandridge of Norfolk State University (Milwaukee Bucks), Cleo Hill of Winston Salem State University (St. Louis Hawks). While star players like Evans

"Lefty" Belton and Al Glover of STA and Fred Bibby of Fayetteville State University were not drafted into the NBA, they were all excellent players. The CIAA type players excited me. I believed that joining this rank of CIAA greats was within my reach and where I wanted to be.

The first reality of basketball tryouts was the need to readjust my time management calendar. Practice did not start in the gym. It started with rising early every morning to run three miles before classes and three miles in the afternoon. This part of practice did not involve handling a basketball. Following the distance running in the afternoon, tryouts would move to the gym, where running continued. This time running consisted of wind sprints in preparation for drills on playing tight and strong defense.

In a short time, I knew that CIAA basketball was played at a different level than what I experienced in high school. In high school, we had practiced in the gym after school. The practice had lasted from about 6:00 p.m. to 9:00 p.m. There had been no running before school began. Running during afternoon practice in high school had involved running hard wind sprints up and down the ninety-four-foot-long court until we almost passed out. There was usually more running if we did not carry out drills in a precise fashion. If a player could not get it, this might result in the whole team being penalized with more running. Running continued with twenty laps around the perimeter of the court, signaling the end of practice. I would soon learn that trying out for the basketball team at STA was far more intense than I had imagined and was accustomed to in high school.

Another shocker came when I stood alongside some of the veteran players and new tryouts. I had not grown taller than six feet three, but my weight had moved to 201 pounds, and I was considered big and strong. My size had resulted in me playing at the power forward spot in high school, but my height and weight were not as imposing among the STA players as it had been among the players I had played with and against in high school. I noticed early in the first practice session that my height was going to be an issue. I had always played down under the basket and had no experience playing at the top of the key. In other words, I had never played at the semicircle that extends beyond the free throw line toward the half court line. This was the area where the guards played.

I was familiar with the skills needed to play at this position for both the point guard, who directed the floor play as the team transitioned from offense to defense or defense to offense, and for the shooting guard. I knew both positions required a level of ball handling skills I did not possess. All of my playing had been as a power forward. If it really turned out that I was not big enough to play power forward, I considered whether I could become the team's small forward. I was not a rigid person. I had enough basketball knowledge to know my shortcomings as a ball handler. Even with the small forward swingman, I knew the position sometimes switched between playing the shooting guard position and on many occasions bringing the ball down court. Regardless of whether I played in the guard position or as a small forward, I recognized there was no getting around increased ball handling if I were moved from being a power forward.

I was used sparingly in practice on both defense and offense. I had no problems grasping the plays. The coaching staff, veteran players, and tryouts all recognized that I was a player who "had game." This included a high level of basketball intelligence. The coaches were impressed with my constant movement without the ball and my ability to anticipate where missed shots would angle off the rim, making me an effective rebounder among taller players. It was clear that my academic ability contributed to and influenced my actions on the basketball court. The coaches often pointed to a feat by me as the way the situation should have been handled. My quick reaction to floor situations did not go unnoticed.

I had an uncanny ability to "anticipate" on the court. This included a variety of actions, from positioning myself on defense where I expected the offensive team would pass the ball to positioning myself for a rebound. My decision making on the court was exceptional. I did not view my on-court skills as an innate ability when it came to basketball. I was keenly aware of drawing on my academic classroom experiences on the court.

For example, when on the bench, as I watched practice sessions or the game, I analyzed every aspect of what was taking place on the floor. I could often see things developing on the court before they actually materialized. I could tell that the player should have cut to the left rather than the right, that he should have made the cut now, or that he should have moved three feet to the right down in the corner because the missed shot was going to bounce off the rim at a forty-five degree angle to the right.

How did I do this? By observing. Based on the initial contact of the basketball with the rim, I was quite skilled at anticipating the trajectory of the basketball off the rim if the shot was missed. I could also tell which trajectory away from the rim the ball would most likely take by watching the basketball from the time it left a player's hands to its most likely initial contact with the rim. As a basketball player, I used the same eye-ball-rim contact to project the basketball movement after contact with the rim as a tennis player uses eye-ball-racket contact to project a tennis ball's movement following contact with the racket. I was clear that my rebounding ability was linked to my understanding and appreciation of mathematical angles found in geometry.

After the final cut of players who did not make the team, the coach met with me to inform me that I had made the STA basketball team. He continued that because of my rebounding ability I would be used in the power forward position. At that moment, it became a reality in my mind that I would play basketball in the CIAA.

When it was time to receive team uniforms, I got another shock. I requested jersey number 33, only to learn the number was taken. I was told the number was used by a veteran senior player who'd had the number since joining the team his sophomore year. In my years of playing organized basketball starting in seventh grade, I had never worn any other number than 33. Putting on a basketball jersey with a number other than 33 was unthinkable. That was my number. That number identified me on the court. How was I to play basketball without wearing number 33? There had been times when I arrived on the courts of opposing teams and could hear echoing from the stands, "Which one is Shepard?" When spectators in the bleachers were unable to recognize me walking toward the visiting team's locker room, there where times I heard someone respond, "He will have on number 33."

I shared my dilemma with the coach. The coach's response was that number 33 belonged to another player and had been worn by a member of the playing squad as far back as he could remember. I asked if the number could become mine once the veteran player graduated or was no longer with the team. The coach said that if no other veteran player was interested in number 33, it would go to me. But for now he instructed me to meet with the person in charge of team uniforms to select one of the unused numbers.

I left the coach's office, still thinking, *How can I play basketball in a number other than 33?* It did not register. I had never given any thought to the matter of what basketball jersey I would put on. It was always number 33. For the first time, I asked myself if my focus on the number 33 was one of those superstitions held by some athletes. I had heard of athletes who would not attempt to play their sport without a pair of socks or underwear they always wore when playing. Others chose not to cut their hair or to grow a beard. Somewhat dejected, I went to pick up my uniform. On the back and front of my STA basketball jersey was number 41.

I continued to perform well in my academic classes but was feeling the effects of basketball practice. I was happy to make the team but began to wonder if the travel schedule would affect my classwork. It was tough. Basketball games would take me all over the state of North Carolina as well as to Virginia, South Carolina, and other distant locations depending on teams we played outside the CIAA.

I was managing my time well, but classes like chemistry and mathematics increased in difficulty and really required more time than I was giving them. My chemistry professor called me into his office and explained that my classroom performance was slipping. He said I needed to make some adjustments and make them quickly. I responded that I could and would get back on track. The professor's only response was that I needed to do it quickly.

I was equally concerned about my mathematics class. As the chemistry professor spoke, I had been thinking about how I was not up-to-date on my math homework. Neither was my study time as abundant as it once had been. But I had been successful in high school with no problem and was confident a few minor adjustments would handle the problem. I was sure I would make the necessary changes to get my studies and my basketball in balance. I kept thinking about the highest scholastic average award I had received three years in a row in high school and was sure I had all I needed to perform at my best in the classroom and on the basketball court. I rationalized that I could do both now, just like in the past.

The basketball season opened with me sitting on the bench. I remained on the bench for the second and third games. This was a first for me. Never had I sat and watched others play a regulation game I believed I was born to play. Sitting was tough. What made it even more difficult was my constant play-by-play analysis of the performance of players out on the floor. On the bench I would think, *That point should have been made. Wrong defensive move. Went to the wrong spot. Ball delivered a second too late. That foul could have been avoided. That foul shot should have been made.* I was convinced I was ready for floor action in the CIAA.

Five minutes into the start of the fourth game, the coach looked down the bench in my direction and summoned for me to come sit next to him. As I rose and sprinted toward the coach, I wondered if this was the time I would enter a game as a player in the CIAA. The coach whispered to me that I would be entering the game to replace the current power forward. I removed my warm-ups, jogged to the scoring table, knelt, and waited for the signal from the referee to enter the game. After what seemed to be forever, the referee blew his whistle and motioned for me to enter the game to replace a player on the floor. In November 1965, I officially joined the ranks of so many talented basketball players who can say, "I played in the CIAA."

STA was playing Fayetteville State University. When I entered the game, we were behind. When the game was over, the scoreboard showed STA as the winner of the game. As a freshman, I had scored fourteen points in my first game in the CIAA. At the sound of the buzzer signaling the end of the game, a spectator leaped from the stands and ran to me. It was a freshman student at Fayetteville State University who had attended one of the rival schools in the WJCAA, where he had seen me as a star player for Garner Consolidated High School.

The spectator introduced himself and said, "Shepard, I thought I was through seeing you in action on a basketball court when you graduated from high school. I cannot believe I am attending a college where I will have to watch you destroy my college team like you did my high school team."

I simply smiled, shook hands with the student from the opposing school, and briskly left the court to head for the locker room. Upon my arrival in the locker room, my teammates walked over to congratulate me on my performance in my first basketball game at the collegiate level. I thanked them and said the come-from-behind win had been a real team effort. Sitting on the bench to take off my sneakers and undress for a shower, I smiled to myself and whispered inside, *I can play in the CIAA.*

I could not dwell on this victory because there were many more games to play before the season ended. Following this first game, I was used for the remainder of the season, sometimes as a starter and sometimes as a sixth man coming off the bench.

As I stood to walk to the shower, my classroom performance surged through my mind. I knew I was getting seriously behind in my tough science and mathematics courses. What troubled me most was that these subjects were the bedrock of the field I had declared as my major— chemistry—with a minor in mathematics. But like in times long past and gone, I had convinced myself I could and would catch up in all my classes.

As my freshman year came to a close I was certain that my sophomore year would be the year I would come into my own on the basketball court in the CIAA and maintain success in the classroom as well. I did not expect to be the sixth man on the team coming off the bench; I expected to be a bona fide starter.

That summer, I needed to work but was not interested in returning to the tobacco fields I had worked in from seventh grade to my senior year. But I needed to earn enough resources to pay for expenses not covered by my scholarship. I decided to ask the coach about student employment on campus. Being on campus would allow me to practice my game when I wasn't working and studying. I could have approached my professor in the chemistry department about summer employment but wanted to have more freedom to devote to my basketball game than would have been possible working in a science laboratory.

The coach said I might qualify for the campus work-study program but that he would need to check with the financial aid office since I was on scholarship. The coach called the financial aid office and informed them that I would be coming over to inquire about a work-study job on campus this summer. He closed the conversation by saying he had work I could do in his department.

As I stood looking at the coach and listening to how he phrased the conversation on my behalf, I thought of how many people in my life had gone before me to help open the doors I had walked through, changing my life forever. Here it was happening again—a stranger I had met only a limited number of times was stopping what he was doing to help me. I did not equate what was happening in front of me as God working in my life, but I sure felt something in my heart as I stood watching the coach speak into the telephone receiver on my behalf. When the coach hung up, he instructed me to go see the person in the financial aid office. She would be waiting for me. This definitely helped to shape my destiny.

I left the coach's office and walked toward the administration building located on the other side of the campus. As I walked, I thought of the many summers I had worked in the hot tobacco fields and how different it would be not doing that this summer. I also thought about my high school chemistry teacher, Mr. Sanders, and how much of his personal time he had invested in me. And a total stranger had just joined the ranks of Mr. Sanders and so many others who stopped to assist me on whatever path I was walking. I thought of how many times my mother had told me that if I worked hard and always tried to do what was right, God would bring people into my life who would try to help me and not harm me. I reflected on how true this had been in my life up to that point.

As I strolled toward the administration building, everything that came to my mind revealed that people, and sometimes total strangers, had reached out to me. I remembered the shoeshine job I was fired from as a boy. I thought about how the white barber had given me an opportunity in a totally white establishment that had not had a Black person working there before or any getting haircuts there. I thought about how again this had been someone helping me. I thought about the white librarian who hired me to tend to her yard. I recalled Mr. Lloyd, the white drugstore owner who had given me the job of cleaning his drugstore at night. The recollection of persons who had helped me on my journey became overwhelming. I arrived at the administration building and went to the Financial Aid Department. Just as the coach had said, they were waiting for me to arrive.

The counselors had already gone over my financial information. They informed me that I did not qualify for federal work-study because my scholarship was covering most of my educational costs, but they shared that the college had some funds that were used for work-study and that I qualified to be compensated from those financial resources. One of the counselors asked if I had talked to the professors in my department about summer employment in a laboratory to tutor students or something in the science area? I said I had not because I wanted a break from the laboratory this summer. The counselor chuckled when I shared how often I had missed dinner meals in the dining hall because of being stuck in the laboratory. I said I wanted to eat more regularly during the summer.

I was assigned a job with other athletes on campus for the summer. This group consisted mainly of members from the football team. They were all pretty big and quite strong. The job involved moving furniture around campus. This included setting up offices and classrooms with new furniture and transporting old furniture to STA's off-campus storage facility. I found the work just as hard as fieldwork in tobacco. The main difference between what I was doing on campus and the work in tobacco fields was that we worked inside on campus, sometimes under air-conditioning, and not outside in the hot sun.

The campus work required sitting down and riding from one location to another, which provided the opportunity for the workers to catch their breath between the take-down process at one location and setup at the next. This often allowed for a brief respite that was not possible working in the fields, especially when mules were exchanged for tractors. Another added benefit of the work was how the heavy lifting helped keep my muscles toned and me physically fit for the basketball court. The one negative about the campus job when compared to tobacco fieldwork was that I received pay every two weeks after taxes were taken out, not daily without anything taken out.

Summer was soon over, and it was time to get back to the grind of both the classroom and the basketball court. I was feeling good about both of them. I tackled my classwork like a man trying to prove something either to myself or those around me. The hard work paid off. My professors were pleased with the outcome of my hard work and encouraged me to keep it up the entire semester. With my classwork back on track, I was ready for the basketball court. I started the first four games but shared more time than I intended with another player. In the fifth game of the season, STA traveled to Winston-Salem, North Carolina, to play a rival school, Winston-Salem Teacher's College (now Winston-Salem State University). The gym was packed to standing room only. Spectators had come from near and far to see one player and no other players on either team. The crowd was in the place to see a basketball phenomenon named Vernon Earl Monroe, who was getting to be known around the world, on and off court, as Earl "the Pearl" Monroe.

It is difficult to go through your team's basketball warm-up drills when all eyes, including yours, are searching the opposing end of the floor, looking for one person. That was the case for the entire STA team that night in Winston-Salem. It seemed that a lifetime passed, and with the roar of the crowd, it was clear that the Pearl had entered the gym. I cast my eyes on him and was quickly taken by his nonchalant attitude on the court but comforted myself somewhat with the fact that he was about my height. There was total difference in the Pearl's movement on the court when compared to all the other players that night. Even his stride during warm-up reflected a basketball confidence I had never seen before. All I could think about was the write-up in *Sports Illustrated* (SI) about the man at the opposite end of the court.

Warm-ups were over. It was now time for the game to begin. I was called on to start. My teammates and I walked to the center of the court and waited for the Winston-Salem players to join us at midcourt to start the game. In what appeared to be at least a minute of waiting, no Winston-Salem and no Pearl. The crowd was on their feet, and the noise was deafening. The moment STA's coach was about to signal for us to return to the bench, the Winston-Salem team broke huddle and strolled to center court. There was one player missing from the floor. Yes, that's right; the Pearl had not entered the game. He was in the building but had not joined his teammates at center court. The noise level rose to yet another decibel level. My heart was racing in what felt like an uncontrollable manner. In all my days of playing the game of basketball, I had never witnessed anything even close to what was going on in that arena. Even the referees were standing, looking, and waiting in an awelike manner.

At his own appointed time, Earl "the Pearl" Monroe slowly walked onto the same court I was on. It was a night I would remember for the rest of my life. The screams and chants were like none I had ever heard on or off a basketball court. "The Pearl is in the house" thundered from the stands. It did not stop there. Others soon followed. "The Pearl, Black Magic, Black Jesus, Earl 'the Pearl' is the best in the world."

The Pearl walked slowly to the center of the court with every eye in the arena, including mine, fixed on him. He shook hands with the STA players closest to him, one of whom happened to be me. The referee tossed the ball high into the air, and the game began. We won the ball toss, and our guard collected the ball. He immediately raced to the right side of the court toward me. I quickly shifted slightly to my right to avoid a defensive trapping situation. The guard passed the ball over to me. Realizing that the defensive guy was giving me too much room, I launched a jumper from the right corner that hit nothing but the bottom of the net. The game was on, and I had scored the first two points in the contest.

Moving back to the opposite end of the court to take a defensive position, I observed the Pearl heading across the half-court mark. The high dribble, the twists and turns, and the fake handoff were nothing like I had ever seen on a basketball court. The Pearl continued his dribble, and once inside the free throw line, he did a 360 degree reverse spin move and launched what appeared to be an off balance shot that went high into the air and then in what looked like slow motion dropped through the basket without the nets moving. The screams and chants in the arena went off the sound level chart.

After the breathtaking move and score by the Pearl, the coach signaled a timeout. Huddling with the team, the coach said to the player supposedly defending the Pearl that it was going to be a long night. He then said to the team to stop standing around worshipping one player and to play our own game. Before sending the players back on the floor, the coach signaled for a player off the bench to take my place. I was shocked beyond words and confused by the coach's move. I had not been assigned to guard the Pearl, so why was the coach taking me out of the game and so early? Without protesting or saying anything that night to the coach, I took a seat on the end of the bench as my replacement went back out on the court in my place.

The game continued to be a show worth sitting and watching, and watch I did. Prior to seeing the exhibition of greatness unfolding before my eyes I had thought my basketball game was pretty good. During my high school playing days I even had distant thoughts of possibly being good enough to play in the National Basketball Association (NBA). But what I witnessed in the performance by the Pearl sealed in my mind that night that I needed to rethink what I would do with my life because the NBA would not be a part of my future. On the bench that night, I assessed that if what I was seeing was a prototype of the players who would enter the NBA in the future, I needed to apply more of my time to my studies. When the game was over and the Pearl had scored 62 points in a one-sided victory for Winston-Salem State University. I made up in my mind that when I returned to campus I was turning in my uniform and spending more time in my books.

On Saturday morning following the Friday night slaughter in Winston-Salem, I washed my basketball uniform, ironed it, and took it with me to the gym to meet with the coach. The coach had not yet arrived to his office, but just as I was turning to leave, he got out of his car in the parking lot. He spoke to me and asked what I thought about the game last night. I responded, "I've never seen anything like that before." The coach only said that Earl Monroe was a basketball talent. After the talk about the game, I asked why he had pulled me so quickly. The coach responded that my replacement had more of the Philadelphia-type playground experience like the Pearl, and he had felt he would fare better in the game than I would. I did not agree with his statement one bit but did not voice my disagreement. I simply said, "Such experience was not shown in the game last night."

I proceeded to explain to the coach that I was getting off the team. The coach seemed somewhat shocked and quickly asked if I was making the decision because I had been pulled from the game. I said no, not at all. I told him my decision was based on a desire to be included as one of the best at whatever I took on, and at my current skill level I knew I could never become one of the best players in the CIAA. I said that if I had come to STA on a basketball scholarship it might have been different. But I was on an academic scholarship instead. I told the coach that to make matters worse, I was in a very challenging major of chemistry with mathematics as a minor and that competing in basketball was causing me to lose ground in my major.

It was clear to me, if no one else, that I was much better at chemistry and mathematics than I ever would be at basketball. Since both endeavors were time-consuming it would be more beneficial for me to spend more time in these subjects. I expressed to the coach that at this point of my basketball development, I did not believe it was realistic to think I would reach a level of performance in basketball much higher than my current skill level, and for that reason I had decided to put my energy in the area I believed would yield me the better return on the time invested. The coach conveyed that there were areas of my game I should work on to improve and that he believed I could become a much better player if I worked at it.

I thanked the coach for his words of encouragement and the opportunity to carry out a childhood dream of playing in the CIAA. I handed him my uniform with the number 41 showing on top. No, there was no thought of whether things would have turned out differently if I had worn number 33. I concluded that the basketball performance displayed by Earl "the Pearl" Monroe was pure talent, and no superstitious act relating to the number on the back of a jersey could have impacted what I witnessed in Winston Salem in November 1966. As I exited the office, the coach's final words were, "You are only a sophomore, and if you change your mind you are welcome to take another stab at it." I smiled and left the gym.

As I walked from the gym back toward my dorm room I wondered how a player could become as talented as Earl Monroe. I wondered if he had been born with such basketball skills or had acquired his talent in the hometown environment he came from in Philadelphia. I thought about the many years I had played organized basketball and the players I had gone up against who I thought had exceptional talent and basketball skills. I thought of how just two years ago my high school team had been playing at a level that had almost resulted in them being the North Carolina State 2A Division champions. I then wondered if any player in the whole state of North Carolina could play at the level I saw on the court in Winston-Salem from Earl Monroe.

As I approached the dormitory and was thinking of how abruptly I was hanging up my sneakers after being a part of the sport for most of my life, I remembered something my mother said that came from the Bible. When any of us mentioned what might have been, Mom always came back with, "He who puts his hand to the plow and looks back is not fit for the kingdom." She never explained the statement, just used it as a warning to us when we looked more behind than ahead. Having never seen this in the Bible for myself, I simply wondered now if her statement applied to me as it related to basketball and if I should as of that day stop thinking about basketball (looking back) and focus more on other possibilities ahead of me in life.

I walked up the steps to my dorm room, opened the door, walked in, sat down at my desk, took out my chemistry book, and began to read. My complete focus shifted to my academic performance and what I needed to do to elevate my studies, especially in my chemistry and mathematics courses. In a short period of time, I returned to a level of performance in my studies that was sufficient enough for me to maintain my academic scholarship.

I was so sure I would become a teacher like my uncles that I became teacher certified before graduating. I was not particularly fond of teaching, but that was all I knew. Not being fond of teaching was confirmed when I taught chemistry, mathematics, and physics during my student teaching assignment at a high school in a county about forty-five miles from Garner. I could not see myself standing in a classroom for the rest of my working life.

Student teaching proved something important to me: there is a difference between a career and a job. Reporting to the classroom for nine weeks was long enough for me to determine that for me, teaching was a job. I was convinced that teaching could never lead me to a career.

One afternoon following a full day of student teaching, grading papers, and interacting with students, before packing up to head back to STA, I sat down at my desk and looked out the window. I began to remember the speech given by the speaker at my high school baccalaureate service. As I gazed out the window and into the distance, I remembered what Dr. Johnson said about the graduates starting their journey toward taking their places in the larger world. I thought about how he stressed the need for us to move from the comfortable but shallow shoreline of the river out into the deep parts that are more hidden, concealed, and unknown.

At that very moment, a large body of water somewhere in a distant land vividly formed in my mind. I could see myself move farther and farther from the shore toward the horizon that sat off a far distance from where I was sitting. I reflected on Dr. Johnson saying there was a need to continue to hold on to God's unchanging hand and take education beyond my high school diploma. My mind went back to the days I spent on my grandmother's front porch, dreaming about one day going wherever US Highway 70 would take me. I had worked hard at every task I started, and now I had added student teaching to my portfolio.

While sitting and reflecting, it felt like something began pulling me away from the student teaching classroom experience. I finally rose from the chair at my desk, gathered my books and papers, and headed for the door. I exited my classroom and entered the hallway. As I walked down the long hallway, my mind once again returned to Dr. Johnson's message. Upon exiting the building, I stood on the small porchlike area looking out over the field of tobacco surrounding the school and thought of the many years I had spent working in those fields.

I looked up at the blue sky, and my mind began to compete with which thought to hold on to about my future. I wanted more but could not settle on what that might be. All I knew was an unsettling feeling weighed on my spirit. Finally, I descended the steps of the building, walked across the parking lot to my car, opened the trunk, neatly arranged my books and papers in a box, closed the trunk, got in the car, and drove away. At the end of my student teaching assignment, I received high ratings from the high school leadership team as someone with talent to become an exceptional teacher and leader within the North Carolina education system. My mother was delighted that I graduated from college qualified to teach science and mathematics in high school.

During my undergraduate days at STA, I did not think about going to graduate school. Like my uncles, teaching science and math in a local high school in North Carolina would be next for me. But a month before the graduation exercise in May 1969, STA hosted an event that would change my path. The event was the twenty-sixth joint meeting of the National Institute of Science (NSI) and Beta Kappa Chi (BKX). The national meeting brought to the campus students and faculty from many of the nation's HBCUs.

The meeting provided an opportunity for the participants to exchange information and present scholarly research papers in science and mathematics at the undergraduate and graduate levels. The meeting provided a platform for practicing scientists to encourage and advance scientific education through original investigation, dissemination of scientific knowledge, and stimulation of high scholarship research in pure and applied science. In addition to presenting research, the joint meeting gave undergraduate students at HBCUs like STA an opportunity to interact with graduate students and professionals currently working in science-related fields.

The graduate students from the Department of Chemistry at Howard University in Washington, DC, were noticeable and dominant participants in the meeting. The meeting was organized so that presentations did not conflict, providing the opportunity for students to hear all speakers. I became mesmerized each time a graduate student from the Department of Chemistry at Howard University made a presentation. Their youthful look and their chemistry knowledge almost took my breath away.

I could not believe that individuals who looked my age were only months away from receiving PhDs. When I learned that some of the graduate students would be twenty-five, twenty-six, and twenty-seven years old when they received their PhDs, I was stunned. One Howard chemistry professor I chatted with looked to be in his twenties. The youth of this group of Black scientists fascinated me. Up to this point I had equated having a PhD only to persons much older, in their fifties and beyond. I thought this because that appeared to be the age range of the STA professors who taught me. At no time did it occur to me to ask how old any of the professors were when they received their PhDs. Never had I seen as many Black scientists in one location as I saw at the meeting. I wanted to become a part of this group. Sitting and listening to the presentations of these Black scientists was as mesmerizing to me as when I had sat and watched Earl "the Pearl" Monroe. The only difference was that the two events had played out in different arenas.

The graduate students and faculty from Howard placed a flier in the hands of every undergraduate student attending the meeting. The flier announced a gathering on Friday night that Howard was having in the lobby of the hotel where they were staying in downtown Raleigh. The flier announced that Howard representatives wanted to talk to all undergraduate students who might be interested in attending Howard University to pursue graduate studies in the field of chemistry. I carried the flier around in my pocket and pulled it out to glance at periodically. I remembered that the young-looking chemistry professor had mentioned the gathering to me before the flier was passed out. Looking at the flier, I figured that my overall grades were not good enough to be accepted into a graduate program. On that note, I put the flier back in my pocket.

Later that day, about ten minutes before the start time of the activity, I called my supervisor at the bank where I was a check sorter to say that something critically important to my school work had come up and I would not be in tonight. He said it was fine and asked if I thought I would be in on Monday evening. I said yes. I then jumped in my car and headed to the Sir Walter Raleigh Hotel downtown. When I arrived, many students were standing and chatting with the Howard faculty and graduate students. The young chemistry professor I had been chatting with earlier in the day spotted me and motioned for me to come over. I made my way through the crowd. The professor shook my hand and immediately asked what I had thought of the meeting. I said it was great. I explained what impressed me most was how young the graduate students looked, even the ones who were just a year or two away from earning their PhDs.

Looking intently at the professor, I said, "When you were introduced as Dr. Jesse Nicholson, professor of chemistry at Howard University, I could not believe it was true. You look as young as me." I asked how old he had been when he received his PhD.

He said, "I was twenty-seven years old when I received my PhD from Brandies University, and I completed my postdoctoral studies at the University of California at Los Angles at age twenty-nine."

At twenty-one years old already and poised to turn twenty-two at the end of the year, I was convinced that it would be absolutely impossible for me to accomplish something as lofty as receiving a PhD over the next five or six years, even if the opportunity presented itself.

As the evening moved on, I circulated among the people. I was somewhat comfortable because two of the graduate students from Howard were former students from STA. I spent time conversing with them about graduate school. I had no reference point relative to graduate school, so any question I posed was viewed as valid. I asked about workload and the makeup of the various classes relative to where the students were from. I wanted to know about living arrangements. I asked everything imaginable about graduate school because, first, I had never considered it and, second, I had never been in an environment where such information could be gathered at one time.

I finally asked about the professors and whether they were harder than the ones I had encountered in STA's Chemistry Department. One of the Howard students suggested that I ask the professors in the room about what a professor in Howard's Chemistry Department was looking for in a prospective graduate student.

I made my way back to Dr. Nicholson. Before I could inquire about the faculty in the Chemistry Department, Dr. Nicholson asked if I was interested in going to graduate school. I was not prepared for the question.

Before I could respond, he asked how my STA grades were. I responded that overall my grades were not good but that my grades in science and math were pretty good. I told Dr. Nicholson that, looking primarily at how I performed in science and math, I believed my performance was very good. Dr. Nicholson returned to the original question of whether I had thought about going to graduate school.

I responded this time. I said I had never thought about going to graduate school because none of my uncles had gone or mentioned going to me. I told him my high school chemistry teacher had thought I was good in chemistry but that he had never mentioned to me education beyond a bachelor's degree. I said I would be graduating on May 12 and would be certified to teach high school science and mathematics in the North Carolina school system. I told Dr. Nicholson that the principal of the school where I had done my student teaching had a contract for me to sign as the science and math teacher at his high school.

Dr. Nicholson asked if I thought I would be interested in attending Howard's graduate school. I was speechless and did not answer right away. After a few seconds, I responded that I had never thought about graduate school and that thinking about it gave rise to many questions and fears. Dr. Nicholson suggested that we not put the cart before the horse. He asked if I could have a copy of my transcript sent on Monday via United Parcel Service (UPS) to the dean of the graduate school at Howard University, along with a copy to him. He also informed me that the Chemistry Department would be reviewing graduate school applications within a week and that he would love to have me included among the applicants. My head was spinning. I kept thinking of how the day was Friday, April 23, 1965, and that my graduation was less than two weeks away. What could be done in such a short period of time that could potentially be so impactful? *Nothing,* I thought. Nevertheless, I promised Dr. Nicholson that he would receive my transcript first thing Monday morning.

All I could think about over the weekend was how the graduate school selection committee would view my transcript. I remembered that I had been straight with Dr. Nicholson about the shortcomings of my transcript in some areas but the strength of it in science and mathematics. I thought about calling him and discussing my transcript in greater detail before sending it on Monday. I concluded that would not be a good idea. So I waited over the weekend until the registrar's office opened on Monday morning. I sent my transcript on April 26, 1969, to the dean of the graduate school at Howard University, with a copy to Dr. Jesse Nicholson, professor of chemistry. I had made an official application to become a graduate student at Howard University in Washington, DC.

I was going through the close-out exercise seniors were required to do. My books, clothes, and other items accumulated during my four years on campus were packed and ready to be taken home until I could find my own place to live. In the afternoon, the senior class reported to the gymnasium for a walk-through and practice exercise for the graduation coming up on May 12, which was now just one week away. In the middle of practice, the coach I had once played basketball for came to the doorway of the entrance to the gymnasium and announced that I had a telephone call in his office. I was stunned and could not imagine who was calling me. I stepped out of line and went into the coach's office to take the call.

On the other end of the line was Professor Jesse Nicholson from Howard University with this announcement: "Robert, because of your high level of performance in science and mathematics at the undergraduate level at Saint Augustine's College, the selection committee at Howard University Department of Chemistry has unanimously voted to award you a National Aeronautics and Space Administration (NASA) Pre-Doctoral Fellowship to attend Howard University in Washington, DC." He continued by stating that a response was needed within a couple of days. The prestigious fellowship was being awarded for me to continue my education by pursuing a doctor of philosophy (PhD) degree in chemistry at the nation's comprehensive, premier, and world-renowned HBCU, Howard University.

I was speechless. I remained speechless and motionless for at least five seconds. I fixed my eyes on the window of the coach's office and looked out across the parking lot into the distance on the other side of the street. For what seemed to be an eternity, I reflected back over my life. Again I thought about the days I spent on Grandma's front porch, looking at eighteen-wheeler trucks go up and down US Highway 70, wondering where they were going. I thought of the jobs I had worked as a young boy, especially in the tobacco fields. I thought of my mother's educational push and my father's strong work ethic. I thought of all this and more but soon answered that I would accept the offer but really needed to get back with him after speaking with my mother and father. I then hung up the phone.

I had been silent so long on the phone and after hanging it up that the coach asked if everything was all right. I looked at the coach and said, "Coach, you will be the first to hear this. I was just offered a fellowship to attend Howard University in Washington, DC, to work on my PhD degree in chemistry." The coach rose from his chair and extended his right hand to congratulate me. He declared, "This could not have happened to a finer young man. It is obvious that you came to the right conclusion that night in Winston-Salem when you decided to move on and concentrate more on your academics. Shepard, I am very proud of you, and I know you will go up to Howard and do a great job." I thanked the coach but said I was still trying to figure out if the call was real. The coach assured me that it was.

Accepting the fellowship would open a new chapter in my life, and what went with it was the inability to look back even if I wanted to. Instead I now needed to focus on what was ahead. Along the way, I found in the Bible the statement Mom often quoted about putting your hand to the plow. It came from the book of Luke: "But Jesus said to him, No one, after putting his hand to the plow and looking back, is fit for the kingdom of God" (Luke 9:62 NASB). I returned to the line for the graduation practice and realized that God had truly orchestrated what had just transpired over the telephone in the coach's office. Another giant step toward me fulfilling my destiny was about to take shape.

Chapter 9

THE SURPRISE SECRET

I NEVER REALLY DATED WHILE IN high school. Basketball was my constant companion and friend. Adding in my academic studies left little time for having a steady girlfriend. With the expectations on me from my family, my church, and my entire community, I did not feel any pressure but had the personal desire to not disappoint those who thought so highly of me. I simply maintained a cheerful attitude and was a friend to all I encountered, both my male and female classmates. There were some classmates I was closer to than others. For example, during my junior and senior year in high school there were a couple of female classmates who I seemed to like and could at times be seen walking them to their buses and appearing to be in very deep discussion as I carried their books for them. Nothing serious ever developed between my classmate friends and me.

I had the ability to deflect potential harmful situations without causing dissension, turning them into more light-hearted moments. During these moments, my friends and acquaintances said to me, "Robert, you seem to be able to change the direction of a situation and not let it bother you." My response was simply to smile and quickly move the conversation in a different direction. On occasions my presence helped eliminate friction from the situation. The ability to bring about such reactions was not something I thought about or was even aware of.

It could be said that I was good at bringing about cooperation as opposed to confusion and conflict. I consciously pursued such a course in my interactions because that had not always been the case in my home when I was younger. At times, when I was in my room I could hear my mom and dad in argumentative and heated verbal attacks. There was not a single time in these instances when as I listened and processed the discussion I could not think of a way to deal with the situation in a more positive manner. So it could have been that I made a conscious effort to try to move encounters in a constructive direction, knowing that if things continued on a more negative path they were sure to result in discord and confusion. Whatever brought me to the point of choosing a more lighthearted approach, friends and acquaintances seemed thrilled to be around me.

I was home from STA for Christmas break during my freshman year one week before my eighteenth birthday. My cousins called to ask if I was interested in going to a high school holiday basketball tournament that involved some of the schools I played against while at Garner. I jumped at the opportunity. Two reasons to say yes quickly came to my mind. First, I loved the game of basketball, and the tournament site was only a short drive away. Second, I realized it was only nine months ago that I had been playing in the conference regional tournament against the team that went on to win the North Carolina 2A Division State Championship. I figured there would be players and fans at the tournament who would recognize me. Sure enough, there was, and some came up to ask if I was still playing.

As I looked over the audience, my eyes fell upon a beautiful young girl. She was standing behind a chair with her right foot on the bottom rung while her left foot remained on the floor. She was wearing a mixed green-and-pink tweed skirt and white blouse. Every time I tried to concentrate on the game being played on the floor, my eyes kept returning to the young girl. There was never a time before in my life that I had seen anyone who without saying a word captivated my attention that way. What was going on here? I had seen attractive and even beautiful girls before, so I surmised in my mind that what was happening to me was a normal thing for someone turning eighteen the following week.

Neither of my cousins was aware that for the first time while watching a game, my mind was not in the game doing my usual mental play-by-play analysis of what was happening on the floor. At the conclusion of the game, my cousins were ready to go, but I asked them to wait a few minutes. I approached a young man who was a student at the opposing school, introduced myself, and described the beautiful young girl I wanted to meet. The student knew who I was describing and said for me to wait and he would go tell her I wanted to meet her. The student dashed off in search of the girl

In less than five minutes the student returned to where I was waiting with word that the girl said she would be there shortly. I asked if he told her who was waiting to meet her. He said yes. I asked what she had said. He said that she just said she would be out shortly. I was somewhat surprised that the girl did not immediately come out to meet me and did not do so at a fast pace. After what seems like five minutes passed, my cousins came over and asked what was going on. I told them I would be ready to go in a few minutes.

They asked, "What or who are you waiting for?"

As I was about to answer, or not answer, their question, I looked in the direction of the gymnasium and gazed upon the beautiful young girl strolling toward where I was standing with my cousins. When my cousins realized the girl was coming over to me, they turned and left.

The girl walked up and asked if I was the person waiting to meet her. I said yes and quickly followed with, "And do you know who I am?" The girl said, "No, I do not, but should I know who you are?" I asked, "You are telling me that you have no idea who this is talking to you?" Her response was the same. Before telling the girl my name, I asked if she attended any Wake-Johnson Athletic Association (WJAA) Conference basketball games over the past four years. She said, "No." I said, "Then that accounts for why you do not know who I am," and went on to introduce myself. I told her I was Robert Shepard, a freshman basketball player at Saint Augustine's

College, and that I was home on Christmas break. I dared not tell the young girl that I wasn't eighteen, thinking age seemed young but college presented the picture of a more mature individual.

The girl said her name was Alzonia (Zonia for short) and that she was a junior in high school. I told her I had friends waiting for me so I had to go but asked if I could see her again. She responded that she was seeing someone and was not sure about that. I smiled, said it was a pleasure meeting her, and left. I felt good because I figured that the young girl and I were close in age, since she would graduate in 1967 and the class I supposed to have graduated in was 1966. Once in the car, my cousins asked who I had been talking to. I just said she was some girl I noticed during the game and wanted to know who she was. The cousins looked at each other and smiled. Then we drove away without any further discussion of the reason we left the game later than originally planned.

The Christmas break was over, and I was back in school, or supposed to be. January 1966 opened with a big blast of weather. Shortly after school opened for the spring semester, classes were cancelled because of severe weather conditions. About an inch of snow was on the ground. Trees suffered broken limbs that were all over the ground. Electricity was out for days in some communities. After the short delay, classes resumed, and I went back to my studies.

I did not have much contact during the first half of the New Year with the young girl. I was able to get her telephone number and moved quickly to contact her. She received my call, and we talked for a short while. I did not want to be too aggressive. The call ended without me making a request like, "Can I come see you?" or "Would you like to go out?" I made no such requests. I simply asked if I could call her again, and she said that would be okay. The calls continued.

In early spring, Zonia informed me she would be coming to Raleigh over the weekend with some friends. I asked if she could drop by. I wanted to show her the campus. She said she didn't know but would try. She visited the campus, and I gave her a brief tour. At the conclusion of the tour, I could tell she had enjoyed it. Her friends arrived to pick her up, and the visit was over.

School was over in May, and I had completed my first year of college. Being a member of the basketball team in the fall had interfered somewhat with the overall performance of my first year. I was not satisfied with my ending GPA. It was high enough, however, for me to pledge the fraternity requiring the highest GPA, Alpha Phi Alpha (AøA). To ensure my GPA would be high enough to be accepted into the AøA fraternity at the start of the fall semester, I went to summer school in 1966. At the conclusion of summer school, my GPA was more than high enough for me to join the fraternity.

Summer school soon ended. I had worked hard in summer school but still found time to speak with Zonia a few times by telephone. The conversation was usually lighthearted and consisted of talking about what the day was like. We lived only sixteen miles apart, but there were no home visits. I had my driver's license, but with only one car in the family and my dad needing it all the time, I had no way to visit her anyway.

At the start of her senior year in high school in the fall of 1966, Zonia was busy on many fronts. She was a member of the girl's basketball team and on several senior class committees. She was excited about her final year in high school. As the second oldest in a family that would soon consist of nine children and with a family background of domestic workers, Zonia had never thought much about education beyond high school even though she was a very bright student. She knew money was tight and that she would need a full scholarship for college to become a reality for her. Even with the distant thought of college for herself, she liked that a boy in college was talking to her. And while she had expressed it to me in only a casual manner, she was giving more thought about college during her senior year in high school than at any time in the past.

In the fall of 1967, my sophomore year in college started well, as did Zonia's senior year in high school. She and I began to talk more often by telephone, but because of transportation limitations, the visits were very infrequent. One weekend a fellow high school basketball player was on campus and spotted me. We briefly engaged in small talk before I asked if he had anything to do when he left campus. He said nothing was going on. I asked if he would take me to see someone who lived about sixteen miles away from campus. The player asked me who I knew who lived that far from Garner. I told him I had met a young girl and wanted to visit her. The player said he would take me but that we could not stay very long. I said fine and that I would be right back but needed to call her to see if a visit was okay.

I dashed off to the dorm to call her. You guessed it. There was someone on the telephone in the hallway of the dormitory, with two others standing in line waiting to use the pay phone. I ran up the stairs to the second floor. That phone was also being used. I climbed one more flight of stairs in hopes of finding a phone not in use. The third floor was free. I called, and she said she was not doing anything and that she needed to check with her parents to see if it would be okay. She asked me to hold just a minute. She returned and said it would be okay for me to visit. I said I would be there in about twenty minutes. I hung up the phone and ran back to meet the person who would become the driver for me to visit Zonia for the first time at her home.

Upon arrival, I met her family for the first time. She had four sisters and three brothers. Her mother was pregnant and due to deliver her ninth child in June 1967, around the same time the young girl would be graduating from high school. I enjoyed the visit and found the family warm and pleasant but soon left with my driver. On our way back, my driver asked where I had met the young girl. I told him at last year's Christmas holiday basketball tournament, which had been held at her high school. The driver told me the whole family seemed to be very nice people. I felt the same way and looked out the passenger side of the car window as we cruised the highway back to STA.

I thanked my driver. Having only three dollars in my pocket, I offered him a dollar for gas. I said I hoped to repay the favor if he needed a ride in the future. He took the money and left. As I exited the car, I wondered how I would make good on my statement of repaying the favor when my use of the family car was so limited. I thought about my being eighteen years old and needing someone to drive me to visit someone who was becoming a female friend. While I was appreciative of the ride, I was somewhat disturbed that at six feet three and 201 pounds I was not able to drive myself on my first visit to see someone who was becoming a close friend. I knew I needed to ask my father about using the family car on some weekends if I was going to visit her again.

Zonia began to talk more about going to college and possibly attending STA. I recommended she apply and do so early. She was not afraid to apply but did have concerns about her application being rejected. We concluded that either way it turned out, by applying she would gain from the process; not being accepted would provide information useful for making alternative decisions if she did not go to college.

Zonia submitted an application to attend STA in the fall of 1967. To prove she was serious about college, she moved quickly to complete and mail her application before the end of January 1967. Now it was time for her to pull together her family's financial picture and other materials needed to support her application. As she got deeper into the process, it was a great relief to her to be able to bounce questions off me, and I was always eager to help her whenever I could.

My second year of college was coming to an end. With basketball now in my distant past, I continued to balance my classwork and the time I was spending talking and visiting with Zonia when I could. I had done really well in my major subjects but again not as good as I could in the other areas. When I could have access to the family car, I enjoyed introducing her to things that were novel and new and she had not been exposed to before.

For example, a part of my freshman orientation had been a trip to the Ambassador Theatre in downtown Raleigh to see the long-running movie *The Sound of Music*. I loved the movie, which played at the Ambassador for an amazing sixty-one weeks from its initial run starting in the spring of 1965 through midsummer of 1966. I delighted in taking her to see the film, and she loved it. Knowing that much was to be done in preparation for college, I also used our time together to check in about her application to STA. I encouraged her to believe that she would be accepted and to be ready when it happened.

There was one more high watermark event before she graduated from high school—her senior prom. She had not participated in her junior prom, so it was important to her to see what a prom was like and not depend on classmates to describe it to her. She wanted to invite me but was not sure a college guy would be interested in escorting a high school girl to a prom. She thought about it for a few days but asked me about it one evening as we talked on the telephone. I was beginning to like Zonia a lot, so escorting her to her senior prom was fine with me.

I kept everything in perspective. As a nineteen-year-old college sophomore, I had no problem escorting a seventeen-year-old high school senior who would soon turn eighteen before I would turn twenty. Anyway, many of her classmates, both the guys and the girls, knew me from my basketball playing days in Garner. And since none of them had seen the two of us together, I knew they would be surprised to see the two of us now as friends.

After being asked to escort her, I started months in advance of the prom, alerting my father of my need to use the car. I explained that he had to let me use the car on this special occasion. My family had met the young girl when I brought her home one evening in the early part of the year. They really liked her, but she did not sense that my family was too fond of her, especially my mother.

My mother assured me that I would have transportation to take my friend to her prom. Following the prom, a good deal of our conversation continued on the subject of college. Now that her application was submitted, she was not sure how she could attend since nothing would be available from home. I suggested not to put the cart before the horse but to wait.

With the investment of time leading up to the prom, during the prom, after the prom, and through graduation, I grew to like Zonia even more and wanted the best for her. The prom, coupled with the application to STA, seemed to link us toward common goals. Our conversations seemed to converge toward a common interest more than in the past. After the prom, transportation remained an issue for me. This dilemma caused me to still only see and visit her occasionally.

A month after prom, Zonia called me with excitement and joy in her voice. I had to calm her down so I could understand the news she was giving me. The information she shared would change her life. She had just received in the mail a letter from Saint Augustine's College that read,

"Congratulations. You have been accepted to Saint Augustine's College as part of the incoming freshman class for the fall of 1967." In addition, to her delight, I also could hear jubilant voices in the background from other family members. It was truly a day of celebration and thanksgiving for her and her family. She was the second oldest of eight children, soon to be nine with the birth of her baby brother, Reggie, in another month, and the first of her immediate family to head to college.

While there was much to celebrate and for which to be thankful, I soon learned of the great fear that had set in on her. As we continued to talk over the summer about a college experience, I assured her that she would be just fine. I assured her that I would be there to help her navigate the details and complexities associated with the college experience.

The acceptance letter acknowledged what fraction of her educational expenses would be covered by federal loans and informed her that she would receive a follow-up letter describing the portion of the expenses her family would have to pay based on the financial information that had accompanied her application. This concerned her. She knew her family would have nothing to pay toward her college education. When I heard her discussing the subject of how her college expenses would be covered, I told her not to worry and that a way would be made when the time came.

Zonia finally received the follow-up letter describing the portion her family would have to pay for the first semester and the amount required for the second semester. The letter described that in addition to the resources she would earn from an on-campus work-study job, it was clear that she would need to take out a personal loan to cover the balance of the cost.

She worked in tobacco the summer after graduating from high school. She continued to work and save as much as possible. Every day as she worked, her thoughts were about becoming a freshman in college. Just the thought of it brought joy and wonder to her spirit. Since college was not something she had thought much about in her early years, she had a lot to undergo in preparation for the transition to college. She was ready for the challenges ahead.

As the summer of 1967 rolled on and we were seeing more of each other, there was tremendous unrest in the nation and in the world. The

Vietnam War was getting worse with what appeared to be no end in sight. US soldiers were being shipped to the war by the thousands. Media coverage on the war clearly revealed a troubled and divided nation. Vietnam week was established in April just before the tax deadline that year. Demonstrations were staged in cities across the country with one of the largest taking place in New York when more than 400,000 people marched from Central Park to the United Nations headquarters.

There were other sizeable protests against the War everywhere. Anti-war and anti-draft demonstrations occurred in many large and small cities and towns throughout the nation. In addition to the anti-draft demonstrations, draft cards were being burned. As the number of young Black boys and Black men being shipped off to the war and their death tolls rose, Dr. Martin Luther King Jr. began to speak out forcefully against the war in Vietnam. Dr. King was not the only voice protesting the war. Stokely Carmichael and many other activists added their voices of disapproval.

When the number of US troops in Vietnam reached 448,400 in June, the nation's leadership seemed paralyzed by what the war was doing to the country. Many colleges and universities were engaged in the war dialogue.

Great concern was expressed by leaders in the Black community and those on the front line of the civil rights movement. The major concern was that Blacks, who made up only 11 percent of the nation's population at that time, made up nearly 13 percent of the soldiers in Vietnam, with a large number of the young men in infantry. With the casualties among Black soldiers on a steady rise, coupled with delays in racial progress in the United States, Zonia and I were center stage watching high levels of frustration and tension mount right before our eyes. Radical leader Stokely Carmichael visited STA and eloquently spoke about the details of the war and the negative impact it was having on Black male students during his keynote address at one of our Afro-American Festival Week activities. We closely watched these events unfold, as did the rest of the nation.

World Heavy Weight Boxing Champion Muhammad Ali refused to be drafted and go to Vietnam. This became a major high point of the news. This action by Muhammad Ali inspired all of us. His action even motivated civil right leader Dr. King to elevate his speech and intensified his opposition to the Vietnam War. His messages began to link the high casualty rate among Blacks in Vietnam to long-standing racism in America. There were accusations of the Vietnam War having brought about a new form of discrimination against Blacks. Civil rights leaders, including Dr. King, described the Vietnam conflict as racist—"a white man's war, a Black man's fight."

Race riots and demonstrations mounted and continued throughout the summer of 1967. In major cities like Chicago, Philadelphia, Newark, Cleveland, Baltimore, Detroit, Los Angeles, and Washington, DC, the riots ended in more death outside the war and great damage to property. The race riots were not confined to major cities but were also on ships and military bases.

I recognized that the Vietnam War was devastating on the Black community. I often thought of my freshman classmate and close friend who in the summer of 1966 was drafted into the army. He was soon killed in action in Vietnam and never returned to college. The war was real for me, and I appreciated having a student deferment. I never feared serving in the military. As noted in earlier chapters, I had close relatives—including uncles, cousins, and brothers-in-law—who were all proud military men. I loved hearing them talk of their time serving our country and the pride they expressed in carrying out their duties. The racial tension surrounding the Vietnam War gave it a more disturbing composition.

I worked on campus during the summer following my sophomore year. Being on scholarship allowed me to save the money I earned and add it to my savings account. Another key thing happened to me during that summer. One of the local banks had an announcement on the bulletin board in the administration building on campus for an opening for a sorter operator on the nightshift. I figured this would be a perfect job for me and would allow me to buy my own car.

To see how much I would need to save for a car, I visited used car lots during the summer of 1967 to investigate prices. I admired three muscle cars during this time: (1) the Pontiac GTO, (2) the Chevrolet Chevelle Super Sport 396, and (3) the Plymouth Road Runner. Upon researching the prices of these cars, I discovered they were too costly for me to consider. But I did strike gold when I visited a Ford dealership to look at used Ford Mustangs. Over in the back corner of the used lot sat a yellow 1966 Chevrolet Chevelle Malibu reasonably priced based on what I already had in my account.

The salesman approached me and asked if I was interested in the lone 1966 Chevrolet sitting back in a hidden corner in the midst of Ford cars. I said I was but being that the car was only two years old, the cost might be a little too high for me. The salesman quickly responded, "You are standing in Ford country, and I do not give a damn about a Chevrolet on my lot." He continued by saying that if I was really interested and could handle a payment plan, he was sure a deal could be worked out. I had not expected the conversation to go in the direction it did at such fast pace. I asked if I could test-drive the car. The salesman went to retrieve the keys, returned, and handed them over to me. He then walked around the beautiful automobile and climbed into the passenger side as I moved into the driver's seat behind the wheel. I sat for a moment and then cranked the engine and cruised the baby across town and onto a thoroughfare where I could test the speed.

The fine automobile drove well, and with the low mileage of only 24,000 miles showing on the odometer, it was better than the family car my father allowed me to drive on several occasions. I was hooked but knew help would be needed for me to get the car that was priced at $2,800, including all taxes and North Carolina license tags. The salesman told me he needed to move quickly because he did not believe the car would remain much longer on his lot. He shared that a potential buyer had shown interest in the car several hours before I arrived. I informed him that I had to speak to my parents about purchasing a car but hoped there was some way the car could be held twenty-four hours. The salesman said he could not do anything without a down payment. I had no down payment and could only hope the car would be there when I returned. The salesman said something I would later categorize as another event where divine involvement was at play in directing my life. The salesman said he would bury the car deeper behind other cars to further lessen its view on the lot.

I thanked him for that and promised to return as soon as I talked to my parents.

When I hinted to my parents the notion of buying a car before I graduated from college, they were not in favor of the idea; my mother was more against the idea than my father. I explained the job prospect I planned to look into that would involve working at night. My mother strongly objected to such an idea. In addition to expressing the need for me to be in my dormitory room studying, she said it would be much too dangerous for me to be out in Raleigh late at night when bad things were happening. I explained how I had demonstrated good time management over the years between my studies and various other things. I specifically pointed to how much basketball kept me away at night but never interfered with my academic performance. And as far as safety was concerned, I said I would be all right. I knew I would need additional steady income to pay for a car. If my parents consented, I planned to pay $300 down on the car and get the Ford Company to finance the $2,500 balance. My parents consented that it was okay for me to get more details about the job and to apply for it if I felt comfortable about the nighttime hours.

I called to inquire if the job posted on STA's bulletin board was still available, and it was. I said to the bank personnel representative that I would complete an application and return it to the bank. The representative asked if I had time to come to the bank that day, fill out the application, and sit for an interview with the hiring official. I paused momentarily as I tried to figure how this could be done without transportation. I asked how much time I would have to get to the bank. The bank representative said the hiring official would be there until six o'clock and if I got there about one hour before he left, that would be good. I said I would be there. The time was 2:13 p.m. With no more classes for the day, I knew I had plenty of time to walk from my dormitory to the bank downtown.

I arrived at the bank in plenty of time to complete the application process and be interviewed by the hiring official. The job description stated that the bank was seeking to hire a sorter operator. Like most people, I was acquainted with bank tellers who cashed checks and processed deposits and withdrawals. At the desk to my left, I saw the bank employee whose job was to handle existing customer issues and open new accounts and new loans. Also like most people, I was not aware of the bank employees who worked behind the scenes. A bank sorter operator was such a position.

In the interview it was explained to me that the various payments mailed to the bank every month were opened by a machine that sorted and processed the mail. The bank had hundreds of thousands of customers who made such transactions, and it was essential that mail sorting be done quickly and accurately. The final step of the sorting process involved making sure the sorted checks balanced with the internal bank statements. It was explained that the last step could be tedious and time-consuming. The hiring official said the standard working hours for the job were set from 6:00 p.m. until 10:00 p.m.; however, he warned that the ending hours could sometimes be much later depending on the level of difficulty of balancing the sorted checks with internal statements.

He asked me if I thought I would be interested in the job. Without hesitation, I responded yes. The hiring official said he would let me know within a few days. I rose from the chair, extended my right hand to shake the hand of the hiring official, and left the office and the bank. As I walked back to campus, I replayed the interview and felt good about my prospect of being hired for the position.

The bank hiring official called me the next day to inform me I had been selected to fill the sorter operator position if I was still interested. I said yes, I definitely wanted the job. The hiring official said I could start on Monday of the next week. He explained that hiring forms could be completed when I arrived on my first day of work. I thanked the bank hiring official and hung up the telephone. With transportation having been an issue from the time I got my driver's license three years ago, excitement overflowed within me as I thought about the possibility of purchasing my first car.

I got up and walked back to the Ford dealership where I had seen the yellow 1966 Chevelle Malibu. The salesman with whom I had met earlier was not working and would not return to work for two days. I walked around the lot and saw that the beautiful automobile was still there and did not have a sold sign posted on it. I wanted only to speak to the kind man I met on my first trip to the lot. I left my name and telephone number where the salesman could call me. The salesman called me during his time off and worked a deal. When he returned to his office two days later, I met him and completed the final paperwork that resulted in me driving the beautiful car from the Ford lot to the parking lot at my dormitory.

Later that day when I knew my father would be home from work, I got in my yellow 1966 Chevelle Malibu and drove home to show my father, mother, and siblings my car. Not sure if the deal was going to come through, I had waited before sharing the news. Now that it was a reality, I immediately called Zonia to share the news with her about my new car. She thought it was outstanding that I had purchased my own car. I asked when could I visit her. She said Saturday would be fine. There was no doubt in my mind that the beautiful car had been kept for me. I drove by the Ford lot many more times to see if another car like my 1966 Chevrolet Chevelle Malibu was on the lot again, and it never was.

Summer was soon over, and Zonia was ready to register to enter STA as a freshman. She would have to commute nearly thirty-five miles round-trip to and from her home to campus. Her father dropped her off at the campus in the morning on his way to work and picked her up in the afternoon on his way back home. When I caught up with her, she was standing in the freshman line, going through registration. We talked briefly as the slow line inched along on a time scale that seemed to take forever. As a junior, I had gone through early registration prior to the close of my sophomore year. Therefore, registration for me had been a breeze. I kept saying to her that after her freshman year the registration process would get easier.

I soon left but said I would check back later to see how it was going for her. She finally arrived at the head of the line, where all paperwork and financial matters were handled. She was able to arrange a monthly payment plan that would come from the personal loan after the federal grant and work-study monies were paid. When I checked back, Zonia met me smiling and saying that she was officially a student at STA. I congratulated her, and we went to the corner grill to get something to eat.

At the start of the fall semester classes, I bumped into Zonia from time to time. I always inquired how she was doing. She would say, "So far so good." As a chemistry major and mathematics minor, I had little time to spend outside the science building. When no classes were scheduled, there was always laboratory assignments to complete or a lengthy series of mathematic problems to solve. Adding my new sorter operator job at the bank to my class work really kept me pretty busy. I was able to handle all of what was required both at work and in the classroom. I sometimes found balancing the sorted checks with bank internal statements complex, causing me to not leave the bank until after midnight. This happened several times. I never mentioned the late-night work to my mom.

Since it could not be predicted when a difficult out-of-balance situation would arise and cause a long work night, I decided to alter my daytime school schedule so that my homework could be completed before I left my dorm room for work. At least that was my intent. Sometimes when I left for work, my homework was completed. There were other times I could not get assignments completed, and I returned to the dorm late with lots of homework to complete before class the next morning. My job at the bank left me with little time for anything other than studying and working. I found time, however, to check with Zonia to see how things were going with her. On several occasions, I found that she needed support. We usually were able to talk the issue through and move on.

Other than the continued massive build up of US troops in Vietnam, the first semester started off uneventful on the world stage. By the end of 1967, the number of US troops in Vietnam continued to rise along with the death toll. The opening of the spring semester in 1968 was a different story. Robert Kennedy announced his candidacy for president.

Then April 4 hit the nation and world with a feeling that mimicked November 22, 1963. Dr. Martin Luther King Jr. was shot and killed in Memphis as he stood on the balcony outside a hotel room. With the war raging on the country seemed to be at complete chaos when Dr. King was killed. His assassination brought about a tipping point for just about every college campus in America where there was a significant population of Black students. This included the students at STA.

When word reached the campus around 8:00 p.m. that Dr. King was dead, every dormitory at STA emptied into the streets of Raleigh. Students were running and screaming in disbelief of the news coming over the airways. Riots ensued. Car windows were smashed. Cars were turned over. Cars coming in the direction of STA that saw the eruption from a distance turned around in the street and went in a direction away from campus. I joined in the chaos that was on and off campus the night of April 4.

There was a call by the STA student government leaders for students to meet in what was known as "the Angle." The student leadership had come together in the night to prepare for a more ordered march in Raleigh the morning of April 5. This action brought about a high degree of calm on the campus for the remainder of the night. The calm did not signify sleep but rather a focused degree of organizing. In the dead of night, city officials were notified of St. Augustine students' intent to march from the campus entrance to the front of the state capital in downtown Raleigh in protest of Dr. King's death. The leaders stated that the march would be orderly and asked if there was any information the group needed in preparation for its intended march. City officials responded that there would be police escort to ensure the march would be peaceful and void of violence. Being an off-campus student, Zonia missed the pandemonium the night of April 4. But when she arrived for classes on the morning of April 5, the two of us joined with other students in the protest organized by the student leaders.

Early on the morning of April 5, nearly three hundred students and others from the surrounding community moved from the Angle to the entrance of STA to take part in the protest march. The crowd size obviously caught the city officials off guard. They had not expected the large number that had assembled. Additional police escorts were ordered to assist with the march. For the most part the march came off without any major incidents or confrontations. Upon arrival at the capitol, the STA student government president delivered a strongly worded proclamation describing how long-standing racism and injustice had led to the death of Dr. King and how the nation needed to enforce laws governing equality of all its citizens.

The protest in Raleigh was mild compared to the uprisings by Blacks in more than 125 major cities across the United States. Six days following the marches and protests across the nation, President Lyndon Johnson signed the Civil Rights Act of 1968 on April 11, prohibiting discrimination in the sale, rental, and financing of housing.

Just as the nation was trying to return to some degree of normalcy from the assassination of Dr. King, Robert Kennedy was assassinated on June 5 after winning the California Democratic presidential primary. Once again the nation was numb. On top of this tragedy, the US troop buildup in Vietnam reached more than half a billion young men and women. Richard Nixon and Spiro Agnew won the Republican nomination during confrontational riots in Miami while the Chicago police, National Guard, army troops, FBI, CIA, and other federal service agents clashed with antiwar demonstrators at the Democratic Convention in Chicago, resulting in Hubert Humphrey and Edmund Muskie winning the Democratic nomination on a pro-war platform amid violent anti-war protests.

With the country so decisively divided over the Vietnam War, I often paused to consider what it meant to have a student deferment. With so many young men being killed in Vietnam and Black boys dying in disproportionate numbers, I was torn by it all. During this time, I found myself sometimes thinking about how my father was treated more like a child than a man when working in the yard of the white lawyer. This led to me replaying in my mind how after placing an order in the drugstore I was required to step back away from the counter until I was summoned. I sometimes wondered what would happen if as much attention was given to fixing the unequal treatment of Blacks in the United States as was being given to whatever the issues were in Vietnam. When would concentrated emphasis and attention be given to the race problem in America?

I was pretty happy with my life at this time. I was still in college and had received a student deferment from the military. Obviously during my youth I had not worked hard in school in hopes of one day avoiding going to the military. I didn't think of such things. Working hard was just part of who I was. But as life usually works out, your path is charted based on small, incremental decisions made along the way. When the small, incremental decisions are joined together over time, the outcome emerges. The hard work I had put forth in school had resulted in me winning a scholarship to attend college, and a student deferment had come with the package. I had worked equally hard on the basketball court but suffered an injury, and a military doctor decided I should continue my education and not be considered for military service. The result was that when I reached military age and the Vietnam War was at its height, I was not drafted to participate. This was part of my destiny.

Zonia and I realized we were receiving our education during a turbulent time in history like no time in recent memory. We soaked it all in. The nation and the world were undergoing change, and we were a part of it. She had an excellent first year of college. In addition to the book knowledge she obtained, she ended her freshman year like I ended my junior year. We both closed the year with new perspectives of the potential significant difference between our futures and the lives experienced by our parents and grandparents.

One lesson learned during Zonia's first year of college was that staying home and making the daily thirty-five-mile commute to the campus took a big toll on her. She discussed the negative aspects of the travel to and from campus with her mother and father. Her father's sister lived close to the campus, and she believed the aunt would allow her to move in with her family.

Her father and mother checked with the aunt to see if she had any available space. Zonia hoped her aunt had an extra room she could rent at the start of her sophomore year. When she mentioned to me the possibility of moving close to the campus next year, I hoped it would work out for her. I had not expressed it before, but now that she had brought the subject up, I shared my concern regarding the lengthy travel, late-night studying, and sometime having to miss key evening events because she was not on campus.

In the midst of such turbulence, there were moments of beauty and excitement, and I continued to look for opportunities to explore new activities with Zonia. Now that I had transportation, it was easier to visit parks, go out to restaurants and fast-food establishments at the mall, and visit with each other's families more. These activities were possible only when we had a little extra money. Remembering how much Zonia enjoyed *The Sound of Music*, we used a lot of our free time going to the movies. I enjoyed taking her to see Sidney Poitier in *Guess Who's Coming to Dinner*, Sean Connery in *You Only Live Twice*, Dustin Hoffman in *The Graduate*, Warren Beatty and Faye Dunaway in *Bonnie and Clyde*, and some others.

A special outing I desired was to take Zonia on a three and a half hour road trip from Raleigh to visit Grandfather Mountain in the mountainous region of Boone, North Carolina. I had taken a course in geology in the spring semester of my sophomore year; the class's final examination had been in the outdoor classroom in this region. Grandfather Mountain and the rocks surrounding it revealed a long history of rock formation and mountain building that dated back billions of years ago. We had learned in class that the mountain was formed more than 300 million years ago and the erosion that had been ongoing for more than 100 million years continued to that day. In the class it was interesting to learn that the mountain got its name from pioneers who recognized the face of an old man in one of the cliffs. It was great seeing the mountain after looking at pictures and talking about it in class. I passed my examination.

When I visited the mountain during the spring semester of my sophomore year, I thought the area was the most beautiful landscape I had ever seen. I immediately wanted Zonia to see it as well. Permission from her parents had to be granted for a trip of this scale. I was sure that taking the trip would take a full day, including driving and sightseeing. I explained the trip to her parents, and they granted permission.

We left Raleigh at six in the morning. We stopped to view many beautiful sights along the way and did not arrive in Boone, North Carolina, until about 11:30 a.m. Zonia was fascinated with the magnificent splendor of nature in and around Grandfather Mountain. We enjoyed the day so much that we failed to realize how high up in the mountains we were and lost track of the time it would take to return to the main highway. When we arrived back to the highway leading to Raleigh, it was after seven in the evening.

We started the long journey back home. I became very tired and realized that I was putting both of us in danger by continuing to drive. I stopped several times and exited the car to stretch and get some fresh air. On my final stop at around 10:15 p.m., I called Zonia's father to tell him the situation and that we would be arriving later than I thought. Her father asked how far away we were from home. I said about an hour. He said for me to take my time and be safe coming in. Before hanging up, her father thanked me for calling to give him an update on where we were. I said I was sorry for not being able to have Zonia home when she was supposed to be. Her father again said for me to take my time coming in.

Shortly after our trip to Grandfather Mountain, it was time to return to school. I was returning for my senior year and Zonia for her sophomore year. It was appearing that 1968 would end on a much quieter note than it had opened. There was still unrest in the United States. Women's liberation was on the rise. Women were expressing their disgust at being viewed as objects in things like beauty pageants and not as intelligent contributors to society. The feminist movement launched with the removal and burning of bras. There was still tremendous discontent and anxiety in the Black community.

The Summer Olympic Games opened in Mexico City with the boycott of thirty-two African nations in protest of South Africa's participation with its Apartheid government regime. Four days after the start of the games, Black US athletes and medalists in the 200-meter dash, Tommie Smith and John Carlos, further disrupted the games by performing the Black power salute during the playing of the "Star-Spangled Banner" at their medal ceremony. Their fellow Olympian Peter Norman of Australia wore a human rights badge on his shirt on the Olympic ceremony stand to show his support of the two Black Americans. The event was one of the most overtly political statements in the history of the modern Olympic games, and it was protesting the treatment of Blacks in America and in other parts of the globe. Their action made headlines around the world.

It was important to me to continue both my studies in preparation for graduation in the spring semester of 1969 and my work as a sorter operator at the bank, which provided me the ability to make my car payments and have some spending change. Zonia started her sophomore year with new determination. Being able to walk back and forth to campus was a dream come true for her. In inclement weather, I would drop over to transport her to campus.

It was at her aunt's home that she and I grew even closer. Her aunt's family loved me. They enjoyed seeing me come. They would say to her, "Robert is headed somewhere special." I felt very comfortable in the presence of Zonia's aunt and her entire family. I liked the times when she and I would leave her aunt's house and walk along a main street heading to campus, just chatting about whatever.

My graduation from STA was held May 12, 1969. A great deal was going on when I graduated that was of interest to me. Being in science, I was aware of some of the technical and nontechnical advancements underway in 1969. Since the war cast such a large shadow in my life during my college years and potentially could be impacted at any time, I was interested in the peace talks that were underway between the United States and Vietnam, along with the beginning of troop withdrawal. This action brought a sense of hope that the many lives that were being lost in the war would now cease.

I heard about the test flight of the gigantic 747 Boeing aircraft and wondered if the plane was too big for safe air travel over long distances. When astronauts Neil Armstrong and Edwin Aldrin Jr. aboard Apollo 11 recorded the first footprints of man on the moon, at first it was hard for me to believe it had happened, but I realized that this was a major moment in history. While thinking about these events excited me, I had grown close to Zonia and felt a sense of sadness and loneliness at the idea of not seeing her as much once I graduated.

The summer of 1969 opened for us just as the past two summers had. I continued my sorter operator job at the bank, and Zonia found work in the evenings with a company her aunt worked for that cleaned office buildings around the city of Raleigh. With both of us working, Zonia had no problem and was eager at times to pay for some of the activities we participated in. We visited our families all the time. Her family always expressed kindness toward me. She had grown on my family to the point that one of my younger sisters asked if I was going to marry her. I told my sister she shouldn't be asking such questions.

An event took place during the summer of 1969 that at the time was unknown to us, but it changed the course we were on. Having won a fellowship to Howard University in Washington, DC, to pursue my doctor of philosophy degree, I wanted to gain real-world experience working in a science environment before I entered graduate school. My uncle Bobby's wife was an employee with the General Electric (GE) Company in Philadelphia, Pennsylvania. After discussion at a family gathering of my desire to gain work experience in a science organization before entering graduate school at Howard, my aunt offered to check in her company to see what might be possible over the next three months. She acknowledged that it was late to find a summer position for college students but said she would do her best to help me out. My aunt called me and said a laboratory manager wanted me to come for a position he believed would open up in a few days. She recommended that I head to Philadelphia with the possibility of working in a summer position at GE. Such experience would be invaluable when I entered my graduate studies.

I shared the news with Zonia. She congratulated me and said she knew I would do a good job. I said I would do my best but would miss her. I promised to stay in contact. My current workload at the bank would not allow me to leave for at least a week. I shared with bank officials and my coworkers of the potential opportunity at GE. They expressed their well wishes and adjusted my workload to coordinate with my date of departure.

I was ready to go. My colleagues at the bank thanked me for my fine work and acknowledged that a position would be waiting for me if things did not work out at GE. I thanked them for the opportunity to gain the computer work experience at the bank. After handling all of my sorter operator job responsibilities, I dropped by to see Zonia—really, now my girlfriend—and then was off to Philadelphia in my 1966 yellow Chevrolet Chevelle Malibu, heading to work at GE.

I arrived in Philadelphia in anticipation of working for the next three months at the GE company. The position did not come open the first week. The GE laboratory manager was very apologetic for the delay in the job but assured me the start date would be soon. During my time away from my girlfriend, I realized how much I missed her. When we talked via telephone, I learned of her desire to see me. I wondered if I was really in Philadelphia to work at GE or to determine how fond I was of Zonia.

After three weeks of waiting and loads of regrets from my aunt and from the manager at GE, I left Philadelphia for home. The GE manager expressed deep disappointment of the job not coming through. I returned to my sorter operator job at the bank and to the cherished opportunity to spend time with my girlfriend.

The first person I saw after my family was Zonia. We sat in the swing on the front porch of her aunt's home and talked about many things. One thing was the possibility of getting married. We wanted to stay true to our dreams of her graduating from college and me continuing graduate studies at Howard University. Also, we were sensitive to our families' expectations that we reach our dreams and aspirations. I was concerned because of the high esteem and constant bragging the community did on me, expressing how I was going places and accomplishing a great deal. My siblings were proud of the path I was on. Knowing that I would be getting married would sadden them, and Zonia felt the same about her seven younger siblings.

After much thought and discussion, we decided to secretly get married. Three weeks before I was to leave for Howard University, our marriage plans were complete. I was twenty-one, and we decided not to move on our plans until Zonia's twentieth birthday, which was August 20, 1969. On Friday, August 22, two days after her twentieth birthday, we drove to Dillon, South Carolina, where marriages could be completed the same day. We had our blood work done some time earlier and had the results with us. We would go to the nearest justice of the peace office in Dillon, complete the marriage application, have a judge perform the ceremony, and return home.

Our plans were seriously interrupted when we arrived in Dillon and were told the state's marriage laws had changed. The law that once allowed marriages to be performed on the same day no longer existed. The new law required that marriage applications be on file for twenty-four hours before a ceremony could be performed. This posed a big problem for us. We had planned to be back home the same day.

We stopped long enough to get lunch and discuss our next move. I was aware that newspapers carried marriage information from distant locations. Dillon, South Carolina, was a good place for us to get married because none of the North Carolina newspapers in the Raleigh area would pick up the information that far away. I suggested that on our trip back to Raleigh, we should stop at towns that looked large enough to have a justice of the peace office and inquire whether we could get married. I believed that a town fifty miles or more away from Raleigh would be far enough that the Raleigh newspaper would not pick up information about the marriage. On the other hand, if we could not identify a marriage site that was at least fifty miles from Raleigh, I recommended that we continue home without getting married and put together an alternative plan.

We finished lunch, stopped at a station to gas up, and headed back toward the North Carolina state line. Lumberton was the first major town we got to after crossing into North Carolina. The town was twenty-seven miles from Dillon and ninety-five miles from Raleigh. We stopped at a telephone booth and looked in the telephone book to see if there was a justice of the peace office. There was one with an office located very close by. I called to inquire of the possibility of a couple getting married. To my surprise, the response was yes, if we had official results of a blood test. I replied that we had the results with us. The person at the justice of the peace office gave instructions on how to get to the office.

I returned to the car with a wide smile and shared that we could get married in Lumberton, North Carolina. We sat in silence for a while. I asked Zonia if getting married was something she still wanted to do. She said with a smile, "This is what we both want to do."

After driving to the justice of the peace's office in Lumberton, we sat in the parking lot and went over the notes we had jotted down. We were familiar with the South Carolina state marriage laws because we had studied them before embarking on our trip. We were less acquainted with the state laws governing marriages in North Carolina. We felt some requirements were the same for the two states. This included age requirement, identification requirement, blood test results, and a fee. We had these documents as required by South Carolina, so we felt everything would be okay. We stepped out of the car, locked it up, and went into the justice of the peace's office.

A receptionist who had a big smile on her face warmly greeted us. She asked if I was the person who had just called inquiring about getting married today and asking for directions to the office. With a sheepish grin on my face, I said, "Yes, I am." The receptionist continued to smile as she instructed us on how the ceremony would be performed. As she went down the checklist of requirements, we learned that we had been right in thinking that the South Carolina and North Carolina requirements were similar. When the receptionist asked if I had brought along my best man or Zonia her maid of honor, we said no. We were not aware that the marriage certificate would need the signature of a witness.

The receptionist smiled broadly and said for us not to worry; her office would produce a witness. She said couples did not always bring a third person with them when they were using a justice of the peace to get married. We smiled broadly at her statement. We handed her all of the required documents. She said the ceremony would start in about an hour and walked us to a room where we could change and freshen up before the ceremony.

Sitting in the room on this beautiful summer day, we were very happy and excited about getting married. I mentioned how I hoped that Lumberton was far enough away from Raleigh for the newspaper not to report on our marriage. We wanted to keep it as our secret and at some point surprise our family with news that we were married. It was time for us to return to the car, get our change of clothes, and get dressed for our wedding. On that note, we left the room, went to the yellow 1966 Chevrolet Chevelle Malibu, opened the trunk, took out the suitcase, locked the car, and returned this time to separate dressing rooms to change.

The ceremony was beautiful. It consisted of the reading of a scripture, a short prayer, and the performing of the vows by the justice of the peace. Some people, none of whom we knew, stood and watched, while others simply filed by. I whispered to myself, "Only God could make this day turn out so wonderfully."

At the end of the ceremony, the receptionist came over to us with the marriage certificate in her hand and a big smile on her face. She shook our hands and congratulated us. Standing to the right of the receptionist were Emily Taylor and Troy Tyler, who the receptionist introduced as the persons who would be the two witnesses to sign our marriage certificate. These two unknown individuals also smiled and shook our hands. They signed the marriage certificate, shook our hands again, congratulated us, and wished us well on our new journey together.

With their signatures on the document, the two witnesses we had never seen before became forever linked to us, and we have a marriage certificate to prove it. The receptionist then handed us a marriage gift basket of toiletries, candies, and a reading and sent us on our way. We returned to the room to change clothes. Then we returned to our car and headed for Raleigh.

I hoped and prayed that the ninety-five-mile distance from Raleigh was sufficient enough not to see our marriage reported by the marriage commissioner's office in the newspaper and that it would remain our surprise secret for now.

We had exactly three weeks to spend together before I would begin my graduate studies on September 15, 1969, at Howard University. We saw each other as much as possible without giving away our secret. I still made sure Zonia—now my wife—was in her aunt's home by a respectable time.

My wife registered and started the fall semester of her junior year the last week in August. She continued with her work-study program on campus and her evening job with the cleaning company. I terminated my sorter operator job at the bank the second week of September. We promised to stay in touch. I set the time of my first trip back home from Howard. I had received enough early information from the Department of Chemistry at Howard University and talked via phone with some of the graduate students to schedule a good time for me to make a trip back. I set my first trip back the last weekend in September. My second trip would be the weekend of October 18. Between my two visits back to North Carolina, my wife would make a quick trip via train to Washington, DC, in early October. I hoped these schedules would work.

Before leaving for Washington, DC, I got the shock of my life. As I was leaving my parents' home one evening in the second week of being married, my grandmother called for me to join her on the front porch (this was the same porch I had spent many hours sitting on looking at big eighteen-wheeler trucks traveling up and down US Highway 70, wondering where the trucks were going). I pulled my yellow 1966 Chevrolet Chevelle Malibu into my grandmother's driveway, got out, walked to the porch, and sat in the seat right next to my grandmother, who was seated in the swing.

She asked if I was ready to go up to Howard University. I said I was, but I had some trepidation about it all because I did not know much about graduate school. My grandmother said for me not to worry and not to be anxious. She said everything would be all right and that I would do a fine job. She said for me to keep my hand in God's hand and God would not let me down. She encouraged me greatly that day on the front porch I knew so well and had so many fond memories of—family gatherings and long times just sitting by myself and pondering my future. There were so many things going through my head as I sat there with my grandmother.

As I stood to leave, my grandmother said, "Robert, I have a question for you. Aren't you married?" I froze with fear but had to think quicker than ever before and do it in a manner that would produce two results: (1) not give away in my reaction that I was married and (2) not lie to my grandmother. For a split second that seemed like eternity, I just laughed out loud and responded by saying, "Mother, what are you talking about, and why would you ask me such a question? Do you know something I do not know?" Not answering my grandmother's question and hoping she would not press the matter was the only hope for me. Inside I was confused and shocked all at the same time. Had my grandmother found out that I was married? But how could she know?

Still frozen in position, I thought about how I had checked the newspaper daily for two weeks and seen nothing about my marriage. Feeling that time was running out for me to answer my grandmother's question, just as I turned to tell the truth, my grandmother answered the two questions that had popped into my head. She said, "I can tell you are married because I do not see that yellow car running up and down the driveway like it used to." I felt some relief. My grandmother's response gave me the opportunity to laugh even more and joke around with her statement of how she knew I was married. We continued with our small talk compared to the heavy question my grandmother had posed before. I hugged her, said good-bye, got in my car, and drove off. My head was spinning in disbelief that my grandmother felt in her spirit, and felt correctly, that I was married.

As I drove to visit my wife, I did not know how much longer our surprise secret would remain just that. With excitement in my heart of seeing her, my mind drifted back in joy to the night in December 1965 when I first fixed my eyes upon the beautiful young girl who now, after four years of moving toward falling in love, was my wife.

Figure 1. My grandmother on my mother's side and her house with the front porch where I often sat as a boy, dreaming about the future. (Courtesy of the author)

Figure 3. My grandfather on my father's side was a man of the land. (Courtesy of the author)

Figure 4. My family gathered at my sister Maxine's graduation from Saint Augustine's College. Mom is in the center wearing the white dress, and Dad is standing on the far right. That's me standing behind Mom. (Courtesy of the author)

Figure 5. The house my father built. (Courtesy of the author)

Figure 6. The house built by my father sits directly behind my grandmother's house. (Courtesy of the author)

Figure 7. US Highway 70 is a major travel route for eighteen—wheeler trucks. (Courtesy of the author)

Figure 8. Working in a tobacco field was hard work. (Courtesy of Tobacco Farm Life Museum)

Figure 9. Temperatures while hanging tobacco in a curing barn reached in excess of one hundred degrees and required endurance. (Courtesy of Tobacco Farm Life Museum)

Figure 10. Some tobacco fields were vast with very long rows. (Courtesy of quiltville.com)

THIS BOOK WAS PRESENTED
AS A WRITING CONTEST PRIZE
BY THE
WAKE COUNTY ASSOCIATION OF
Phi Beta Kappa
TO

DEC. 6   ΦΒΚ   1776

Robert L. Shepard

ROBERT SHEPARD
GARNER CONSOLIDATED H.S.
APRIL 1965

Figure 11. My paper titled "The Day America Slept" was judged a winner by the Wake County Association of Phi Beta Kappa. (Courtesy of the author)

Figure 12. As a writing contest winner, I was awarded a copy of President John F. Kennedy's book Profiles in Courage. (Courtesy of the author)

	North Carolina High School Athletic Conference State Men's Basketball Champions			
Year	*Champion (2A Division)*	*Score*	*Runner-Up*	*Score*
1963	Clayton Cooper	69	Elm City Douglass	68
1965	Method Berry O'Kelly	84	Kinston Woodington	76
1967	Whiteville	87	Garner	71

Figure 13. Over a five-year period, the all-Black basketball conference I played in sent three of its schools to the North Carolina state championship game. Two of the teams won the state title, and my high school was the runner-up in 1967. (Data from website http://www.nchsaa.org/pages/15//#.UPIfZI7AVic)

Figure 14. I was a scholar-athlete and was inducted into my school's Alumni Sports Hall of Fame in 2009. (Courtesy of the author)

Figure 15. This is me shaking hands with NBA Hall of Famer Earl "the Pearl" Monroe at a CIAA tournament. (Courtesy of the author)

Figure 16. That's me, number 41 for STA, but not guarding future NBA Hall of Famer Earl "the Pearl" Monroe. (Courtesy of Saint Augustine's University)

Chapter 10

HOWARD UNIVERSITY

M Y MOTHER WAS THE FIRST family member to hear the news of my being awarded a fellowship to continue my education at Howard University in Washington, DC. Mom was thrilled at the opportunity but less excited about me going to Washington, DC. I had only lived in the small town of Garner, and the thought of me living in Washington, DC caused her some anxiety.

She shared so much with me as we talked about my heading away to another university. While I had lived on campus the entire four years at STA, moving away would be different. STA was across town so to speak. Washington, DC, was not across town where I could return home any evening. I was thinking about how often I might come home to see my new wife but not come to see my family. Just thinking about it at times made me sad, almost to the point of tears.

I had always enjoyed being in the presence of my family, and now I had to deal with the thought of first leaving and then returning from time to time without a visit. I did not know which was worse—keeping my marriage a secret or telling my family. I still believed that telling them and leaving would produce the most pain. So I kept to what my wife and I had agreed on. I enjoyed the sound of the term, *my wife.*

I knew my mother was delighted that I was being rewarded for the hard work I had put forward over many years now. But being the ever forward-looking individual that she was, she quickly moved from thinking about where the opportunity was and thought more about what possibilities were ahead of me. As she stood over the kitchen range, she spoke of purpose and focus. She talked of God charting a person's footsteps while individuals planned mostly with their heads. She talked to me about learning to follow my heart. That day in the kitchen with just the two of us, my mom poured her entire self into me.

She talked of how God had ordained what was happening to me from the foundation of the world even before I was born. She told me that God had great plans for me and for me to remember that I was really a representative of God and not man. She said my brothers and sisters would be watching me and desiring to follow in my footsteps and how it would be important for me to do my very best. My mother said I was not to feel pressured about the advanced level of study that lay ahead of me or to worry about the expectations of others but to look at this opportunity as a continuation toward achieving my destiny. After all, I had earned the opportunity to study at a prestigious institution like Howard University, who was paying me to study and train rather than me paying them for the opportunity.

I did not have much to say that day. After my mom talked about what was happening and the investment that was being made in me, she suddenly turned around from the kitchen range, looked directly at me, and said, "And you better go up there and not make anything but As and A++s. What an honor this is for you and for all those you represent."

I felt the joy coming from my mother as she continued to talk about the prospects that lay ahead for me. My mother's voice now sounded as if she was dreaming way into the future, even beyond Howard University. After a while, there was silence. When I rose from the chair at the kitchen table and went toward my mother, I could see tears forming in her eyes. Based on the conversation we had that day in the kitchen, I knew they would soon flow as tears of joy.

It was Saturday morning, September 13, 1969, and I finished talking with my mother and told her I was going to pull together some last-minute items and spend some time with Alzonia. My mom called her by her shortened name, Zonia, and said for me to tell her hello. I said I would. It felt strange to know what I knew and keep it a secret from my mother. I could not think about that too much because it seemed to confuse the issue. I left the kitchen and headed to my car.

I arrived at Zonia's aunt's home at about eleven o'clock. The two of us sat in the swing on the porch and talked about our marriage ceremony and how much we wanted to be together. I asked Zonia how she was doing with our secret. She said that at times it was hard and did not feel right. I was somewhat glad to hear her say that because it gave me the opportunity to express to my new wife that the same thing had happened to me at times. We continued to talk when I said I had to go shopping for a couple more items in preparation of packing and leaving Sunday morning. A sad look came across Zonia's face when I spoke of leaving tomorrow morning. She said she wished I did not have to go but quickly followed with, "But I understand."

Zonia went in to tell her aunt that I was there and that we were leaving to go shopping but should return within a few hours. An acknowledgement came from somewhere in the house: "Do not go anywhere until I come out to say hi to Robert." Her aunt dashed out the front door, wiping her hands on her apron and reached to give me a hug. She asked when was I leaving to go to Howard University. I said tomorrow morning.

She said, "We are all going to miss having you around. It felt like you were part of the family, and we really enjoy having you come by."

I responded that I would be back from time to time because I had to come see Zonia.

She responded, "Come by any time, Robert. You are always welcome to visit and act like you are home when you do."

Zonia and I walked down the steps from the porch to the landing on the sidewalk next to the street where the car was parked. We got in the car, waved, and drove away.

We went shopping for just a short while. Classes had begun at STA, and Zonia had settled into her junior year and was already tackling the subjects in her major, which was English. As we drove to the shopping mall, I expressed my concern of having her walk the mile or so from her aunt's home to and from campus, especially on rainy days. Zonia assured me that she would be all right.

We finished our shopping and were soon off to the park to sit, hold hands, and talk about her junior year at STA, my unknown trek to Howard University, and our future together. Always the serious one, I was already twenty-five years into our future when Zonia slowed me down so I could catch my breath and focus a little more on the here and now. I smiled as the conversation returned to the immediate needs facing us from now until the end of the academic year. We left the park, and I dropped Zonia off at her aunt's home. Zonia got out of the car, and I told her I would come by early in the morning so we could spend some more time together before I headed to Washington, DC.

I was packed and ready to go early Sunday morning, September 14. My family gathered in the living room in a circle holding hands to send me off with a word of prayer. I cannot remember who prayed, but it was not me. It must have been Mom. After I left, they would head to church. I made my rounds in the neighborhood beginning at my grandmother's house. I walked up to her house where again prayer was offered for me to have a safe trip and a blessed new experience. My grandmother asked why I was leaving on Sunday and not going to church? I told her the friend I would be staying with while I looked for my own place to stay told me to come on Sunday since he would not be at work. I promised her I would find a church when I got to Washington. She looked at me like she had when I sat on her porch dreaming or daydreaming when I was supposed to be working.

From there I went over to my aunts' and uncles' homes so they could add their sendoffs and well wishes. I knew at each stop there would be prayer because growing up it seemed like my folks did a lot of praying, and I had seen it before when my uncles and brother-in-law were coming and going while they were in the military. In fact, there was always prayer, even if someone was going only a few feet from home. My grandmother did not believe it should be taken for granted that you would arrive safely where you were going. So she instilled in her children, who instilled in their children, that you pray and ask God to keep you safe as you go from place to place. Those who prayed always added somewhere in the prayer, "Lord willing."

As I grew up, I added the phrase, "Lord willing, I will do this or that." Every prayer prayed for me on Sunday morning, September 14, 1969, ended with, "Lord, if it's Your will, take 'em to Washington, DC, without any hurt, harm, or danger coming to him." Now that I was all prayed over, it was time to go. Since there was only one church service that started at 11:00 a.m., I was convinced that praying for me would continue at Wake Baptist Grove Church in a couple of hours.

I walked back home, said my good-byes, got in my car, drove up the driveway away from the house I had helped my father build, passed my grandmother's home and the front porch upon which I had sat so many times as a boy, passed my uncle's home, and finally passed my two aunts' homes. I was soon out of sight of the family community where I had spent my childhood. Many tears were shed in the household I was leaving. Driving away, I was convinced that even more tears would have been shed that morning if my family and community had known I was married.

After leaving my family, I went to spend time with Zonia before finally heading to Washington, DC. I arrived at her aunt's house around 9:00 a.m. When I arrived, the family greeted me as they ate breakfast, and Zonia and I left. It was a beautiful 85-degree summer day. So we drove to downtown Raleigh and walked and talked. I had planned to be on my way to Washington, DC, not later than noon. It was now after 2:00 p.m.

Driving to Washington, DC, for the first time, I wanted to get to the city in daylight. I realized now that I would not arrive until after dark. I had traveled through the area last summer when I went to interview for the potential job at the General Electric Company, but I still was not as familiar with the drive as I would like. I dropped Zonia off back at her aunt's home.

We needed a moment to talk about our finances, and this was the only time we had before I left. I had saved nearly $1,500 from working both in tobacco and as a sorter operator at the bank. The amount would increase when I received my last paycheck from the bank. We had enough money until I started getting my graduate stipend at Howard.

Zonia had preregistered at the end of her sophomore year for the start of her junior year in the fall of 1969. She had done exceptionally well, especially as an off-campus student. Staying off campus was hard for her. Now that cold weather was about to set in and with me not around to help support with transportation, I desired for her to be on campus. Her work-study position as a freshman and sophomore was in the college's health unit. Before leaving campus for summer break, she inquired about other work-study opportunities. I asked if she heard of any possibilities. She said not yet. On that note, I got in my car and drove off, heading toward Howard University in Washington, DC.

The day remained beautiful as I drove along Highway 85 North to 95 North into Washington, DC. Night had fallen. I could see the tall Washington Monument and the bright lights of the city in the foreground as I crossed the Potomac River on the 14th Street Bridge. It was a view like none I had seen before. When I traveled to Philadelphia to inquire about a job at the General Electric Company, the daytime view had not been the same. I had seen the Washington Monument on that trip but not as it glowed in the darkness of night. It was a beautiful sight.

As I crossed the 14th Street Bridge, I witnessed a sight I had read about and heard about but never seen. The city of Washington, DC, or at least the 14th Street section of the city, was alive and functioning as if it were a weekday and not a Sunday. All of the stores were open, the neon signs of nightclubs and liquor stores were on, and people were walking and parading everywhere. To me this was a sight from a movie and could not be happening in real life.

At home where I had just left, people spent most of the day in church, and stores being open was unheard of. Do not even mention an ABC Package store open on Sunday in North Carolina. In the minds of North Carolinians, this kind of activity on Sunday was considered blasphemous at the highest level and would lead one straight to hell. So based on my interpretation of what I was exposed to in Sunday school and church, every person I saw on the streets of Washington, DC, at around 7:00 p.m. on Sunday, September 14, 1969, was hell bound. This had to be the destination of these folk. With what I was taught from my childhood, how could hell not be where these folks were headed? I wanted to get out of the midst of such corruption as fast as I could.

Apparently I was not moving fast enough because it seemed that every car that came in close proximity to my yellow 1966 Chevrolet Chevelle Malibu was blowing its horn and motioning for me to get out of the way. To make matters worse, a Washington, DC, police officer pulled up beside me and motioned for me to roll down my window. When I did, the officer shouted, "What kind of driving are you doing? I just scraped up one body off the street and do not want to have to do another one tonight before I get off. Get off the road if you cannot drive." The officer then sped away, blending in with the other wild drivers on 14th Street.

At this point I was so confused by the commotion and police officer's comment that I thought the only safe place to drive was on the sidewalk. That option would not work because the sidewalks were jam-packed with the people who were hell bound. And if these early episodes weren't enough of a shock to my system, a car traveling right in front of me heading north on 14th Street suddenly made a U-turn in the middle of the street and went back in the south direction, a move on a US roadway I had never seen before. This action caused me to slam on my brakes, fearing I was going to hit the car. You guessed it. Persons in the cars behind me began blowing their horns even louder. I was becoming a nervous wreck on 14th Street in Washington, DC.

As soon as I arrived at my fraternity brother's home, I did not want to do anything but get to a telephone and call home to report my safe arrival in what seemed to be Sodom and Gomorrah. That had to be what this city was like because folks in North Carolina were living in what appeared to me to be a pristine environment on Sunday compared to what I was observing, and they did not engage in such sinful activities. Everything was closed on Sunday, and the North Carolina Blue Law made certain of that. I surmised that the law kept North Carolinians from living like what I observed when I entered Washington, DC.

In North Carolina the Blue Law kept Sunday as a day to avoid vices like alcohol, sex, gambling, working, and other worldly trappings that tended to lure people to antisocial behavior. I believed the Blue Law was good since it kept people from some of the harmful practices on Sunday that took place the other six days of the week. All I was accustomed to on Sunday was getting up early enough to have breakfast with the family in the kitchen. This was followed by dressing and waiting for the church bus to pick us up and take us to Sunday school. After Sunday school was sometimes church. Sunday school was every Sunday, but we could sometimes return home after Sunday school on the Sundays when Mom did not make it to the 11 o'clock church service.

Since our father was not a churchgoer and my mother had no transportation, she would need to catch a ride with other family members. On some of those Sundays, we stayed for church but walked home because there was not enough room in the car to transport everybody back home. I enjoyed the Sundays I had to walk back because it gave me time to think and sometimes laugh about what had gone on in the church service, especially the times when women got happy and broke out in uncontrollable shouting that sometimes resulted in them losing their hats and even slapping the persons seated next to them.

Once home I would change out of my Sunday clothes and get ready for dinner, which was usually prepared the Saturday before. After family dinner with everybody seated at the kitchen table, we would walk the twenty yards or so to my grandmother's front porch, which was the gathering spot for all the families living in the Sally Whitaker subdivision. So I was used to Sunday being a day of rest and family conversation, and what I observed upon entering Washington, DC, was an absolute shock to me.

I parked on the street in front of my fraternity brother's home, walked up the steps onto the front porch, and knocked on the door. There was no response. I continued to knock, but no one came to the door. I looked down at the address on the piece of paper in my hand and back up at the house number on the door; the two matched. I came down off the porch and walked to the end of the street to look up at the street sign to see if I was on the right street. The street sign matched the street shown on the piece of paper. I returned and knocked some more and then sat down in one of the chairs on the porch.

After thirty-five minutes passed, it was getting late and turning much darker, and still no one had returned home. I became concerned. Since the houses along the street were all connected in a row, I felt that my fraternity brother and his family would definitely know their neighbors on both their right and left sides. So I decided to go first to the house connected on the left of where I was standing to inquire if the persons there knew the whereabouts of my fraternity brother and his family.

I needed to call home before it turned any later. I did not want my family and especially my young wife back in North Carolina to be worried. Having not spoken to them since about midday, I needed to contact them. I knocked on the door several times and got no answer. I figured the people living there were out mixing with the crowd on 14th Street that was hell bound.

I then went to the house on the right, and an elderly man opened the door and asked if he could help me. I gave him the background of who I was and why I was there but said that the person I was to meet was not home. The elderly gentleman said the family usually attended a late evening church service that lasted into the night. He recommended that I go to the Anthony Bowen YMCA to see if I could get a room for the night. The elderly gentleman told me that the Y had sleeping rooms and was located at 1816 Twelfth Street NW between S and T Streets. He told me I would easily recognize the building because it would be the tallest brick building in the residential area.

I explained that I had not spoken to my family in North Carolina since early morning and asked if I could use his telephone to call home. The man said sure. I made the call. Since I was standing in the house and using the telephone of a complete stranger, I decided not to share with my family the details of what I saw upon my arrival in Washington or that the people I was to stay with were not home. It would only worry everybody, especially my wife and my mother. I said that I left Raleigh a little later than expected but arrived safely and was okay and would talk to them tomorrow once

I got settled at Howard University. I hung up the telephone and thanked the elderly gentleman and told him to please tell my fraternity brother that I was in town and that I would call him tomorrow. Stepping down from his porch, I thought how wrong I probably was about the whereabouts of the persons in the house on the left. They could have been at a late-night church service. The thought caused me to remember my teaching to not judge.

I arrived at the YMCA. The men there looked to be from the rough side of life. Some even had the look of being violent if wronged. I had learned when growing up in Garner to never show signs of fear. This was important when boys from Garner would go to Chavis Park in Raleigh to compete in basketball and baseball games. I was aware that boys from the city could tell when a boy not from their neighborhood felt uneasy. So I checked into the YMCA on 12th Street, NW and looked every bit as if I had been in and belonged to an environment equal to whatever experiences the other young men were accustomed to. I went straight to my room, entered, closed and locked the door, got a good night's sleep, and did not come out of the room until daylight Monday morning. Confident I would not be returning to the place for a second night, I gathered my belongings, carried them down to my car, and headed to Howard University.

I arrived at Howard University and, per instructions received in the mail, reported to the administration building. The letter I had received from Howard mentioned the difficulty of finding parking in and around the university, but I had no clue it would be as difficult as it turned out to be. I was immediately challenged to find a parking space. I rode all over campus looking for a "legal" space to park but was not successful. This was one of those times I appreciated arriving early for an appointment. Even after arriving early, I began to come close to the time I was to report to the proper unit in the administration building. I decided to go a short distance off campus out on Georgia Avenue to look for somewhere to park. I was successful. I found a legal metered parking space. The parking space was just one block off campus and west of the administration building.

I stepped from my car, locked the doors, and looked at the instructions on the meter. The sticker on the meter read, *Park for a total of two hours and insert twenty-five cents for a total of one hour.* I reached into my pocket, pulled out two quarters, and inserted them in the meter. The meter registered two hours.

I turned, walked to the administration building, and met with the representative from the Graduate School of Arts and Sciences who was waiting for me. The representative introduced himself and congratulated me on being accepted as a graduate student and receiving a NASA Fellowship to Howard. I had in my possession all the necessary paperwork for me to complete registration and become a full-time graduate student in the Department of Chemistry at Howard University in Washington, DC. I read and signed all documents, especially the one showing my monthly pay after my fellowship covered all of my graduate school expenses.

At the completion of my registration, the graduate school representative shook my hand again. I then turned to leave, but the representative called to say he had forgotten the last thing that needed to be completed. He informed me that since Howard did not have on-campus housing for graduate students I was to go down the hall and meet with the director of off-campus housing. He said appointments had been set up for me to visit the home of two families. The representative motioned for me to return to his desk for the room number of the director down the hall. I again thanked the representative. I continued down the hall to the director of off-campus housing, who already had the addresses and telephone numbers of the two families and gave them to me. He said both families were expecting to hear from me before the end of the day. I thanked the director and left.

I slowly walked across the huge campus toward where my car was parked on Georgia Avenue, but I paused for a moment to take in the massiveness of the campus compared to STA. When I arrived at my car, I had four minutes left before the parking meter expired. I unlocked the car door on the driver's side, got in, cranked up the engine, and drove off. When I reached down to retrieve a tape from my once-extensive collection of tapes to put into my under-dash mounted eight-track tape player, I noticed the unit had been yanked out and stolen. The only thing left to remind me of the classic unit I had loved so dearly was the many dangling and now exposed electrical wires.

Robert Louis Shepard, PhD

I pulled over to the curb in a space that said, *Loading Zone, No Parking*, to ponder my options. I had never experienced anything like this in North Carolina, even when I left my car doors unlocked. It was at this point that I realized the break-in had happened while I was in the administration building, registering to enter graduate school at Howard University. All I knew to do was to drive back to the administration building at Howard and report the incident to the representative I had just met.

When I reported the incident, the representative said things like that happened around Howard and asked me if my car had been locked. I said yes. The representative said how sorry he was that this had happened and called the district police office closest to the administration building. When an officer came on the telephone to ask specific questions about my yellow 1966 Chevrolet Chevelle Malibu, the representative handed me the phone for me to answer them.

The officer instructed me to go outside and stand next to my car and an officer would meet me to record an incident report. I thanked the representative for his help and left his office to go stand next to my car to wait for the officer. After I waited for about forty-three minutes, an officer finally arrived and took down the details of the car break-in. Needless to say, after giving the information to the officer, this was the last time I heard from the police department and certainly the last time I ever saw my beloved eight-track tape player. I returned to playing the old car radio with its boring tunes that could not compare to my collection of classic music that was now gone forever

After being in Washington DC for only two days, I had experienced things I never had before. I was hungry and decided to stop in a fast-food shop to get a sandwich and make a call to the families about housing. My plan was to call the first family on the off-campus housing list to schedule a time that I could come by and then order my food and eat it as I headed to my first appointment. I ordered a cheeseburger, fries, and a vanilla milkshake to wash it down.

After ordering, I changed my mind. I took my tray of food to an empty table, sat down to eat, and began to replay in my mind the events that had happened to me during my first two days in the city. I reflected on the Sodom and Gomorrah sight I saw Sunday night when I arrived in the city on 14th Street. I thought about the police officer's comment when he yelled out his window about my driving. I mulled over having to find a place to stay Sunday night at the YMCA because my fraternity brother was not home. I mused over the cost of parking and about the break-in of my car. All this had happened in less than two full days, and I had not set foot in the Chemistry Department at Howard University.

I concluded that I had to regroup and do so fast. I reasoned that it would not be good to call my wife or my family every time I experienced a challenging event. And if the whirlwind of events that had come upon me in just less than two days would be the norm, I decided that daily reports of such information to those who could do absolutely nothing about the situation would be counterproductive.

The second day in Washington, DC, I made up my mind that I would not view myself as a victim and would not report every new incident to my young wife or my family. I vowed to be observant while moving around in the big city but to not move in fear. I believed all of the horn blowing and gesturing directed at me the night I arrived, and especially the comment by the police officer, had been the result of drivers noticing my license plate from North Carolina and using that to try to distract me. That day sitting in the fast food restaurant I also resolved that no such actions would distract me from focusing on why I was in Washington, DC, which was to attend Howard University.

With a smile on my face and excitement in my heart, I slid my chair away from the table, emptied my tray, went to the pay telephone outside the eating establishment, and made an appointment to meet with the first family regarding housing.

The visit to the home of the first family did not go particularly well. Upon answering the door, it was obvious that the homeowner was expecting me by name but not by sight. He looked out his front screen door and decided I would not be a good fit as a rooming guest. I was a little shocked by the homeowner's reaction. For a few seconds I wondered if the homeowner had reacted to my long afro hairstyle and full beard. I quickly remembered my recent promise to myself at lunch of not allowing any unfavorable action to distract me from focusing on why I was in Washington, DC. I simply thanked the homeowner and left to find a pay telephone to schedule an appointment to visit the second listing of off-campus homeowners.

I arrived at the second home. When the lady of the home answered the door with a big smile on her face and said to me, "The director of off-campus housing at Howard called to tell us you would be contacting us," I knew this interview was going to be as different as day and night compared to the place I had just left. The lady continued, "Come on in. We were waiting to meet you."

The lady informed me that she and her husband had listed their home on Howard's off-campus housing roster for twenty-seven years and had enjoyed the many students they helped through the years. She shared that they owned a second home in another section of Washington that had been on the university's off-campus roster even longer. From the pictures she showed me, the second home was equally as beautiful as the one I was sitting in. The lady offered me something to drink as we sat in the beautifully decorated living room of the fine home, and she continued the interview process.

She began to talk about how Washington, DC, was before the 1968 race riots. I listened intently because my wife and I had been part of the struggle that took place in Raleigh by the students at STA and Shaw University the day Dr. Martin Luther King Jr. was assassinated, apparently at the same time unrest was going on in the nation's capital. The lady spoke about how bad the destruction had been in a section of the city where Blacks shopped and ate at fine restaurants. She said it was now in shambles, with every aspect of the beautiful area forever changed.

I could almost feel the lady's pain as she continued to describe what had happened on April 4, 1968, near where we were sitting when word came that Dr. Martin Luther King Jr. had been assassinated in Memphis. She caught herself and said, "I know you have got to go, but it was a time that I will remember." She closed this part of the interview by saying that if I was around on Sunday after church, she and her husband would love to drive me through the area. I jumped at the opportunity and said I would love to see it.

The lady of the home was delighted to hear that I was from the great state of North Carolina. Her husband was also from the state. She said she knew that being from North Carolina I was well trained, a Christian, and a good person. She said she had only met good people from North Carolina. She said as soon as I got settled in she wanted me to attend church with her and her husband. I said I would like that.

After a few more minutes of small talk and a slice of the homemade cake she had baked over the weekend, the lady of the house took me on a tour of the area of the home that housed graduate students. I was absolutely stunned by the upstairs living quarters and the downstairs fully equipped kitchen with a washer and dryer. These items were separate from the homeowner's kitchen and washer and dryer.

Following the tour, we returned to the living room. She asked how I liked the space. I said it was beyond anything I had expected as a place to live while pursuing my graduate studies and that while I loved the space I did not know if I could afford it. The lady said she and her husband had always wanted to help students who were about improving their lives.

She said, "So the monthly rent for the entire space is forty-eight dollars per month, and if you would like to have it, you can move in today."

I was speechless. In a split second I thought about the polar opposites between the first homeowner I visited and the home I was about to secure as my new living space. The lady asked me if I wanted the space. I said yes.

She said, "Okay, and you can move in right now if you like."

With all my items from North Carolina in the car, I moved in. I went outside to retrieve my two suitcases and various other small items tucked away in the car. Upon returning with my items, I sat back down and said there was another important thing we needed to talk about. I told the lady I was married and asked if there would be a problem if my wife visited me sometimes. The lady said not at all and admitted that she had been going to ask if I was single because she had two daughters away in school who were not married and I looked like a fine young man to introduce to them. I smiled and went outside to get the rest of my items to move into my new living space in Washington, DC, as a student about to enter graduate school at Howard University.

When I came back inside, I asked about having access to a telephone. The lady said she had forgotten to go over how the telephone worked for student tenants. She said the telephone in the hallway upstairs was on a separate line from the other house phones but was in her name. She said the phone line was set up that way years ago so that students staying in their home would have privacy and the feel of being in their own home. She also noted that all charges to that telephone line would be the students' responsibility.

When the other two upstairs bedrooms became occupied, she explained that she would divide the basic phone bill charges by the number of students, meaning all would have equal monthly payments for telephone service and fees. For long distance charges, it would have to be an honor system in which the responsible students would pay for their individual calls. She told me that this method of providing telephone service had worked well over the years. I liked the way the telephone was set up, and I knew I would not have a problem because the bill would reflect the location and numbers to where long distance calls were made.

With these instructions, I took my items to my new home (one of the upstairs bedrooms) and made a call to Zonia to let her know I had found a place to live in a very beautiful and upscale neighborhood of Washington, DC. I followed this with a call to my mother and informed her of the same. When I came back down the steps to head over to the Chemistry Department at Howard, the lady said she knew how busy I was and that I would not have time to find a grocery store, so she invited me to have dinner with her and her husband when I returned. I was somewhat taken aback by the loving hospitality of the lady. I had not met her husband, but to be married to such a gracious person, I figured that he too was very good and kind.

I responded that I did not know what time I would return. The lady said it did not matter and if I got back late she would leave the food in the oven for me to help myself but asked me to return what was not eaten to the refrigerator. I thanked her and dashed out the door to my car to head to the Chemistry Department at Howard University for the first time.

Sitting down in my car I could not help but once again reflect on the night-and-day experience between the first house I visited and the second home that took me in. As I started my car and headed to Howard University, I wondered if what I was experiencing was what I heard in church all my life about doing good and good coming back to you. I was not at a point in my life when I could put together in a spiritual sense what was happening to me, but I dared not take it for granted. That was enough for me right now.

I was far enough in my understanding to know that you cannot control outcomes. The two house hunting experiences had proved the point. I wore the same afro hairstyle and full beard on both housing interviews. I was not considered by one location but was by the second. I figured that was the way life unfolds.

I was finally inside the walls of one of America's renowned academic institutions—the Chemistry Department at Howard University. The place where names of great Black chemists like the renowned Dr. Percy L. Julian and Dr. Lloyd N. Ferguson were posted behind glass cases with their biographies for every entering graduate student to see. And there was Dr. Moddie D. Taylor, one of a few Black chemists to work on the Manhattan Project that produced the first atomic bomb during World War II, who was the current chairman of the Chemistry Department.

I had known before leaving STA that one of my fellow alumni, Dr. Lloyd A. Quarterman, was another Black chemist to have worked on the Manhattan Project. I sat down in a chair along the hallway to reflect on what it meant to be embarking on a path that had been traveled by giant Black men of science. I did not want to fail. I knew next to nothing about what was before me, but I understood what hard work had produced for me up to that point. At an early age I had embraced the idea that work was my friend. When everything else failed, I could depend on the outcome of hard work.

As I sat in the chair and waited to enter the Howard University Chemistry Department's front office, I vowed to let work be my friend in this new venture and leave the outcome to those who would judge my performance. I made up my mind that I would work like there was no tomorrow. In fact, based on the paperwork I received from the department, there would be no tomorrow if I was not able to demonstrate an ability to do the work and do it well. The time for thinking about the journey had ended, and it was now time to take a step on the new road. The receptionist said the chairman was ready to see me. I rose from the chair I was seated in and entered the front office of the Chemistry Department to meet the chairman and register. As I was leaving his office, the chairman asked if I had been to administration to complete the paperwork for my fellowship. I said yes. He said, "Then welcome to the Department of Chemistry here at Howard University." I shook his hand and said, "Thank you."

I had successfully registered, left the front office, went around the department, and saw some of the graduate students I had seen at STA when they attended the National Beta Kappa Chi/National Institute of Science meeting back in April. The graduate students congratulated me and welcomed me to the department. All of them were quick to note that it was no cakewalk getting through Howard University's Chemistry Department with a PhD. But they each pointed out that if I was ready to roll up my sleeves and work hard, I would be fine.

They also emphasized for me not to try to go it alone. They encouraged me to join the Graduate Student Council without delay. The seasoned graduate students said it was never too early to join, but they could point to many cases when students had joined much too late and were in trouble and failing. One graduate student asked if I knew which discipline I was going to concentrate on. I said I was thinking about organic chemistry but would give myself time to learn more about the other disciplines.

After about eight hours inside the chemistry building on my first day, from 2:00 p.m. until 10:00 p.m., I was hungry and tired. I exited the building, got in my car, and went back to my new home. I noticed several things I thought I should pay attention to early on: I left the building alone, left late, and probably would at times leave even later, and my car was parked in a poorly lit area that was extremely dark. As I drove Georgia Avenue toward home, I looked to my right and left to observe people's movement. On the six-mile drive from the chemistry building to my home, I saw a variety of activity. My car had been broken into, so I concluded that I needed to be observant at all times. This was as important to me as working hard and sometimes late at night as a graduate student.

I arrived home and upon entering the door to the kitchen noticed the oven light reflecting the containers of food in the oven. With a smile, I put down my carrying case, opened the oven, and prepared a plate of delicious food. After eating, I cleaned up, put everything away, and went upstairs to call my wife. We talked for a while. Zonia said all was well with her. I was glad to hear that. Before hanging up, she said she had some exciting news to share. She had received a letter saying she had been selected as a dormitory counselor (DC). A room in the dormitory was a prized benefit of being a DC. The first thing I thought of was Zonia not having to walk to campus from her aunt's house. This was a major blessing for us.

Highly regarded students were selected for these prized work-study positions. Students selected as DCs were required to be enrolled full time and in good academic standing. Also, DCs were required to be proficient in the basic skill areas of writing, speaking, and studying, and they had to be exceptionally good listeners with a strong sense of personal responsibility and fairness. Most DCs at STA demonstrated remarkable leadership ability, and this was true of Zonia. At times she had to be reminded how good her leadership skills were.

As a DC, Zonia would assist the dormitory manager in helping the female students in a wide range of areas. DCs were required to attend training sessions to keep abreast of their roles, duties, and responsibilities. Zonia was a conscientious and serious student and was not going to let this blessing get away. Listening to her share the information, I was confident I was on the path to fulfilling the destiny God had set for my life and that Zonia was an integral part of my destiny.

I shared a little of my first day inside the chemistry building but not the entire day because it was late and both of us needed to prepare for tomorrow. I said I was going to try to make it home to see her in about two weeks as we had planned before I left. As I hung up the receiver, I wondered if it would be possible for me to get back so soon with all that was before me as a first-year graduate student who needed all my time to start well.

I got ready for bed but decided to take out the graduate school material I had received in the mail titled *Requirements for the PhD Degree, Department of Chemistry*. I read it again for what seemed like the hundredth time. But I wanted to make sure I had a full grasp of the program requirements.

The material described how the program components were designed to promote disciplinary competency, professional development, interdisciplinary training and research, and much more. Emphasis was placed on the integration of professional development into the curriculum and the opportunity for interdisciplinary training and research.

I looked at the total number of hours of coursework and research required to receive a PhD. I moved on to the section that described how divisions or advisory committees could require that more hours of coursework be taken. When I came to the section that said students were expected to complete the PhD degree within seven years of the date of initial registration in the program or be dropped from the program, I laid the material to the side and wondered what was ahead of me.

I wondered if the few students I had met who were in their seventh year and had not moved beyond a bachelor's degree would be leaving the program without a PhD. I did not want this to happen to me. I asked myself what would keep me from completing the program in three years. I figured I would need to keep a close watch on this section. I picked the material back up and kept reading.

I was skilled in organic chemistry, but that would not be enough. Taking at least two graduate courses outside your subdiscipline was required. I needed to think about this carefully. It was not clear what other chemistry courses or courses from other departments such as engineering, biochemistry, or related disciplines I would take. I needed to make sure I took something I could handle. I kept reading.

I soon came to another section that concerned me. The requirement said a cumulative average of 3.0 (B) was required for graduation. Only two grades below B would be permitted. Receiving a third grade below B would result in the student being dropped from the PhD program. This caused me to swallow hard and pause for a while.

I viewed the next sections as critically important. They reflected requirements that could have the greatest impact on my life going forward. This included the comprehensive examinations, specialty examination, research proposal development, departmental seminars, foreign language requirements, additional requirements, and dissertation defense.

The requirements were time-sensitive, meaning completions were set to lengths of time within the program. For example, a student was given four opportunities to pass comprehensive examinations in three of the four chemistry areas (organic, inorganic, analytical, or physical). The first attempt came the first week of the student's first semester. The second attempt came the first week of the student's second semester, the third attempt at the end of his or her second semester, and the fourth and final attempt the first week of his or her third semester.

Some of the information was not clear, and I would need to talk to someone to get clarification. I thought about the distance I would need to travel to satisfy all the requirements I had just read over once more. I rose from my bed, went over to look out my window into the dark of night, and replayed over in my mind what I had read.

I was left with fear and trembling each time I read the requirements for attaining a PhD degree in chemistry from Howard University. The possibility that it could take up to seven years to obtain a PhD was something I could not accept. Looking out the window, I thought about being married and the need to get on with building my life and family. So once again, my mind was flooded with the thought of needing to work extremely hard in my classes and my laboratories. The comprehensive exams sounded intimidating to say the least. And there was the specialty exam, where the flexibility existed for division leaders to add more requirements to the already tough examinations requirement. If the student failed, he or she would have a second opportunity to pass the exam during the first week of the fifth semester but no more attempts after that. The night outside seemed to become darker.

The part that mentioned having to stand in front of the department to give two seminars—one in my research area and one in a technical area unrelated to my own research— caused me to take deep breaths. This requirement suggested that not only would I have to be proficient and good in my chosen discipline, but I would also have to show technical depth in a broader sense.

Standing there thinking about the foreign language section, I saw no way possible for me to meet the requirement. It frightened me beyond description. Passing Princeton-administered foreign language examinations—one in German and one in French—with a score of 450 or better on each to remain in the PhD program seemed a slim possibility. I was so afraid of the foreign language requirement that I told myself that if anything sent me back to Garner, North Carolina, without a PhD, it would be my inability to pass the German and French examinations administered by Princeton. I agonized over this requirement.

I moved from the window and sat back on the side of my bed. I took out my notebook and made a checklist with a time line based on the schedule of when activities must be completed. The checklist would become my Bible, so to speak. I would check it on a daily basis to make sure I was on track and remained on schedule. I got into bed.

To avoid becoming totally traumatized by the language requirements, I lay in bed that night and decided that because I had recently finished taking German as an undergraduate at STA, I would sit for the German examination during my first semester while the language was still somewhat fresh in my mind. With that positive thought, I closed my eyes and went to sleep.

I settled down somewhat as my studies got underway at Howard University. One setback was that I was not able to travel back to North Carolina to see my wife the times we had agreed I would return. That really bothered me, but we both understood and let our long phone calls serve as an adequate substitute for not being able to be together. My wife was working hard in her junior year, and I was working hard in my first year as a graduate student. I was able to go see her and visit with other family members the second weekend in October.

It was a beautiful fall day when I left Washington with my radio playing the newly released number-one song by the Supremes titled "Someday We'll Be Together." How appropriate the song was to the feeling I had in my heart about my wife and me. We spent quality time together talking about our campus experience without each other. Although we were apart, it was clear from our focused conversation that we were making progress toward a common goal that one day we would be together just as the Supremes' song suggested. During this visit together, we made a decision that Zonia would come to Washington to visit me for Thanksgiving. We also made a major decision that day in October 1969. We agreed that before Zonia traveled to Washington, we believed it was time to expose our secret by telling our families we were married.

The family took the news better than either of us thought. My mother immediately asked if we planned to stay on track with our goals and complete our education. We said of course. My mother asked no more questions but said she thought it was nice that we were married. My siblings were quite surprised and somewhat saddened. I assured them they had not lost a brother but had gained a wonderful new addition to the Shepard family. They all knew this was the case, but they told me that somehow it just did not seem like it would be the same with me married and not able to spend as much time with them in the future. In the end, it was about our new life together, and the family was happy.

When I told my grandmother the news, she said the announcement did not surprise her one bit because she had known I was married the day I got married. I asked her why she had thought I was married. She said she noticed a change in me and that I was more serious than I used to be. I asked my grandmother if she remembered that I did not tell her a story when she asked if I was married. My grandmother said she had been so sure I was married that she had purposely moved the conversation on so I would not have to tell her an untruth. I smiled, embraced my grandmother, and told her I would soon bring Zonia by to meet her. The reaction my family displayed when learning of our marriage was the same reaction Zonia's family exhibited when we told them. It was a great relief for us to know that our families expressed joy about our decision to make a life together.

By the time Thanksgiving 1969 rolled around, I had settled down and was feeling comfortable in my classes. Zonia came up by train, and we really enjoyed our first Thanksgiving together. The lady of the home where I lived was happy to meet my wife. The time seemed to pass much too fast, and before we knew it Zonia was on the train heading back to STA. We agreed that Christmas would be coming soon, and I would be home. This time we would have the chance to spend a longer time together.

All of my advanced classes were going very well. I was running second in the very difficult advanced organic chemistry class and at the top in my other courses. The student leading the advanced organic class demonstrated knowledge of the subject matter beyond anything I had ever seen in a fellow classmate. On every exam, the student from Japan would have the top score, with me coming in second. After this pattern continued through the fourth examination, I reconciled to myself that in the advanced organic class it would be okay to end the semester second to the student from Japan. Anyway, the difference between first and second was such a narrow margin that in my mind I looked at it as two students in first place. However, I did have an "aha moment" in the advanced organic class.

When the professor of the class administered his exams, he would return the results at the end of the next class lecture as the students exited the classroom. As the professor returned my results from the fifth examination, he held the exam results out for me to take but pulled the paper back before I could take it. He looked at me and asked, "You are the student from the small school in North Carolina, right?"

I responded, "If you are referring to STA in Raleigh, North Carolina, then yes."

I thought the professor was going to congratulate me for my top-level performance in his difficult class; instead, looking directly into my eyes, the professor said, "I do not understand how you are performing at such a high level on my exams coming from such a small school."

I was shocked. My mind tried to figure out what had just happened. I was speechless and somewhat stunned. In a flash, I collected myself and simply said, "I am from the small school in North Carolina," as I took my paper and walked away from the classroom.

As I continued to walk down the hall, I tried to process the deeper meaning of the professor's comment. Was he saying, "Shepard, you must be cheating on these examinations?" Was his comment an indictment on the education of Blacks from the South? Or was his comment genuine and simply limited to the smallness of STA? My next class would not start for three hours, and I was grateful that I did not have to immediately go to my next class with the professor's comment fresh on my mind. My plan had been to go to the corner store to get a sandwich, but all that had changed now. I was not hungry anymore.

I decided to go home and think some more about my dialogue with the professor. What gave me the greatest difficulty was that the first year of graduate school marked the beginning of my seventeenth consecutive year of education, and the professor who made the comment to me was the first white American teacher I'd ever had. At STA, I was taught physical chemistry and the German language by German professors. I knew race relations were trying to change in the education system in North Carolina. As I drove up Georgia Avenue, I asked myself, "Is what I just experienced with the white American professor what Black students will be faced with as the South moves from the K–12 segregated school system I was trained in to the new desegregated systems that's coming?"

When I got home, I was still trying to process the statement the white American professor made to me. I would have occasion to deal with the issue of the perceived ability of Black students in the years to come. For now, I addressed the issue the only way I knew how. Sitting on the side of my bed, I made the decision that I would take every course offered by the white American professor and make sure he was a member of my dissertation defense committee. The encounter with the white American professor motivated me to the point that I no longer entertained the idea of not succeeding in graduate school but instead knew I would receive a PhD in chemistry from Howard University.

Before I left Howard for Christmas break, I was informed that I had passed the first two of my three comprehensive examinations on my first attempts. One was in organic chemistry, and the second was in analytical chemistry. I also learned that I had been successful in my decision to take the Princeton-administered German foreign language examination. A passing score was 450, and I had made 453. It wasn't by much, but I passed. With these early successes, I was at the top of the first-year graduate students in chemistry at Howard University. The time off for Christmas break was good, but I took all my books with me to North Carolina so as not to lose momentum.

My wife and I spent a lot of time together and with our families. Soon Christmas break was over, and it was time for us to think about returning for the second half of our educational journeys.

The second semester of my graduate studies continued where they left off the first semester, or so I thought. I constantly reviewed the departmental information, making it a checklist to ensure I was up-to-date or ahead on all of the PhD requirements. With the success of passing my first two comprehensive examinations, one in organic chemistry and the other in analytical chemistry, as well as the German foreign language exam on my first attempt, I decided to sit for the other two comprehensive examinations.

I had one more comprehensive exam to pass. The subject matter choices were either advanced inorganic chemistry or advanced physical chemistry. I was not particularly good in either of these courses but felt I would do better in physical chemistry rather than inorganic chemistry. This would be my first attempt at the physical chemistry comprehensive exam and the second attempt of the four-attempt maximum to pass them all. And while everything was moving along so nicely, I figured why not take the French foreign language exam as well.

I completed these exams all in a matter of one week. The problem was that I did not feel as good about either of them as I had felt about the first set of exams. For several days I tried to reconstruct the examination questions and my responses but was not successful at either remembering the questions or my answers. So this exercise was short-lived, and I moved on with my studies.

I received word two weeks after taking the series of tests that I had failed them both. A degree of fear settled in on me at hearing that I had not only failed, but I had failed miserably. The question in my mind was what could I do to overcome my lack of knowledge in physical chemistry—an area of chemistry I did not feel comfortable with in the first place. Failing the French foreign language exam did not surprise me at all. I remembered when I took French in high school that the teacher had wanted to give me a D, but both the counselor and principal had intervened and asked that I be allowed to do a special open-book project in order to obtain a B, which would keep my GPA up and not jeopardize any scholarship offers I might receive.

I remembered the French teacher being infuriated by the request and threatening not to do such a thing. After more convincing that it was the right thing to do in this unique case, the French teacher decided to allow this exception. She immediately called me into her office and said, "You do not know any French at all, and I want you to promise me that you will not embarrass me by taking French in college or anywhere else." In her office back in 1965, I promised that when I got to college I would take a foreign language other than French. That was why I took German at STA.

As I now sat alone in the graduate student lounge, thinking about how I should have been more serious in my high school French class, I knew nothing else to do but seek some guidance.

Until this setback, everything for me had progressed smoothly based on my own initiative and hard work. But now I felt somewhat lost and confused. I had never experienced this kind of failure in my academic pursuits. I seemed unable to separate the failing of the chemistry exam from the failing of the foreign language exam. My mind kept returning to the material describing the requirements for receiving a PhD in chemistry from Howard University. I had failed a major portion of what had to be passed, no exception. It all seemed somewhat overwhelming.

Before leaving the lounge, I took a deep breath to try to collect myself. I once more processed the chemistry part of my failure. At this point, I was able to deduce that for the chemistry part I would have to seek help from graduate students who had chosen physical chemistry as their specialty. I thought about how well I had performed on both the advanced organic and advanced analytical comprehensives. It became clear to me that I had failed the other exam because I had not had as much interaction with physical chemistry to be able to pass the test. This rationalization seemed to calm me down and bring about some clarity of thought on how to proceed. Just as I calmed down, my mind raced forward again because I was already into my second semester, and the requirement clearly stated that all comprehensive exams must be passed by the first week of a student's third semester.

The first week of my third semester in the Department of Chemistry was the fall of September 1970, and this would be the last time I could attempt to pass the remainder of my comprehensive examinations. If I was not successful, I would be dismissed from the PhD program at Howard University. The thought of failing seemed to haunt me in a way I could not easily shake off, although I had said in my mind that I would not fail. Since I had not chosen a research advisor, I decided to meet with the professor who was responsible for me being at Howard to see if he had some ideas as to what might be a good next step toward preparing to take the physical chemistry and French foreign language exams a second time.

Several days passed before I was able to get in to see the professor. It was a rare occasion when his office was not flooded with undergraduate students standing around in need of extra help and graduate students discussing research results.

When I started to describe that I had failed my last comprehensive examination, the professor interrupted me and said, "Youngster, I know you were successful on your first two chemistry exams and the German foreign language exam, but you are aware that you only got the rest of this semester and the first week of the fall semester to pass one more chemistry exam and the French exam, right?"

I responded that I was very concerned about the short time frame left to pass the exams. I then shared with the professor my idea of trying to get graduate students in the field to tutor me before I took the advanced physical chemistry exam again. The professor agreed with my strategy and gave me names of graduate students I might approach. He asked what was I going to do to pass the French foreign language exam. I told him I did not know. He suggested that I work into my schedule a time when I could audit an undergraduate French class before taking the language exam again.

An excellent graduate student in physical chemistry tutored me for six weeks. The light came on for me in the tutoring sessions, and in March 1970 I passed my last comprehensive exam. This was a major step toward becoming a PhD scientist. Just as I was experiencing joy from having successfully passed all of my chemistry comprehensive examinations, I received word that I had failed the French foreign language examination a second time. The emotional roller coaster of the pass-fail cycles was about to get to me. Once again it was time to pull out the PhD requirement checklist and assess my progress.

A breakthrough came on the French language situation in a way I never could have imagined and certainly could not have brought about. One day, I completed my day of classes and laboratories and went home to get some rest before returning to the building. When I arrived at home, I walked in and was introduced to a female student who was being interviewed for one of the rooms upstairs across the hall from where I was staying.

The lady of the house introduced the student to me and said with a smile on her face that she believed the student was going to move in. Shaking the student's hand, I interjected that I believed she would enjoy living in the space of the home of such a warm family. I then went upstairs to my room to take a nap. Since it was around 4:30 p.m., before lying down I decided to call Zonia to see how her day had gone. Although I became excited, I was utterly shocked at what my wife said before we hung up the phone.

Zonia said her day had gone very well. She had gotten the results back from a major test and had aced it. We talked about how much we missed each other. As the conversation continued, I shared the positives going on with me and made no mention of my struggle to pass the French examination. Thinking Zonia may worry and not remain as focused as she needed to be on her studies, I mentioned passing the last chemistry comprehensive exam but did not bring up that I had failed French for a second time.

As the conversation was about to close, Zonia said, "Robert, I have something very important to tell you." Without stretching it out, Zonia said she was pregnant. There was a long silence on the line. Zonia asked if I had heard what she said.

I said, "Oh yes, I heard."

At that moment my mind was flooded with a bit of fear. I was trying to form the right words to respond to the girl I loved so much, but at that moment I was agonizing internally a great deal—the commitment the two of us had made to our families of not letting our marriage interfere with our education, my wife with less than one year to go to have her degree in hand, the difficult journey of pursuing a PhD in chemistry, the hurdles I needed to overcome, the fact that I wasn't sure if my student health insurance would handle the medical cost of bringing a child into the world, and then the financial needs on a daily basis after the child was born.

I was taking too long to respond, and Zonia asked again, "Robert, are you all right?"

This time I quickly responded that everything was fine but that I was shocked and excited at the same time by the news and just trying to decide how things were going to work out in the future. We had only been married seven months and had not discussed beginning our family so soon. In her calm way, Zonia said it would be fine and that we would take it one day at a time.

I collected myself enough to ask sensible questions like how far along she was. When would the baby be born? And how was she doing? It wasn't until later after hanging up that I realized all of my questioning had been out of sequence and that my last question should have been the first (i.e., I should have asked Zonia how she was doing first).

Zonia's responses demonstrated to me that my young wife had done a lot in my absence and done it well. It was amazing to me that while I was here in Washington, DC, struggling with failed examinations but not wanting to worry my wife about my problems, she was in North Carolina dealing with being pregnant, going to school, working, and going to doctors appointments but not wanting to worry me. Now the two of us were on the phone, trying to see how to proceed. I was impressed by my young wife. I figured that was why she could easily say in such a calm manner, "We will take it one day at time."

I settled down, but with my mind moving at the pace it was, I could not remember asking Zonia if she knew when the baby would be born. So I called back. She said that based on her last visit the doctor said she was already one month and the baby would be born toward the end of December or early in the New Year.

I decided to use this as a time to lighten the conversation. I told Zonia that without either of us having full-time jobs, having our baby born as the first baby of the New Year in Raleigh would be a great windfall for us. Zonia asked what I meant. I said that the first baby born in a jurisdiction in the New Year received enough support from the business community that the parents did not have to purchase anything for about the first six years of the newborn's life. Zonia said that would truly be a blessing from God. I said God just might make it happen for us.

Moving the conversation back to the issue at hand, Zonia said it would probably be necessary for her to drop out of school her last semester, get situated with the baby, and then return and finish. I responded to the statement with an emphatic "no way." I said, "Let's not go there. You said let's take it one day at a time, so that is what we're going to do." I said for Zonia to continue her classes and keep an eye on her physical condition. I asked if she would be able to keep her work-study job as a dormitory counselor. She said she did not know but would check on it. I suggested that she not check on it but leave things as they were and continue the path she was on until her pregnancy began to show … and to remember to just keep moving forward.

Finally, I asked whether I should come home. Zonia said she was fine and for me to stay focused on my studies. I had one more question before we hung up. I asked if our parents and siblings knew of her situation. She said not yet. I asked when we should tell them. She said we should do it together the next time I was home. I agreed.

It was now 8:45 p.m., and I had not rested my eyes as I had planned. I got up, packed my books, and came from my room to go back to the Chemistry Department, where I was certain to find many other graduate students hard at work. When I got outside the doorway of my room and into the hallway, the new student tenant was coming up the steps with her last items. She had officially moved in diagonally across the hall from my room. We introduced ourselves once again but this time took a moment to ask what degree each other was pursuing.

I said I was pursuing a PhD degree in chemistry but might take a master's degree as a precautionary measure and then continue on. The female graduate student told me she had gotten her undergraduate degree in foreign languages and had entered Howard to pursue a master's degree in French. I stood for a moment in total disbelief, but I was smiling on the inside with a deep assurance that something special was happening in my life even at that moment. I was convinced that the French major now living across the hall could help me pass my Princeton French examination, and that was exactly what she was able to do.

I passed my French foreign language requirement after two weeks of intense tutoring in the evenings and on weekends by the female student who had moved in across the hall from where I was staying. Every time I think of the incident in the upstairs hallway of the house where I was living back in 1970 with the new graduate student studying French, there is no doubt in my mind that my destiny was planned for me before I was born.

With the passing of my last foreign language exam, my enthusiasm returned. I had selected my research and thesis advisor and sat with him to bring him up to speed on the pregnancy of my wife. I explained that the pregnancy would not impact her completing her junior year, which was ending in another couple of weeks. I discussed with my advisor that I would be in and out of the laboratory during the summer but that it would not impact my moving forward on requirements.

My advisor simply said, "It is all up to you, youngster, how you use your time, but remember this: it will all be revealed when you stand to defend your research." He congratulated me on my progress but warned that over the next several years it would not get any easier.

I wanted to ask my advisor about the rumor floating through the department of him leaving to take a research leadership position at the Proctor and Gamble Company in Miami Valley, Ohio. But our dialogue went so well that I decided to reserve the conversation about the possibility of his leaving Howard for another time.

My first eight months in the graduate program had gone well, and I was feeling good about my ability to not just complete the program but to possibly stand out in the program. Unlike undergraduate school, in graduate school I only studied something I truly loved—chemistry—and learned that I could really excel when my mind was dedicated totally to the subject and not split on subjects I was not that interested in.

My advisor and I had talked many times before about how my research project was designed in such a way that I could pick up a master's degree and continue on for the PhD with minor impact on the length of time it would take to complete the requirements for a PhD. In his office on this occasion, he made clear the impossibility of being able to predict research outcomes prior to establishing and testing the hypothesis. On that note, he warned that the connection between what would be enough to constitute the awarding of a master's degree and continuing the work toward a PhD would not be known until the research began.

I was fully aware that it was impossible for either of us to sit in his office and predict the actual length of time it would take for me to complete the PhD requirements without even having started my laboratory research project. Before leaving his office, I decided it was a good time to ask him if he was considering taking a research position with the Proctor and Gamble Company. He said he had not given a definitive yes or no to the offer but told me that while there the previous summer he had enjoyed the work and the challenges it afforded him. He said he would let me know if he was leaving Howard.

I thanked him, left his office, and returned to my laboratory. I pondered my advisor's response to the question of him possibly leaving Howard University and seriously considered the idea of writing a master's thesis. I had enough data to write the thesis; however, it was a matter of whether I should do it if the next phase of science questions left to answer would meet the requirements of PhD and fit the realm of PhD quality work.

I continued to make much progress in the laboratory and was looking to make greater strides in the summer before my third semester started in the fall of 1970. As if I didn't have enough on my mind with my wife's pregnancy, another major issue surfaced in mid-March that would slow me down some more.

The front office secretary called my laboratory to tell me that my mom was on the line but for me not to worry because everything was fine. I instructed her to tell my mom I would call her from my laboratory phone in one minute. I immediately stopped what I was doing and made the call home. My mom told me I had received a letter from the Raleigh Draft Board, and it was marked urgent. I asked her to open the letter and read it to me. The letter read,

Dear Robert Louis Shepard,

Based on your number in the Vietnam lottery held on December 1, 1969, to determine the order of call for induction during calendar year 1970, you have been reclassified from 2-S deferment because of collegiate study to 1-A, available for unrestricted military service. In about thirty days the Raleigh Draft Board will notify you of when you are to report to take a physical examination to determine if you are acceptable for military duty.

There was a long silence. My mom asked, "Are you still there?

There was still no response from me. I finally said, "Yes, I'm still here just trying to process what's going on in my life right now." I asked when the letter was dated. She said March 14, 1970. I quickly realized that I would be receiving information on taking my physical examination around April 14. I told my mom I needed a moment to think but promised I would call her back soon.

I sat in a chair next to the window in my laboratory and quietly looked out at the beautiful tall oak tree off in a distance. This was something I had done so many times before. The tree and its beauty seemed to help me focus on what to do when I got stuck in my research progress. I sat there for more than an hour, just looking out my laboratory window. I was known to sometimes go into a trance like this when I was trying to work it out first mentally and then verbalize what was already crystal clear in my mind. This was one of those times. During these times, my family knew to give me space and that in time I would surface and sometimes share and other times not share what I had worked out. Zonia had not yet figured out this part of me.

After what seemed like an eternity, I finally got up, went to my thesis advisor, and informed him of my new classification for military service. This time I simply said that I was putting my classes and laboratory work on hold. I promised I would return but explained that I needed some time to deal with the draft board letter. My advisor said that with my wife's pregnancy and the military draft matter, it seemed I had a lot on me right now, but he said he hoped everything worked out for me.

In an oh-by-the-way manner, he then mentioned that when he was either a graduate student at Brandeis University or a postdoctoral student at the University of California at Los Angeles, he had heard that showing evidence of a medical condition could be taken into consideration during a military physical examination. I thanked him for that information, shook his hand, and promised him I would earn a PhD from Howard University with him as my advisor.

I returned to my laboratory to shut down my experiments and packed up what seemed to be a ton of books and papers in preparation for my trip to North Carolina. I left the laboratory not knowing when I would return.

I left the building with a couple of boxes of papers and books and packed them in my 1966 Chevrolet Chevelle Malibu. Then I went to my room at the boarding home. Entering my room, I continued to think about the draft board letter and what was coming in the next thirty days or so. I lay across the bed, staring at the ceiling. I stared so long and intensely that I fell into a deep sleep.

I woke in the early morning hours and believed a solution had come to me. What I needed was a little time. My mom had read the letter to me over the telephone on March 17. Now with just twenty working days before April 14, the question was if there was enough time for me to implement a solution.

The clock was ticking fast, and I needed to talk to somebody and gather more information about what was really going on. I decided the best place to start was with the head of the Raleigh Draft Board. I presumed that the draft board was not aware that I was still a student and now one working on a PhD in the field of chemistry. I rehearsed in my mind the speech I would make. When the sun rose, I called my mother and explained that I was going to call the draft board and explain my situation. She advised me to do what I thought needed to be done.

I called and asked to speak to the head of the Raleigh Draft Board. The receptionist asked the subject of the call. I said to request a meeting about the military letter I received dated March 14. She said the head person was not in but that it would probably not be a problem and that she would get back to me.

I told my mom that I planned to leave Washington, DC, and travel to North Carolina in anticipation of meeting with the head of the Raleigh Draft Board. In a more cheery conversation so as not to worry Zonia, I called to let her know that I had put things on hold in my laboratory to come home and meet with the Raleigh Draft Board.

Chapter 11

PLACED ON HOLD—VIETNAM

I LEFT HOWARD UNIVERSITY NOT SURE about my immediate future. I was confident of returning to complete the PhD program. I just did not know when.

I did not have any time to waste. When I arrived home I shared with my mother the details of what I was doing. My mother asked if the head person at the draft board knew I was Black. I said he must know and asked her how he would not. My mother said she just did not know of anyone, let alone a Black person, who had met with the head or any other person at the draft board. After rehearsing to my mother, though she did not know it, what I would say to the persons at the Raleigh Draft Board when I met with them, I felt confident that my informing them of my student status would result in the Selective Service Agency reinstating my 2-S student deferment classification.

At 8:00 a.m. on the morning of March 18, 1970, I received a call to see if I could meet on March 19, 1970, at 9:30 a.m. with the head of the Raleigh Draft Board. I said yes. When I arrived the next day, the look from the secretary made me revisit my mother's question: "Robert, do they know you're Black?" It was evident from the look on the secretary's face that she definitely had not known I was Black, and I now believed that none of the persons I would meet were aware of it either. Sitting out front, waiting to be called into the meeting, I rehearsed in my mind again and again what I was going to say. Halfway through the second pass on my rehearsal, the head of the Raleigh Draft Board opened his door and summoned for me to step into his office.

I thanked the officer for allowing me the time to meet with him. The officer asked how he could help me. I explained that I was a student at Howard University working on a PhD degree in chemistry. The officer asked what I planned to do when I finished. I said work as a research chemist in industry. I continued explaining what I had rehearsed in my mind over and over again. I said that up until March 14, 1970, I had a student deferment classification of 2-S but that after the draft lottery I received the letter stating that I was reclassified to 1-A, making me draft eligible.

The officer said my classification was correct based on my lottery number of eighty-four, which put me in the "highly likely to be drafted" category and that this would not be changed. The officer then repeated the contents of the letter I was holding in my hand: "In thirty days you will receive a letter from this draft board instructing you to report to the Veterans Hospital in Raleigh for a medical exam." The officer then asked if there was anything else he could help me with. I said no. The officer stood. I stood, thanked him for his time, and left the office.

Leaving the draft board, I recalled how I had watched the lottery on television on December 1, 1969, as had every other young man in America. I thought hard about that evening. As I left the building, my number eighty-four kept flashing in my mind and how the officer had used the phrase *highly likely to be drafted*. Two things had now moved to the front of my mind: (1) finding someone else to talk to who could move me from the "highly likely to be drafted" category and (2) not taking a medical exam in Raleigh, North Carolina.

Time seemed to pass quickly during the rest of March 1970, and I had made no progress as to what I was going to do. I settled into the bedroom

I had grown up in as a boy and tried catching up on some of the graduate school requirements, which I was steadily falling behind in. I found that reading through the requirement checklist reminded me even more that I was losing ground. My mind could not focus. I had difficulty calling and talking to my wife in a cheerful and confident way with any sense of certainty about anything.

While in my room in my parent's home around late March, what occurred to me was the last statement my thesis advisor had said about evidence of a medical condition helping in a military medical examination. I remembered the fall on my knees I had taken during a basketball game in high school back in 1963 and how my mother had taken me to the top orthopedic clinic in Raleigh. I did not want to get too excited because it had been seven years ago and they probably did not have any record of the visit. I left the room and walked to the kitchen to tell my mother what I was thinking. I told her what my thesis advisor had shared with me about evidence of a medical situation sometimes helping in medical examinations for the military. My mom liked the idea and said I should drop by the clinic and ask.

She said, "You have nothing to lose."

I left and drove to the Raleigh Orthopedic Clinic to inquire about a medical incident involving me back in 1963. To my surprise, the doctor who had attended to me was still there and remembered me. More importantly, he provided me with a copy of the medical diagnosis he did on my leg. The doctor asked what I was doing, and in a matter of minutes, I was able to give the doctor the speech I had planned to give to the officer at the draft board. The doctor said he hoped things turned out in a way that would allow my education not to be interrupted. As I was leaving, the doctor said to me that it was good to see me after so many years. We shook hands, and I exited the office.

As I slowly walked out the clinic, I was in tears as I read the letter, thinking how unlikely it was for me to be holding the seven-year-old letter in my hand. When I reached my car, the tears flowed more freely as I thought some more about those in my path who had caused me to be holding the letter. I first thought of my mother, who had driven me to the clinic and how persistent she had been about my seeing a second doctor, and not just any doctor but one who specialized in legs and limbs.

Next I thought about what had seemed at the time to be a casual statement by my thesis advisor of what he had heard about evidence of a medical situation. I thought about the doctor in the clinic who had maintained a record-keeping system that was able to quickly retrieve a seven-year-old document so important to me now. As I sat behind the wheel of my 1966 Chevrolet Chevelle Malibu, I declared that all of this pointed to God planning my destiny and all the persons in my path who were coaching, mentoring, and encouraging me to fulfill the destiny set for me. Once I collected myself and dried the tears of joy from my eyes, I drove off to share the letter with my mother.

When I returned home and showed the letter to my mother, she said her customary phrase: "Praise God." She asked how the letter was going to help me. I said I did not know yet but that I believed it would, just not in Raleigh. My mother asked what I meant. I explained that after meeting with the draft board official, the feeling I left with was that nothing could help change the course the board had set for me of being drafted into the military, not even the letter I was holding in my hand.

I said that what came to my mind right now was to get my physical moved away from Raleigh, North Carolina. Mom asked how I was going to do that. I told her I did not know but felt that it must be done. My mom responded with another of her customary phrases: "Pray about it." I told her to let me know immediately when the next letter arrived from the Raleigh Draft Board. I left and went to show the letter to Zonia. My wife said her customary phrase: "Robert, it's going to work out." We hugged, and I got in my car to travel back to Washington, DC.

I returned to Howard and checked in with my advisor, who welcomed me back and asked if I would be staying for a while. I told him I hoped so but didn't know yet. Then my advisor returned to what he was doing before I poked my head through his doorway.

Being back in that environment brought renewed energy. I got back into the groove pretty quickly, or as quick as possible still realizing that I would soon receive instructions for taking a medical examination. I checked around and found that young men living in DC received their Armed Forces examination in Baltimore at a US Army post named Fort Holabird. I called the post and asked if I could have my medical examination moved from the Veterans Hospital in Raleigh, North Carolina, to Fort Holabird and what the process was for doing so. The person asked why I was requesting the change. I said because I was in school at Howard and the change would be less disruptive to my research work. He said making the change was not a problem but that it would have to be approved by the Raleigh Draft Board. I was also told that it was something done frequently and that I should make the request to the Raleigh Draft Board the moment the letter arrived.

My mom called on April 11 saying the draft board letter had just arrived. I asked her to open and read it to me. The letter instructed me to report on April 22 to the Veterans Hospital in Raleigh for a medical examination. I had to wait until the weekend was over to call the Raleigh Draft Board and request that my medical examination be moved to DC. Early Monday morning I called. Without any discussion about it, the Raleigh Draft Board approved the move. The person charged with such requests told me that he would arrange for me to join the group of men in DC scheduled for medical examinations on May 4 at Fort Holabird; that the paperwork for the transfer to Fort Holabird would be taken care of by the Raleigh Draft Board; and that I would receive the letter indicating the transfer within a couple of days.

The person speaking with me on the telephone at the Raleigh Draft Board was someone totally different than the head official. The person was someone with compassion. I could tell it in his voice. I thanked him and said I would be looking for the letter. After hanging up the phone, I took a few minutes to reflect on what I had done by way of thinking and moving around to be at this point. I will confess that at this stage of my life, my mother and others—many of whom I will never know—did most of the praying on my behalf. But this day I whispered a prayer of thanks to God for myself.

Instructions were for me to meet at the DC Armory in time to board the Greyhound-style bus that would transport me and other young men from DC to Fort Holabird. What I witnessed when I arrived at the armory will remain forever etched in my mind.

When I arrived at the DC Armory on an extremely hot day to board the bus, there appeared to be a sea of young Black men ranging in age from eighteen to twenty-five milling around in the armory yard, waiting to board twenty buses. I stood and estimated that with a seating capacity of fifty persons per bus, at least one thousand young Black men from DC were scheduled to take medical examinations with me that day. It was an amazing sight to see. If young white men were present in the group, they were not visible to me or I overlooked them. I tightly clutched my letter from the Raleigh Orthopedic Clinic as I boarded one of the buses for Fort Holabird. As I looked out the window on the trip over, I wondered how it would all end for me.

When the buses arrived at Fort Holabird, an officer came aboard and said for those with letters from medical personnel to raise their hands. I was the only one on my bus with a letter. The officer requested that I gather my belongings and come to the front of the bus. As I collected my items, the young man sitting next to me asked where I had gotten my letter. I told him I had gotten it from a doctor who had provided medical service to me seven years ago. With that, I was off the bus and taken with a handful of men from the other buses to a waiting room. The nearly thousand other men were led away toward a different building.

The young men with letters were told that the process would be for us to strip down to our underwear and line up in the section under the sign that said *Medical Examiner*. When our names were called, we were to step forward and give our letters to the examining officer. The officer would read over the medical situation and conduct an examination to determine the current status of the condition. We were told that in most cases conditions that had been determined to be a problem in the past would not show up as problems now, meaning that the condition had been corrected. Persons whose conditions did not show that there was a problem would be ruled as fit for military duty.

Once we were lined up, I counted thirty-one men who had letters. We were led in our underwear to the examination room, where two medical officers sat at two separate desks. There was a big difference in age between the two men. The medical officer on the left appeared to be in his mid- to late thirties while the one on the right looked to be in his sixties. We were ordered to stand along the wall in a single-file line. I ended up in the middle of the line. As each young man reached the beginning of the line, he stepped forward and handed his letter to the medical officer, and then his examination was performed. Then the next young man in line was asked to step forward.

While waiting in line, I observed the difference in how the two officers dealt with the young men. It appeared to me that the younger officer had a smile on his face when talking with the young men. The older officer appeared to have a much more stern, serious demeanor. In fact, much about him reminded me of the officer heading the Raleigh Draft Board. Another key point to me was that the older officer did not seem as comfortable interacting with the young Black men as the younger one did; not once did he look them in the eye. When I say not once, I mean not a single time that I observed.

I had now reached the front of the line. The older officer said, "Next in line." I did not move. I could not move. It was as if I was frozen. The officer repeated the call and this time much more forcefully. I still did not move. I quickly turned to the young man directly behind me and suggested that he go. The young man simply stared at me as he went around me to go to the older officer. Shortly afterward, the younger officer called for the next person in line, and this time I stepped forward.

I handed the officer my letter, and before really looking at the letter I had given him, he looked directly at me and asked what I did. I told him I was in graduate school at Howard University working on my PhD degree.

Still not having looked at my letter, the officer responded, "There are times young men like you come before me to be examined for military service at points when they are doing great things to move their lives forward. I have asked myself on these occasions what would have happened to my dream of becoming a medical doctor if I would have been taken off track in the midst of my studies. Looking at your letter I see that you got injured playing basketball."

He then asked if I had been a good ball player. I said that I was considered pretty good but not good enough to make the NBA. The officer smiled broadly and said I really looked like a basketball player. He then said, "I do not think it's necessary to reexamine you at this time, and I want you to get dressed and return to your graduate studies at Howard. In a couple of weeks you will receive a 1-Y classification indicating that you have a medical condition limiting your availability for military service. There is talk that the 1-Y classification will change to 4-F sometime in the future, which you may or may not ever receive."

I stood up, thanked him, and shook the young medical officer's hand. I returned to the dressing room and this time was fully persuaded that it had been God Himself who led me to choose the young medical officer to examine me.

When the bus returned to the DC Armory, I exited with the others but was physically and emotionally drained. I walked toward my car, looking at the beauty of the day and thinking once again of the amazing thing that had happened to me. I was thankful that the military aspect was at least over for now and that my pregnant wife was progressing well as her junior year came to a close without being stressed by this episode.

I welcomed the summer of 1970. While my graduate studies had been put on hold slightly, I still found joy when I reflected on my accomplishments both in the program and outside the program during my first two semesters. But there would be no rest for me in the coming months.

Chapter 12

PLACED ON HOLD—FIRSTBORN

T HE SUMMER OF 1970 ENDED, and with the military draft matter over, I was hard at work back at Howard and steadily checking on Zonia. I had not realized how taxing the draft matter had been on me, but I was relieved it was over. I was back on track traveling back and forth from DC to North Carolina to check on my pregnant wife.

Her senior year got off to a good start. Her classroom performance remained stellar. It appeared the doctor had accurately pinpointed when the baby would come, which was late December or early in the New Year. I was hoping our baby would be the first child to be born in Raleigh on January 1, 1971. We both were glad when Thanksgiving rolled around. This gave Zonia an opportunity to get off her feet and do less moving around. She continued her classes up to the Christmas break, at which time Zonia was heavily with child, and she was definitely showing it.

I was in the midst of some very important experiments in the laboratory but knew that when I left for the Christmas break in December 1970, it was possible I would be gone for a while and might have to put my work on hold once again. I went to my thesis advisor to explain my situation.

My advisor said he understood my situation but immediately moved to discussing the critical milestones the two of us had agreed I would complete before the end of the first quarter of the New Year.

My advisor said, "Youngster, I hear you, but taking more time off in addition to the time off in April and May will definitely delay other important steps in the process, which could end up lengthening your stay in the program."

I simply responded by saying, "Doc, I understand the potential negative fallout and setbacks from this, but my wife has just about made it, and I want to do all I can to help out so she can, if possible, graduate with her class in May 1971." I explained that my work was at a good point for me to leave because I needed to do some important library research on some references and that I would be conducting this research at the library at North Carolina State University.

My thesis advisor looked at me and said, "That's all I have to say. I guess I will see you when you return."

I had established such a good relationship with my thesis advisor that I was confident of his support of my decision. I went to my laboratory and again shut down my experiments, packed my books, lab notebook, and key papers, and headed home to prepare to go to North Carolina to be with my wife.

With my birthday coming on Christmas day and Zonia looking as if the baby could come at any moment, I began to wonder if the baby would be born on my birthday. I still preferred the baby being the first baby born in Raleigh, or even the first baby born in the United States in 1971.

I was able to conduct valuable library research as I waited for Zonia to deliver. When December 25 passed and the doctor had not admitted Zonia to the hospital for delivery, my attention turned to wondering if we would be blessed with the first child born in the United States in the New Year. This was certainly my thought when December 30 came and went with no admittance into the hospital.

On the morning of December 31, 1970, I spent most of the time in my boyhood room with my wife, and it was still not time for delivery. At 5:00 p.m., my mother came into the room, looked at Zonia, and suggested I take her to the hospital. She advised us to call the doctor and give him an update. The doctor said for me to bring her in. We had icy and freezing weather that day but did as the doctor instructed. We checked into the hospital and continued to wait.

When 9:30 p.m. arrived and no delivery, the doctor said the baby could be born at any moment but that it was possible that if Zonia worked with him he could make the delivery at 12:01 a.m., January 1, 1971. He asked if we wanted to do that. Zonia immediately said in a stern voice, "No way." That killed my dreams of not having to purchase anything for our new baby for the first year of his or her life. So we continued to wait.

Zonia gave birth to a beautiful baby boy on December 31, 1970, at 10:07 p.m. I looked at our new baby boy and then looked at Zonia, who was totally drained. The doctor said Zonia needed rest after her long labor and delivery. I left the hospital to go celebrate the coming of the New Year by announcing the good health of my wife and the birth and good health of our firstborn, a son.

I got up early the next morning and turned on the television to see what the first child born in the New Year in Raleigh was receiving. The child received gifts from a number of businesses. A store gave a year's worth of disposable diapers, a Raleigh bank gave a US savings bond, and the hospital provided a new baby safety car seat. A department store provided a day of free shopping in the baby section of the store. A grocery store offered free shopping for baby food items. I estimated that the freebies for the first baby born in Raleigh in the New Year totaled at least $1,000 dollars worth of items. With my wife being an undergraduate student and me in graduate school, I thought of how much this would have helped us out.

Christmas break was soon over, and Zonia was in no shape to return to school. Up to this point we had not gone to our families to ask for help. But help was needed now. Not yet on her feet after Christmas, Zonia remained with her family after the birth so family members could help with childcare and other needed support. Everyone helped us out in whatever capacity they could.

Once again the conversation returned to the possibility of her taking the final semester off and completing her studies the first semester of the next year. We talked it over briefly, but I wanted to keep to our goal of Zonia graduating with her class. With only four months remaining in the semester, she was too close to turn back. We came up with a plan we believed would work to keep her on schedule with her class.

I contacted my thesis advisor to let him know I needed to remain in North Carolina longer than expected because of the timing of the baby's birth. Zonia did not want me to take any more time away from my graduate work, but more time off from Howard was needed, and I had to take it. Zonia had preregistered for all her final courses before leaving for the Christmas break. Our plan was for me to complete the registration for her and then meet with the professors for those classes to explain my wife's situation.

I met with the professors for each of Zonia's classes and explained that I would drop by class each day to get the reading assignment and pick up lecture notes and class assignments for her. I assured them Zonia would keep to the schedule and would be on par with the rest of the class when she returned in February. They all agreed to cooperate and see how it worked out. The plan worked well.

With support from her family, Zonia excelled in her classwork while studying and working from home. I saw how well she was doing in the first two weeks of the new semester, so we agreed that I would return to DC at the end of January to resume my work at Howard.

She returned to class full time in February and participated in her graduation exercise with her class on May 9, 1971.

I had made up my mind to write a master's thesis because my advisor never clarified with me whether he would or would not accept the offer to leave academia and go to industry. Prior to the birth of my first child, I was on schedule to receive my master's degree in chemistry from Howard University during the graduation exercise on May 12, 1971, three days after Zonia received her bachelor's degree in English from STA. I was delayed because of the lengthy breaks I took in the fall semester of 1970 and spring semester of 1971.

When I returned to my classes and my laboratory, my thesis advisor was pleased. He said what I had accomplished outside my studies without my studies being seriously sacrificed in any way was nothing short of miraculous. I said to him that God had been with me every step of the way since I put things on hold at Howard. Because of my high level of performance in the graduate program, even with the interruptions, my thesis advisor said he wanted me to apply for a fellowship that would take me abroad to study with a mass spectrometry group at the University of Zurich in Switzerland that was conducting some advanced research in physical organic chemistry similar to the direction my research was going.

The idea posed by my thesis advisor was that since I had about four more months to complete my thesis work, research with this advanced research group in Zurich would be a great addition to my training. I was flattered by the offer but turned it down because bringing my young family together was past due, since we had been apart for nearly two years. I moved Zonia and our son to Washington to start our life together under the same roof. My thesis advisor said he understood my decision.

The three of us moved into the same room in the house I was renting from the family in upper northwest Washington. Shortly afterward, I secured a lease to a three-bedroom apartment in Oxon Hill, Maryland.

I received my master's degree August 23, 1971, three months after witnessing Zonia walk across the stage to receive her bachelor's degree in English. The delays in my personal schedule had been worth it to be able to witness my wife reach a goal she had never really thought about as a young child. I was grateful it worked out like it did because if I had remained on schedule, both of us would have graduated the second weekend in May and probably would have missed each other's graduation exercise. This way we were both able to witness each other's great accomplishment.

Chapter 13

RECEIVING THE PhD

WITH MY MASTER'S DEGREE IN hand, the delays behind me for now, and my wife and son with me in DC, I was ready to push toward completing all the requirements for my PhD. I had completed four semesters of coursework and research and needed only two more semesters of graduate work to satisfy the requirements to move to the next phase of the program.

I had only two more graduate courses left to take in my specialty of physical organic chemistry and two additional graduate courses outside of my subdiscipline, which could be in chemistry or courses from other departments such as engineering, biochemistry, or other related disciplines. In consultation with my thesis advisor, I decided that the two courses outside my specialty would be computer basics for chemists and quantum chemistry. I was comfortable with also adding to my schedule a seminar course on advanced topics in mass spectrometry.

Before I completed my master's degree, I had already interviewed individual faculty members to identify common areas of interest. This was how I selected my research advisor and project focus. I had chosen my research advisor but still met with more than five other faculty members to get background on their PhD research interests. The research project I had conducted for my master's degree had required use of research facilities that had not been located in the Chemistry Department or anywhere on campus. I had performed my master's degree research under the joint direction of my on-campus advisor and a highly regarded experimentalist at the US Naval Research Laboratory (NRL). Now that it was time to continue, my research toward a PhD required a write-up of proposed research ideas. My advisor had accepted the proposed research project, but it needed to be accepted by the research group at NRL.

My project was accepted. The expert at NRL said that probing deeper into the research questions not answered by my master's degree had great potential to add new knowledge to the field of science at the PhD level. After a series of meetings with my research advisor and experts at NRL, I wrote up the project design, describing how my master's degree research could be continued to the PhD. My project outline needed to be clearly defined to enable an objective assessment of the quality of my research work. Through a series of more rewrites and additional consultations with my faculty advisor and NRL experts, the extension of the research was accepted, and I started the PhD part of the research at NRL in the fall of 1971.

Another major milestone had been put in place by this action. I had identified three of the five members who would become my PhD advisory committee: my thesis advisor, the NRL expert experimentalist, and the advanced organic chemistry faculty member who had not been able to believe that I was performing at such a high level in his class during my first semester in graduate school. I was aware that the identification of the committee would be in consultation with my advisor and the director of graduate studies. I was very sure that I could and would guide the selection process because at this early stage of my research I had already pulled together the argument for why these persons should be on my committee.

There is an important point to note here. The advanced research facility I would work in at NRL possessed five different mass spectrometer instruments on which I could conduct my research. I had familiarity with only one of the machines from my master's degree research. Now I would be given access to the full suite of instruments. Since my PhD research was probing more advanced science questions, the NRL expert determined that I needed to gain greater theoretical knowledge behind the instruments. Therefore, at the start of my PhD research work I spent more time reading the theoretical background associated with the use of the instruments to produce the expected results and interpretation of the results. The additional reading deepened my knowledge about this powerful analytical tool.

I continued to perform well in lectures, laboratory, and my personal research. At the beginning of my fifth semester, I had a cumulative grade point average of 3.95 (A) and expected it to climb higher. I had received a grade of B on only two occasions: one in quantum chemistry and the other in inorganic chemistry. Conducting my PhD research as I had done with my master's degree research at the world-class NRL national laboratory further strengthened me as a science student by exposing me to practical training in cutting-edge research. The growth and expansion of my knowledge in physical organic chemistry was quite evident when I passed my specialty exam by answering more than 80 percent of the examination questions correctly.

The most important milestone for me was that in the fall of 1971, I had satisfied all eligibility requirements and was admitted to candidacy for the PhD. Admitted to candidacy for the PhD meant I had completed the required coursework, demonstrated proficiency in expository writing, successfully passed all comprehensive examination requirements, successfully passed foreign language examinations, demonstrated competency in the oral presentations, and completed an approved written research proposal. The only requirements left were for me to complete my research, write my dissertation, and undergo dissertation defense in the form of an oral examination before an examining committee.

I received an unexpected call in the fall of 1971. I was notified that I was being considered, along with other PhD students who had been admitted to candidacy, for the Howard University Terminal Fellowship. The prestigious fellowship was given to the graduate student whose performance was classified as superior. The fellowship allowed the successful graduate student recipient the luxury of going into his or her final two semesters of the PhD program without having to teach or serve in any other capacity other than stay in his or her laboratory to ensure completion of the research and writing of his or her thesis.

In late fall of 1971, I was selected as the graduate student to receive the Howard University Terminal Fellowship. The fellowship allowed me the opportunity to spend the entire year of 1972 conducting my research. The monetary value of the fellowship totaled $5,200, paid in four nontaxable lump sums of $1,300 dollars per quarter. I was absolutely overwhelmed at being selected for the fellowship. For a young man from the tobacco fields of North Carolina whose initial stipend was $233 per month ($2,796 per year) as a married graduate student and moving to $3,396 when my son was born, the honor of receiving the fellowship left me speechless.

I knew others could have received the fellowship, but I was the one who had been awarded it. Now I was thoroughly convinced that God was charting my destiny. I could not explain receiving the fellowship in any grand theological fashion, but I had been taught that one does not make it through the world on talent alone. I had been taught about what it meant to say that "favor" had been shown toward an individual. I kept it simple by saying that favor had been sent my way, because I knew that hard work alone was not the sole reason for my making it to where I was now.

As I looked at the fellowship material I needed to complete and return to the graduate school office, I paused for a long time. I reflected on the distance I had traveled. I thought about my mother's wisdom; my father's quiet but strong work ethic; my siblings, who had simply loved me as a big brother and acted as my greatest cheerleaders; my grandmother's unspoken belief in me; all of my Sunday school teachers; the decision to leave my longtime friends in the seventh grade to go to an unknown teacher; my high school chemistry teacher, who had returned after long days of teaching to work one-on-one with me on my science fair project; my high school English teacher, who had decided to move forward with entering my writing project in competition; my high school classmates, who had such faith in me becoming someone great; my mom's determination to take me to visit the orthopedic doctor when I fell on my knees; the soft voice of my thesis advisor saying a medical record could be of value; the doctor who maintained a written record of the diagnosis on my knees for seven years; the day I decided to turn in my basketball uniform after meeting Earl "the Pearl" Monroe; the professor from Howard who requested that I send him my undergraduate transcript; the young doctor at Fort Holabird who told me to get dressed and return to my graduate studies at Howard because he was not going to reexamine my damaged knee to see if I was fit for military service; and so many others. And of course, my wife, Alzonia, who always said, "It is going to work out." To me, all of this simply pointed to favor sent my way.

My laboratory research work was complete and the first draft of my thesis was well underway in the early part of the spring semester of 1973. My thesis advisor and I discussed setting a date for when I would defend my thesis and present a final research review to the committee members and chemistry faculty. Before the date could be set, I had to give a departmental seminar highlighting my research findings.

I had made it through my first seminar in the area outside my specialty but not in spectacular fashion. I faltered in handling some of the questions. This next seminar was on my research. Demonstrating expert and comprehensive knowledge was required this time around. My thesis advisor assured me I would do fine, but the butterflies in my stomach told me something different. It was another hurdle I had to overcome.

My seminar date was set for February 1973. This important milestone required my committee members to be present at the seminar and evaluate my performance before scheduling my dissertation defense exam.

I had concerns about my seminar because on one occasion I had witnessed a candidate having a tough time in his seminar with very difficult questions from faculty who believed the student should know every question asked of him. Not only was this seminar very stressful for the presenter but also to other graduate students present in the audience who had yet to reach that point in the program.

Even after winning prestigious fellowships, performing well in all classes and laboratory exercises, and progressing well in my research, there always seemed to be a next requirement that stood on its own without regard to past performance. The seminar was one of these requirements. It appeared to me that faculty members who had not taught the candidate any courses were the worst. Their questioning sometimes came across to me as punishment to the candidate for not having received any training from them. Unlike earlier points in my educational career in which past performance counted, it just seemed not to be taken into consideration when going from one requirement to the next. This is why every phase of the chemistry graduate program at Howard University was filled with pressure.

I passed my seminar but not without serious challenges. Questions came from everywhere, and at one point one of the faculty members asked if I should be expected to know the answer to a question posed by a certain faculty member. I remember the question dealt with if I knew a certain physical constant of the derivative of a compound I used in one of my experiments. Before I could try to respond, another faculty member responded to the faculty member who asked the question by saying that no one walked around with such information in his or her head. He said most scientists would find the information in the appropriate resource material.

I was glad the faculty member had interceded because as the question was being asked, I did not know the answer and had not thought of how it should be answered. As the two faculty members jabbered at each other, the answer came to me that for such questions an appropriate answer would be that I didn't know but would find out and get back to them. I passed my seminar and was now ready to set a date for the last major requirement—my dissertation defense.

Prior to scheduling a date to defend my dissertation, the final dissertation copies had to be critically reviewed by my dissertation committee, which consisted of the advisory committee, the external examiner, another faculty member, and the chair of the Department of Chemistry. Based on acceptance of my dissertation submission and recommendation of the department, the graduate school would then formally appoint the committee members for my defense. With all requirements met, the graduate school scheduled Wednesday, April 4, 1973, as the date of my dissertation defense.

My final oral examination took place before the five examiners that had been selected. The outside expert was my advisor at NRL. It took five hours to complete my oral examination. My fellow graduate students all milled around throughout the chemistry building, steadily checking the room and the secretary in the office to see if I had finished. It was a nail-biting time when candidates remained in the room as long as I did. This sometimes signaled that things were not going well for the candidate.

In my case, it was just the opposite. The committee was genuinely interested in the research I had done for my PhD and the explanation of my results. My research project was extraordinarily unique at the time. The uniqueness was that while most of the mass spectrometry community attributed the fragmentation of compounds to primary electron impact, I proposed that the only way my results could be explained was that observed fragmentation patterns could not be based solely on the interaction of primary electrons but that less energetic secondary electrons had to be involved in the process. The research was so rare my thesis advisor recommended I send an abstract of the work to be presented in the poster session at the Twenty-first Annual Conference on Mass Spectrometry and Allied Topics. Although no word had been received from the conference organizers, I felt good about the level of interest being shown in my research by the dissertation committee.

Much personal satisfaction came when the white American professor who had said he could not understand how I was performing at the level I was on his exams stopped in the middle of asking me a question. Before completing the question, he turned to the outside expert and asked if the question he was asking me was fair. The expert responded, "Robert should be able to answer the question." The professor finished asking it. Before answering, I said to the professor that I definitely knew the answer to the question but wanted to know why he wondered if it was a fair question to ask. The professor responded, "You were looking at me as if I should not ask the question." I responded that I was looking at him, waiting for him to complete the question. At that point I felt that I had come full circle with the only white American professor I had ever had as part of my now twenty-year academic training program.

I was asked to wait outside the room while the five-member examination committee went into deliberation. When I exited the room, my fellow graduate students crowded around me and asked how I had done and if I passed. I said, "I did my best, but they are deliberating, and I will know if I passed when they come to a decision." I continued by sharing with my fellow graduate students that I felt good and that all the questions had been fair questions for a person with as much exposure to chemistry as I'd had over the past four years—well, eight years when undergraduate training was included. After about twenty-five minutes passed, I was still sharing with my fellow graduate students when the expert examiner came out of the room and requested that I come back in.

The committee stepped back through every aspect of my oral examination. They shared with me where they thought I was very strong. They also communicated to me areas I would need to work on as I moved forward in the field of chemistry. The committee ended by saying that my research was some of the most interesting any of them had seen and for me to definitely continue and expand my research.

The white professor who had questioned my chemistry knowledge a few years earlier stood, shook my hand, and said, "You are a fine chemist." I had grown to really appreciate the professor and his dedication to excellence. I thanked him for all he had personally done for me.

After the feedback, the committee voted unanimously to pass my oral examination and at the same time voted unanimously to accept my dissertation and declared that I had fulfilled all the requirements for the PhD in chemistry from Howard University.

I left the examination room that day with a flood of emotions flowing through every fiber of my body. One moment I was ecstatic. The next I was in a state of disbelief. Another time I felt completely emptied out. When I tried to play back in my head the entire sequence of PhD requirements from the checklist that had become my guidebook for the past four years— well, four and a half years if I counted from the time I received the material in the mail before starting the program—it was too overwhelming. I took refuge for just a little while in the graduate student lounge. I needed the time to try to process the milestone I had just completed. My fellow graduate students understood where I was mentally, and having seen it before, they did not enter the room with me immediately.

My mind drifted back to my mother and her encouragement and motivation and, most importantly, her love. I could not put it all together that day, so when my fellow graduate students opened the door to the lounge, I focused on sharing my experience with them because I could tell it would be a while before what I had accomplished hit me. The department then brought in refreshments to celebrate my success.

Two weeks after celebrating my accomplishment, I received a response from the organizers of the Twenty-first Annual Conference on Mass Spectrometry and Allied Topics. The committee said because of the far-reaching nature of my research, I was invited to San Francisco to make an oral presentation at a session large enough to accommodate many attendees rather than a poster presentation where only a few could gather. I called my thesis advisor to convey the good news. He asked when was my presentation? I said it was scheduled for May 23. He responded that the Chemistry Department would make sure we both were there. I said it would be great if my wife could travel with me since neither of us had been to California. He was not sure how that could happen.

It was now time to celebrate at the university level. Saturday, May 12, 1973, was the celebration day. There was a major conflict that my family had and did not know how they were going to handle it. If there was a time that conflict was okay to have, then it was considered a good conflict. On the same day I was receiving my PhD from Howard University, my brother Roger was receiving his high school diploma. My mother was torn as to what to do. This situation showed how a large family could be beneficial. With a total of nine family members, the siblings and dad all decided to attend Roger's graduation. My mother, along with her brothers Thurman and Howard, traveled to DC to witness me walk across the stage at Howard University to receive my PhD.

What a proud moment that day was for my family. I always knew how special the moment was for not just my family but my extended family in Garner who had invested themselves in me. My high school chemistry teacher said it was a proud moment for him as well. After delivering a powerful speech to the graduates, Dr. Ossie Davis, a writer, director, and actor, was awarded the degree of doctor of humanities from Howard during the same graduation ceremony.

At Howard's graduation ceremony, only the PhDs got the opportunity to walk across the stage to have their degrees conferred upon them. The bachelor degrees, master's degrees, and all the other candidates, including those receiving doctorates in medicine, dentistry, and juris doctor, simply stood up, moved their tassels from right to left, turned to the left, and allowed the person in back of them to place their hoods around their necks. Following these events, the candidates sat down.

It was different for those receiving PhDs. PhD candidates were asked to rise and come to the platform to have their degrees conferred upon them. As I rose to my feet, a sense of unspeakable fulfillment and joy flooded my mind. My last name beginning with *s* placed me toward the end of the line. As I inched my way toward the platform, I knew the trek to this point had been a step-by-step journey that I could not take full credit for charting.

Soon, I was finally standing and shaking the right hand of Howard University President James Cheek and listening to the incredible words he spoke to me and every other PhD graduate: "Having fulfilled the requirements and having been recommended by the faculty for the degree, it gives me great pleasure to confer upon you today the doctor of philosophy degree with all the rights, privileges, and honors pertaining thereto. With this degree, you are not only entitled to teach others but to also teach yourself. Congratulations."

With these powerful words echoing in my mind and deep in my soul, I realized that my destiny was being fulfilled even if I could not fully understand where it would take me. I left the stage and returned to my seat having officially received my PhD. Excitement filled the air even more because the day before this celebration one of the companies recruiting me called to say they would pay for my wife to join me at the conference in California. At this moment, all I could do was look out on the day and whisper a prayer of gratitude in my heart.

Before leaving Howard University armed with my MS and PhD in physical organic chemistry, I needed to pick up some of my materials from the physics building. I had gone in the building many times over my four-year stay in what was called at Howard "the Valley." In fact, I had completed some of my experiments using a new mass spectrometer purchased by the physics department toward the latter part of my research. I did so to compare some of the data collected on the systems at NRL. The data had compared very well.

I went to the front office to say farewell to the physics department chair, who was a fellow North Carolinian. I saw a distinguished gentleman come in, and the chairman asked if I knew him. I said I was sorry but did not know him, but I quickly extended my hand to introduce myself as a friend of the physics department chair. The chair told the gentleman that I had just completed my PhD in chemistry a couple of days ago. He congratulated me on a great achievement. The chairman then told me who the individual was.

The gentleman was Dr. J. Ernest Wilkins Jr., distinguished professor of applied mathematical physics and also founder of Howard's PhD program in mathematics. I had just met another Black scientist and mathematician who had worked on the Manhattan Project during World War II. He had gained fame working in and conducting nuclear physics research in both academia and industry. I said it was an honor to meet him.

As I exited the physics building, saying good-bye to my friend the chairman, I was even prouder of my accomplishment in that two renowned Black scientists who had worked on the highly regarded Manhattan Project were in my midst: Dr. Moddie D. Taylor, who had taught me advanced inorganic chemistry, and now Dr. J. Ernest Wilkins Jr. I was indeed ready to move to the next level of my upward growth path along the journey I was taking step-by-step. This next level was traveling to California to present my PhD research results before a body of international scientists. My presentation was well received and the interest by this audience was on the same level as that shown by the dissertation committee. An added bonus was that my wife and thesis advisor were in California with me.

Chapter 14

THE FIRST ONE

WHEN I WAS A YOUNG boy moving around in Garner, I wondered with regularity what was on the other side of the white-only signs that society would go to the extreme of establishing laws that would say to me, "Not allowed." I would look at signs and say, "Whites only," and I would wonder, *What's inside that I am not allowed to see or take part in?* It was in this early stage of my life that I decided if the opportunity ever came to go behind the doors that had been shut to me as a boy, I would do so. The desire to enter this virgin, foreign territory that was kept from me never left and made it easy for me to decide what was next for me after receiving my PhD.

My undergraduate institution, Saint Augustine's College, offered me the opportunity to return and join the chemistry faculty. This was not in the cards for me. It wasn't that I would not consider becoming a faculty member at my alma mater, but I did not contemplate joining any historically Black college and university (HBCU) because these were environments saturated with people who looked like me, and it was a place I had spent the developmental stages of my life. I made a decision early in my graduate training that upon completion of my studies I would enter the industrial or corporate sector as a research scientist to get a glimpse of the "other world" that was virtually unknown to me. Growing up under the segregated Jim Crow laws of the South had pointed my destiny toward the doors society had closed off to me. It was now my turn to see what was behind the doors that had escaped me in my youth.

As a newly minted PhD, I interviewed and was offered positions with four different organizations. Three were research and development groups within large industry corporations. Two of the three industrial corporations were located in the Northeast, and the other was in the Midwest region of the country. The fourth offer was with a research laboratory located in the South. The two interviews in the Northeast and one in the Midwest went well. It was obvious during the interviews at these three organizations that their employees and staff were accustomed to and familiar with Black PhD scientists because some were in responsible positions within the laboratories. This was not the case for the interview with the organization located in the South. This interview was eye-opening.

The organization had never employed a Black PhD scientist, and if I accepted their offer, I would be the first to work in the analytical chemistry research division at this location within the corporation. During the interview, I was looked at as if I was a man from outer space. The stares and looks I received as I moved through the laboratory were piercing. When I arrived at the laboratory I would head if I accepted the position, the research director introduced me to the two white laboratory technicians. Having experienced the segregated South, I could feel the discomfort of the two fellows, and in my opinion, it was impossible for the research director not to be aware of the uneasiness in the place as well.

As I continued to tour the laboratory, I recognized to a degree what the laboratory was attempting to deal with. When my wife and I were picked up from the hotel later in the afternoon and taken to a restaurant to meet others for a seven-course dinner, I understood the situation very well. This was only 1973, and I was not naïve; I knew unrest and an unspoken tension remained from the 1968 race riots that had brought devastation to the country. In a span of just five years, a company that had no Black PhD scientists in their laboratory workforce now had to adjust to a new day. In many ways they were being forced to by federal mandates rather than coming to the decision on their own. I had diagnosed the state of affairs in America at the time of my interview.

I was aware that the only reason my wife and I were at the dinner was because of federal laws passed by Congress that were now being enforced. I was cognizant of the fact that creation of the Equal Employment Opportunity Commission (EEOC) was part of the historic Civil Rights Act of 1964, which prohibited discrimination in employment on the basis of race, color, religion, sex, or national origin. I was equally sensitive to the fact that because of resistance by industry and the government to promote equal employment opportunities to all Americans that Congress had just passed the Equal Employment Opportunity Act of 1972, which made an attempt to give backbone and enforcement power to the EEOC in an accelerated effort to eliminate discrimination in the workplace.

I sat at the dinner with my wife, interacting with a race of people who throughout my life had been, for the most part, total strangers. I understood the situation, both the dialogue during the interview earlier in the day and now at the engagement at dinner. I believed corporate leadership put forth a genuine effort to make me feel comfortable. I was comfortable, but the expressions around the table gave me the impression that more rehearsing was done to prepare for me than probably was necessary. It was as if company leadership wanted me to know that they knew the world was changing. While eating my dinner and fielding questions about myself, my family, and a few critical thinking questions about the future direction of the laboratory should I accept the job offer, I pondered the experience and found it somewhat disheartening.

I struggled at one point trying to assess if the evening was genuine and the company had decided on their own to do the right thing and was not responding as a result of forced actions by the federal government. At that moment, I remembered a commitment I had made to myself years ago of not using my energy trying to figure out things I had no control over. I had finished with this and moved forward to give my very best wherever I ended up. With this in my mind, I returned to the business of the evening.

My wife and I returned to the five-star hotel in which we were staying and laughed about the experience of a seven-course meal that started with the appetizer of escargot (snail) stuffed in mushrooms. This was definitely a first for us but a second for me. I had experienced such an elaborate dinner setting when I received my writing contest award.

A week after the interview, I received an official offer from that corporation, which was located in Charlotte, North Carolina, only twenty miles from the South Carolina state line. Of the four offers, this company offered the lowest starting salary. The position was one that I found very interesting, but the salary was not attractive. The position was in North Carolina, and that was appealing to me. I had made a decision to not accept any position that was too far from North Carolina. This constraint was linked to my love of family and my roots in the small town of Garner. The love I developed for both as a young boy never left me.

Zonia was not as fascinated with the offer from the company in North Carolina when compared to the offers from the other three industries. With such a low offer from the company, she did not fully understand why I was seriously considering taking the position. I had been in school for eight straight years since graduating from high school, and it did not compute for her that I would start my career at such a low income. Her not being impressed with my consideration of the job offer in North Carolina caused me to reflect on my consecutive twenty-year-long journey from the time I entered first grade at age five until receiving my PhD at age twenty-five.

Thinking about the trek, I concluded that my wife was right. I should not accept the offer. After such a long time in school, there was much catching up I needed to do in all areas, especially in the area of finances, and starting on what the company was offering wasn't going to move us upward very fast, if at all. While not in debt, I was short on financial resources and other things in life that seemed to put me way behind my peers, who had started their families and careers right after high school or after receiving their undergraduate degrees.

But I believed the job offer in North Carolina would net me something in the long term that would be more valuable than financial resources in the near term. I figured that if I joined the organization I would find out characteristics about myself and about life that could not be learned in any of the other three companies. I shared with Zonia the reasons why I wanted to take the position in Charlotte. With it being a Southern city close to the South Carolina border, I could use the location and prevailing culture of Charlotte to introduce something new and different to the company, to the region, and possibly to myself.

During my interview, I had been able to almost feel, and in some cases actually see, the high level of discomfort in many of the scientists I spoke with. Some of the persons had demonstrated great difficulty and uneasiness interacting with me. I had noticed their inability to make direct eye contact with me when they asked questions and when I responded. I viewed it as me performing before an audience of strangers who knew nothing of the play or anything about the performer.

Other than small talk about basketball, UCLA winning the NCAA, and the reigning NBA champions, the New York Knicks, the only other significant question was if we had children, but they asked nothing pertaining to aspects of my culture. Despite the fact that I had come from Washington, DC, where one of the biggest events on the world stage at the time of my interview—the Watergate scandal that led to the resignation of President Richard M. Nixon—was being played out, there was no dialogue on the matter. I thought the overall interactions were awkward and clumsy. Nevertheless, Zonia agreed, and I accepted the offer in the end.

Becoming the first Black PhD scientist in the Analytical Chemistry Division at the Charlotte operation, I was confident I could have a positive impact that would make it easier in the future for other Black scientists and engineers who would join the division I was in and the entire corporation. I also believed the environment would benefit me and help me better understand race relations in a changing world.

In June 1973, I moved my family from Washington, DC, to start employment at the Celanese Research Laboratory in Charlotte, North Carolina. With my strong background in mass spectrometry, I desired to run the mass spec laboratory but instead became head of the corporation's X-ray diffraction and atomic absorption laboratory. As one with a strong background in a range of analytical measurement tools, I had no problem shifting over to a totally different area. What mass spectrometry, X-ray diffraction, and atomic absorption had in common that I knew going into the position was that they all required understanding of the importance of system calibration and interpretation of system outputs. I was adept and meticulous in both these areas of analytical measurements.

I arrived on the job ready to establish my footprint on the laboratory. The two white technicians who were now employees under my leadership eagerly welcomed me. I sat down and had a long meeting with them to discuss my laboratory policies and my expectations of them working under my leadership. After the meeting, I took a short walk down the hall of the laboratory and let persons I had met during my interview know I had arrived. The brief conversations didn't feel as awkward this time around. Soon I was back in my lab and was off and running in my new environment.

An interesting incident occurred just two weeks into my assignment. My supervisor informed me that the annual radiation inspection of my laboratory would occur in a few days. I asked questions about the inspection so I would know what to expect. There was no surprise when a North Carolina health physicist from Duke University dropped in to conduct the check of my laboratory. What was surprising were the results.

The results revealed scattered background radiation in excess of allowable limits for worker exposure in the state of North Carolina. The radiation inspector described the situation in detail and ended by saying the condition needed to be corrected within the next sixty days. He would return at that time to conduct a follow-up inspection, and if the problem was not solved the laboratory would be shut down.

Only two weeks in the laboratory and faced with such a challenge, I did not know what to think. During the exit interview, the radiation inspector was required to give a verbal and written report of his findings to the next upper level management. My supervisor noted that the problem would be rectified. After the inspector left, my supervisor suggested that I go to Atlanta to discuss the problem with the company that had made the instrument to see if they were aware of the problem and had come up with ideas to solve it.

I remembered a fellow Howard colleague who had shared with me on an earlier occasion how, as the first Black scientist in the corporation he joined, he had also been faced with a serious problem within a few weeks of joining the company. In my friend's case, he was asked to lead the investigation and suggest what needed to be done to bring a plant back on line after a violent explosion had ripped through it. He handled the problem and was able to identify and reconstruct the reaction that had led to the explosion. His landmark research had changed the way all of the company's plants would operate in the future.

I wasted no time wondering if the problem predated my arrival in the laboratory. I knew I was now being looked upon to come up with a solution to this critical problem facing my laboratory and the overall company. I understood the important role my laboratory played in the corporation's new product development mission. The raw materials characterized in my laboratory fed into the pilot plant operations being led by the engineering division. The quality of the final product that went to the marketing department depended on the quality of the starting raw materials used in production. Results from my laboratory established raw material quality. The time-sensitive nature of how results from my laboratory flowed into the next phase of the operation was critical to meeting development trials, test runs in batch operations, and ultimately product trials in the field. The bottom line was simply that my laboratory could not afford to be offline for any length of time.

It was good that during my first week on the job I had visited several of the company's plants scattered from Maryland to Georgia to discuss new requirements governing laboratory health and safety of American workers. My knowledge of the new requirements found in the Occupational Safety and Health Administration (OSHA) was causing me early in my position to become the go-to person in the company when there was a question regarding OSHA and the company's workers. I wanted the best for all employees in the company, and now my laboratory employees were personally involved.

I took a flight from Charlotte to Atlanta to meet with the management and technical staff at Philips Electronics, specifically with the developers of the industrial X-ray equipment. I took along the report from the North Carolina radiation inspector. Philips was aware of workers in X-ray environments being subjected to excessive levels of background leaking from their X-ray units, the optics of which were exposed to the atmosphere. I was taken to a laboratory and given a demonstration of Philips's proposed solution to the problem.

The company had developed a heavy rectangle-shaped housing design comprised of clear lead-plated glass. One of the sides of the design had a slide door connected to a microswitch. The housing was to be placed over the exposed X-ray unit, eliminating radiation leaks from the equipment into the laboratory environment of the technicians. To operate the unit with an installed housing, a laboratory technician could not turn on the unit unless the lead-plated door with the microswitch was closed. When the door was opened, the unit would shut down and could not be operated when crystals, detectors, tubes, and samples were being changed and manipulated in various ways.

When the door was open, the entire X-ray unit would become inoperable. With the door closed, a test for radiation leaks measuring five centimeters from the housing unit showed normal environmental background radiation, as would be the case with the X-ray unit turned off. I was told the protective housing unit cost $10,000. I was scheduled to stay a second day with the instrument developers but chose to leave after the first day following the company's demonstration of their solution for my problem.

I returned to my hotel room, took out a sheet of paper, and began to sketch a design I was sure would work to solve my radiation leak problem and at a cost far below $10,000. I packed my luggage, headed to the airport, and took a flight back to Charlotte. Upon arrival, I shared with my supervisor the instrument developer's solution for solving the problem but said I believed their fix was much too costly. His response was, "If $10,000 is what it will cost, you need to go with it." I then described my alternative idea for solving the problem. The supervisor asked how I knew my idea would work. I said that prior to bringing the idea to him, I had blocked the radiation by holding a sheet of notebook paper in the path of the leak exactly five centimeters from the X-ray optics located on top of the unit.

I was convinced that constructing a barrier from material much lighter than the heavy housing developed by the instrument maker would do the job. My supervisor reminded me that I only had about six weeks before the unit was re-inspected. Realizing I was confident of my solution, my supervisor requested that I take my sketch to the graphic design group for drafting of a complete set of drawings and then take the drawings to the machinist to request the cost for fabricating the design, but I was not to get the work done before I reported back to him. We agreed to meet again once I had these items in hand. I secured the drawings and presented them to the machinist, who priced the total job at under $500, including construction, painting, and installation on the instrument. I was given clearance to move forward with my idea.

A shield was fabricated from a 1/8 inch thick aluminum sheet and designed so that no alterations to the X-ray unit itself were required. The design consisted of two parts: one stationary shield surrounding the X-ray optics and one hinged shield above the section of the instrument that allowed the detector to rotate through various angular positions. The stationary section of the shield was secured to the top of the table with four bolts. The hinged (mobile section) was designed to allow technicians access to crystals, detectors, and the sample chamber. With the hinged shield lowered, the instrument operated as intended. When the hinged shield was raised, the instrument would shut down and could not be turned back on until the shield was lowered again. The instrument developer used one microswitch in their design. I used two microswitches in mine. The two were attached to contacts on the X-ray tube that were energized when the hinged shield was raised, switching off the X-ray tube power supply. The microswitches were attached in a way that mimicked water pressure drop to the tube. My measurement results had shown normal environmental background radiation. The design provided maximum protection to technicians and visitors in my laboratory.

I did not think my solution to protect my technicians and others who might enter my laboratory was that big of a deal. This began to change when management began to invite every visitor to the company to stop by my laboratory to see the design developed by one of its scientists. I really thought more of my design when I was called into a meeting that included my supervisor and a representative of the company's legal staff. I was asked what I thought about filing a patent application for my design.

I could not believe this was a serious request. The main reason was because I knew my design was similar to the kind of solutions my father came up with all the time even though he had only attended school up to the third grade. I said I knew nothing about filing a patent. The legal representative spoke up and said most scientists were not familiar with how patents were developed but that his office would help guide me through the application process.

As I exited the room, I once again thought of my dad and how proud I was that after obtaining education to the PhD level, I had used what I learned while assisting him on projects around the house to make my first big impact in my career. A smile came over my face, and warmth resonated deep in my soul.

I worked diligently with the corporate legal staff to write up and file a patent application for my design. I completed and filed the application three months after I started with the company. After eighteen months passed, company management decided it was not in its best economic interest to continue the costly patent review. I was informed that a decision had been made to not continue the patent process. The decision to discontinue the patent effort was based on the company not being an industrial equipment manufacturer and the fact that the instrument was becoming obsolete. My supervisor suggested I publish the design. I submitted it to the journal *X-Ray Spectrometry*, where it was accepted and approved with me as the single author.

I had been very good at approaching science in graduate school, but I had no experience or any coaching or mentoring on how to tackle science in a corporate culture, especially as it related to the intellectual aspects of managing science. Even without such guidance or training in this area, there was one event that took place in my laboratory around my design that did not feel comfortable to me. It was the day the president brought to my laboratory the president of Philips Electronics, the manufacturer of the X-ray equipment, along with two of Philips's chief scientists.

I had met the scientists during my visit to their laboratory in Atlanta when I was in search of a solution to the scattered radiation problem. The president informed me that the Philips Electronics team was meeting in Charlotte and had dropped by his office. While they were in the building, the president wanted them to see the design I had developed for the instrument made by their company.

At first I was excited by the idea. But as I stood in the background watching and listening to the excitement expressed by the president of Philips Electronics about my design. I said in my mind, *There is more to what I see here than I can figure out.* This meeting in my laboratory took place about one month after my design was published in the journal, making it free for public use. Many years have passed since the meeting at the start of my research career, but along the way, I came to understand intellectual property and the power and importance of things created in the human mind.

While I was in Charlotte, there was an event that took place in Washington, DC, that would have a significant impact on my wife and me. The event was the formation of a new federal agency, the US Nuclear Regulatory Commission (NRC). Before the NRC was created, nuclear regulation had been the responsibility of the Atomic Energy Commission (AEC), which Congress first established in the Atomic Energy Act of 1946. Eight years later, Congress replaced that law with the Atomic Energy Act of 1954, which for the first time made the development of commercial nuclear power possible.

The act assigned the AEC the functions of both encouraging the use of nuclear power and regulating its safety. The AEC's regulatory programs sought to ensure public health and safety from the hazards of nuclear power without imposing excessive requirements that would inhibit the growth of the industry. This was a difficult goal to achieve, especially in a new industry, and within a short time the AEC's programs stirred considerable controversy. An increasing number of critics during the 1960s charged that the AEC's regulations were insufficiently rigorous in several important technical areas, including radiation protection standards, reactor safety, plant siting, and environmental protection.

By 1974, the AEC's regulatory programs had come under such strong attack that Congress decided to abolish the agency. Supporters and critics of nuclear power agreed that the promotional and regulatory duties of the AEC should be assigned to different agencies. The Energy Reorganization Act of 1974 created the NRC, which began operations on January 19, 1975. The NRC, like the AEC before it, focused its attention on several broad issues that were essential to protecting public health and safety. The newly created and highly technical agency was in need of staff.

A colleague with AEC who transferred over to the NRC suggested that I look into some of the technical opportunities being developed at the new agency. Although my career was on an upward trajectory and the family had grown with the birth of our second child in 1974, a daughter, my wife and I talked about how different the slow living in Charlotte was compared to the vibrancy we experienced living in the Washington, DC, area. Having used our weekends while in DC to explore the museums and other cultural venues, we desired for our children to grow up in such an environment.

I contacted the agency and was invited for an interview in April 1975. The mission of the NRC sounded exciting. They offered me a position, and I accepted.

I left the company in Charlotte in August 1975 to become a member of the technical staff of the NRC. As was the case when I joined the laboratory in Charlotte, I was the first Black PhD scientist to be hired in the Office of Nuclear Material Safety and Safeguards (NMSS) within the newly created NRC. Upon arrival, I assessed where the greatest need existed in the office. In every meeting I took part in there were discussions of the problem the new agency was experiencing with bias correction requirements for fabricators of nuclear fuel. The problem was one that had followed the new agency from its predecessor, the AEC.

I quickly established myself in the new technical agency by publishing a paper on bias correction in the measurement of special nuclear material used in the nuclear power industry. My paper resulted in me being promoted to the Office of Research, where I would play a key role in the development of the NRC's Safeguards Research Program for the next fifteen years.

My position at the NRC provided me many opportunities to see science and practice science at a higher level than what I had experienced as a bench scientist at the company in Charlotte. Just as bench science often consisted of small-scale studies, I found the opposite end of the spectrum when it came to big science and big federal science budgets when I joined the NRC. I became fascinated with how science was done at the federal level, from design to full implementation.

Through my work at the NRC, I would be introduced to large-scale research being conducted at a network of facilities called national laboratories. My knowledge of these laboratories was limited to the little I was exposed to at the NRL, where I had done my master's and PhD research. I studied tirelessly to understand these systems of laboratories and related research facilities whose research purpose was to advance science and technology and ultimately bring about positive impact on the quality of life of the American citizenry. They all operated on large budgets. I enjoyed learning about these laboratories and their far-reaching impact on every aspect of life, from energy to the environment, health, commerce, space travel, and much more. There was no aspect of life that was not touched by research coming from the national laboratories.

By the time I became fully integrated into the mainstream of the US NRC Safeguards Research Program, my eyes were opened to the bigger aspect of science. Not only was I becoming educated on what was going on in my research area domestically in the United States, but I was also developing an understanding of how my work linked to a more global mission. I was expanding my horizons much according to what had been said to me as I marched across the stage at Howard University to receive my PhD, when the president said to me and the other graduates as he handed us our degrees, "I confer upon you today the doctor of philosophy degree, which not only entitles you to teach others but to also teach yourself."

I spent many hours teaching myself about the larger aspect of my own science program, asking such questions as, "Why is this important? How does it fit in the broader context of science and science policy?" I began to focus on the politics of science by closely following what the various congressional committees and subcommittees were saying and doing about science. Through my own efforts, I delved into the powerful congressional authorization and appropriations (CAA) process. It was here that I began to understand that science did not start with the solicitation announcement emerging from a federal agency but rather from the CAA process that ultimately provided the funding through the federal budget down to the various federal agencies.

The science money trail intrigued me. Based on this knowledge, I believed it was important for researchers, especially those in academic institutions with which I was familiar, not to simply visit federal agencies but to develop relationships with their congressional representatives on both sides of the aisle. I saw little difference in party affiliation when it came to science and its impact on people's lives. I began to develop my own philosophy of why most scientific results remained stored away on shelves in libraries. I believed the primary reason was because scientists were not doing a good job of communicating the importance and the relationship of their science to the quality of life for mankind, starting with those in the local regions where their research was performed. The scientists were being underutilized by focusing too much attention on producing publications that talked to each other and too little on explaining to decision makers what they were doing and its importance.

As I expanded my understanding of science beyond the bench while experiencing tremendous personal growth and development, I noticed there was little growth in the presence of Black scientists in the conversation or engagement in big science research programs. I began mentioning my concerns about the lack of participation in meaningful science discussions to my small group of science colleagues. There was agreement that more serious effort on increased participation of underutilized and underrepresented groups in science was needed, especially among Blacks. As I became more involved in the science dialogue inside and outside my research area and my small network, I shared my ideas with anyone who would listen.

I was aware that everyone was not in agreement with my view of casting a wider net to expand the pool to include talent from segments of society often left outside the mainstream. The pushback for maintaining status quo around the table often came in the form of this statement: "But you are different and capable." I squashed such nonsense early in my career. I never entertained the notion that I was different and that because of my difference I could handle the technical issues placed before me. I never forgot the design I developed for an industry that probably went on to make enormous sums of money from it and that was something my father, with just a third grade education, could do and do better than I had done.

I knew I was not special and that within the community from which I came there were many bright minds that could be used to solve societal problems in a variety of areas. My slogan became, *"The first one, yes; the only one, absolutely not."* With a self-imposed disappointment of being the first and many times the only Black scientist present, I set out to introduce the federal government to a virtually untapped talent pool of capable academic institutions charged with producing top-quality Black scientists. A few years and numerous detours passed along the way before I was able to turn my complete attention to becoming a voice of change in support of the nation better utilizing an underutilized segment of its population.

Chapter 15

THE TELEPHONE BOOTH— A DEFINING MOMENT

S INCE NRC WAS A NEW federal agency and charting its course on a daily basis, I had no reservations during my interview about asking if a position could be found at the young agency for my wife. The human resource (HR) director asked if my wife had a degree in any technical fields like physics, chemistry, mathematics, or one of the engineering disciplines. I said, "No. She has a degree in English."

The director acknowledged that for now the agency was looking mainly for technical staff but would check with the administrative division to see if they were hiring personnel. When I notified him of my start date of August 1975, the HR director said that administration was still in early development but that he would continue to check with them about my wife. On April 5, 1976, Zonia received a call from the HR Department asking if she could come in for an interview. On April 26, 1976, Zonia was hired by the new NRC federal agency.

Marrying Zonia had been the most important decision of my life.

However, in the early stage of our marriage she was faced with a potentially life-threatening condition that resulted in a defining moment in my life. At the young age of twenty-six, while in the shower, during one of her routine checks Zonia discovered a lump in her right breast. As a young couple that had only been married for seven years and had a five-year-old son and a two-year-old daughter, our world was shaken. Even without information from medical experts, uncertainty became my thought process on a daily basis.

Zonia has always been an avid reader. As a result of extensive reading, her knowledge of breast cancer in women was far beyond what I knew about the subject. Based on our discussions about the lump, we decided to consult with medical experts but to not trouble our families with the situation. They had enough to deal with in their own lives in North Carolina.

At this point, Zonia was two months into her first professional job at the NRC. We decided it was not necessary for me to go with her on her first doctor's visit; instead, I remained at home with the children. Her report from the doctor lifted our spirits beyond measure—at least for a short while. This short-lived "good news" was based on the doctor's office examination, after which he concluded that without medical test results the lump in Zonia's body was not life-threatening because of her young age and the fact that she had birthed both children before the age of twenty-six. Zonia explained to me that the doctor said that for persons with her profile, there was nothing to worry about and no need to go further and that a biopsy was not necessary at this time.

But as young people from what would be considered a not-so-highly sophisticated region of North Carolina, our families had taught us common sense, and we knew more was needed than the routine examination conducted by the doctor in his office. Therefore, we pressed for an X-ray examination of the lump. While reluctant to do so, the doctor consented to order an X-ray examination, saying he would notify her when the results came back to his office. With the many questions I posed to Zonia about the routine doctor's visit, we decided that I should join her on this second visit to hear the results of the X-ray examination.

Upon arrival at his office, the doctor placed two X-rays on his light box—an X-ray image of Zonia's right breast and one showing the left breast. In explaining the X-ray results, the doctor noted that his conclusion was correct and that there was insufficient evidence from the X-rays that warranted going further. I stood in the background looking intensely at the two X-ray images and could clearly detect what I considered a glaring difference in one image versus the other. Slight though it was, this difference was a small dense dark spot on the right image in the area where Zonia had felt the lump. This spot was missing from the left image.

At an appropriate point in the discussion and trying to be careful not to go beyond the point of what I believed a doctor would tolerate, I stretched forth my hand and directed my finger to a place on the right image for the doctor to examine more closely, pointing out what I considered to be a clear difference in the right and left X-ray images. The doctor first glanced at me and then at the image but remained steadfast in his opinion that the minor difference was not significant. It was clear from his response and my knowledge that the conversation had to go further.

Still pointing at the two images, I immediately followed up by asking, "Why is this spot on the right image missing from the left image?" This time the doctor provided greater detail in his response by indicating that breast tissue could show some degree of variation, but he said in this case the variation was insignificant. This dialogue continued as I explained my awareness that the degree of dense variation in the image coming from the same person was rare and that the variation appeared much too great to be considered insignificant.

Receiving no response at this point, my final question to the doctor was if he would expect similar results in the majority of his female patients. It was at this point that the doctor consented to bring a specialist into the conversation who would perform a biopsy. The tension in the room was also clear, as was the fact that the doctor was not pleased at all with my probing him on this medical matter. The tension did not bother me the least bit. In my view, this could be a life-or-death situation, and the beautiful young girl I wanted to spend many more years with was at the center of it. How the doctor felt at that point was of no concern to me at all.

Four days following the dialogue in the doctor's office, Zonia was scheduled to meet with a surgeon who would perform the biopsy. The surgeon described the process that would include Zonia being admitted to the hospital for him to collect tissue from the lump for analysis in a laboratory to determine if it was benign or malignant. He said if the results showed the tumor was benign—like the doctor he too expected the mass to be benign—it would be removed, and Zonia would go home probably the same day. On the other hand, if the laboratory results showed malignancy, he would need to perform a radical mastectomy. Therefore, we needed to sign an authorization form giving him permission to continue while Zonia was anesthetized.

Fear set in, and questions began to flow. We knew the process had to continue but became frightened when the surgeon placed the authorization form in front of us that needed our signatures before the start of the process to collect the tissue samples. We signed the forms giving the surgeon permission to continue with the surgery if the biopsy came back showing the tumor was malignant. The doctor shook our hands and said nurses would come in about five minutes to take Zonia to where she would be prepped for surgery. He then left the two of us in the room. We hugged and prayed, and Zonia said her customary phrase as the nurses entered the room and I was leaving: "It is going to be all right."

On Thursday afternoon, August 5, 1976, at Providence Hospital in Washington, DC, Zonia went into surgery for a biopsy of the lump in her right breast. Time seemed to stand still. I stood and walked over to the window and looked out at the beautiful day. Looking at the many people walking and talking, I wondered if any had turmoil going on in their lives like what we were facing at this moment. Time did not move fast. I drifted off to sleep but quickly woke when I heard a person calling for a couple sitting in the waiting room. I did not sleep any more but continued to wait.

After about two and a half hours of sitting and waiting, I looked down the long corridor and saw the surgeon coming toward me. I began to smile because it appeared that he was smiling and coming to bring good news. It looked as if he was walking in slow motion, with the distance he had to walk to reach me getting longer rather than shorter. This sight caused my smile to instantly disappear. Looking at him as his stride seemed to get slower, I thought about how everything in my life had gone so well.

I thought about my moving to the new seventh grade teacher. I recalled my high school science experiment and essay that had both won awards. I reflected on the national meeting of the Black scientists at STA and how that had led me to Howard University. I considered my journey at Howard and becoming the recipient of its highest award for graduate students going into their final year. I even thought of how when I started a semester in graduate school with a list of things to be accomplished within a certain time frame, it had always come to pass.

The job offers that had been presented to me flooded my mind. I even thought about my technical design at one company and the progress and influence I was making at the new federal agency. When my eyes closed from exhaustion, I remember thinking about the young beautiful girl I loved dearly and praying in my mind that she would come out of this okay. These were my thoughts until I looked up and saw the surgeon standing right in front of me.

He informed me that the surgery was going to take longer than he had originally expected and that what he had not believed would take place had turned out to be the case and the worst-case scenario. The surgeon told me that the biopsy had shown malignancy and would require removal of the breast. He noted that because we had given authorization to proceed if necessary, he would proceed with the radical mastectomy while she was still under anesthesia.

He said I could wait but that it would probably be best for me to come back in about four to five hours. The surgeon left to continue the surgery. Up to this point I had been shaken but not frightened to the degree I was now. The statement that the results showed cancer shook me the most and brought on great fear as I began wondering about the future, especially concerning our children's well-being and other internal feelings words were not adequate to express.

Before this announcement, I had not experienced any situation, problem, issue, or difficulty for which I had not been able to come up with a solution. It did not require much thinking. It was different this time, and

I knew it. I had no solution and could not come up with one. It was at this point that it began to sink in that Zonia was in a serious place, and there was nothing, absolutely nothing, I could do about it. Even though I did not know the severity of the situation or the magnitude of the surgery, I became somewhat settled when I thought of Zonia's patented phrase, *"It's going to be all right."*

I held on to those words and continued to tell myself it was going to be fine because every potential setback I had faced up to this point that could have taken me off course—like Zonia not having to leave school and graduating on time, my selection of one military medical doctor over another, my near draft into the military before completing my PhD—hadn't materialized. So based on my successful track record, I thought this had to turn out just fine. The major difference in this situation was that I could do nothing to alter the outcome. There was no registration I could handle, no letter I could show, and no experiment I could run to predict or change the outcome. There was absolutely nothing I could do to influence what I was facing.

I was really afraid. I wondered how the earlier prognosis that because my wife was young and had two children before her twenty-sixth birthday the lump was probably not that bad could turn to her now potentially having to battle for her life. At one point I imagined I was dreaming and the surgeon had not returned yet with the results. But I knew he had just left me with the regretful news.

While I had been raised in the church and believed myself to be a good person, at this time I did not have a personal relationship with God; did not know what it meant to have such a relationship; was not in the house of God as I should have been; and had not given Him credit for working in my life. I was in tune with the training I received as a child to know to call on God in prayer and fall back on my faith, especially when trouble showed up. I had heard these words as a youth, but in my present situation I realized I knew nothing about what to do. I thought of how often I spoke about God, but it was clear now that I had been doing so because everything had always turned out good. Sitting there helpless, I recognized for the first time that my life was more a "head" relationship with God than a "heart" one. As fear continued to brew, I called on God that very moment in prayer while still sitting in the waiting room, which was the only thing I knew to do. For the first time, this prayer was from a completely broken and frightened individual who had no fix for the problem.

As I mentioned earlier, we had not shared Zonia's condition with any of our family members in North Carolina because we did not want them to worry about us in Maryland. Knowing of their own responsibilities, we knew they would feel helpless at not being able to assist us with this crisis and would immediately start worrying unnecessarily. Since both of us came from large families (Zonia being the second oldest of nine children and me being the second oldest of seven children), we had rationalized that we would tell them once all of this was over.

After the doctor left me alone and I whispered a prayer to God, many things were still a blur, and new things began to crowd my mind. Sitting there not really knowing the condition of Zonia and not knowing if she would survive, I wondered if the worst did not happen, how long she would have to remain in the hospital. One thing I did know was that in four days something would have to be done with our children so I could report back to work. There was so much uncertainty surrounding me at that moment. I had previously been able to bring my mind to eventually think positively in panic situations but not this time.

Practical reasons dictated that I could no longer carry the weight of the situation alone. As I sat there trying to figure all this out, I finally got up, went outside to a telephone booth, and called both our parents to inform them for the first time of the gravity of the situation. Both my mother and Zonia's mother responded that they did not know when they could get to us but that they would be there as soon as they could.

As I sat in the telephone booth replaying what my mother and mother-in-law had just said and wondering when either of them would be able to come to my aid, another event took place in that telephone booth on that sunny day in August 1976. My relationship with God changed. The moment the responses from my mother and mother-in-law rung in my ears, I heard what appeared to be an audible voice saying after each of their responses, "*I will never leave you nor forsake you.*" This same message was repeated twice. The voice was so audible that I turned around to acknowledge it both times to see if someone was close enough outside the telephone booth to overhear my conversation. I had thought I had a strong relationship with God prior to Zonia's experience, but this occurrence revealed I really did not, and this manifestation put me on a course that would change my relationship with God forever. I began to recognize the difference between thinking of myself as good and being a person who actually walked with God.

Recuperation was difficult. I cannot remember how many days Zonia remained in the hospital before being released, but there was a very key point I will always remember. The hospital staff had developed a special mandatory counseling program for husbands of wives who had experienced radical mastectomies. Husbands were required to sit in on the counseling sessions before returning home to continue their lives. The thought was that because the wife had lost a very visible and important part of her body, husbands would benefit from education on how to interact with their wives going forward.

Following the instructions of the experts, I attended the first of what was to be a series of education sessions. Two things stood out to me in stark contrast in the initial session: I was the youngest man and the only Black in the room. Based on the discussions in the first session, coupled with the knowledge I had of my relationship with Zonia, which was far more important and real than what was being expressed by the speakers, I attended just one of these sessions and never returned.

Over the next several years, I really became a strong student of God's word. I studied the Holy Bible deeply almost daily with many references and other resources, not able to get enough of what His word taught me. In the early years of my studying, I quickly began to see how much of my journey was God's plan for my life. This understanding made clear my need for Him in all aspects of my life, and I desired to share with everybody inside and outside my network who wanted to improve his or her life and recognize his or her need for Him as well.

After many years of personal study and at the age of thirty-nine, I became a Sunday school teacher for senior high school students in grades nine through twelve. It became evident that God was working with me and giving me the vision for the next quarter of a century to pass His word on to many generations of young people. After my first year of teaching Sunday school in 1987, I was asked to coteach a men's Bible study that met every second, fourth, and fifth Saturday morning of each month. It is a class I now have been teaching for more than twenty-five years. The pastoral leadership of my church also asked me to coteach with the ministerial staff of a churchwide Bible study class I now have been teaching for twenty years. Teaching God's word became a joy and passion of mine. It is truly one of my gifts and one I love.

God promised never to leave me or forsake me, and I can truly say that in both the ups and downs of my life, He has been true to His word. I would like to make one final point on the discussion I had with the doctor when I joined my wife on her second visit. What the doctor had not known at the time was that I had spent the first two years of my professional scientific career heading an X-ray diffraction and atomic absorption laboratory in a corporate setting and was adept at reading the results of X-ray film.

I never got to know the doctor that well, other than in conversation during Zonia's recovery. It was obvious, however, that he respected the role I played at that point as a central figure in the life of our young family. In fact, during a visit to the hospital to check on Zonia following her successful surgery without me being present, the doctor asked her, "Has your husband received his medical license yet?"

The cancer returned in the same spot a year later; however, the small growth was successfully removed with no trace of any residual cancerous growth. Today, Zonia is a thirty-seven-year breast cancer survivor who retired in 2006 following an illustrious thirty-year career with the NRC. When Zonia crossed the line at a Susan G. Komen Breast Cancer Walk for the Cure event and announced at the finish line that she was a thirty-six-year survivor, the crowd burst out with a shout of excitement. At the time of the onslaught of the disease, we did not know anyone else in her family who had the disease; therefore, there appeared to be no genetic disposition.

Since that time, she has another sister who is a breast cancer survivor and another younger one who passed away from the disease in 2007.

Zonia and I have been in love for nearly forty-seven years, married forty-three years, with thirty-six of those years taking place since her surgery. Our two children, who were five and two during this period of our lives, are today forty-one (son) and thirty-eight (daughter). Another worthy point to make is that while the medical profession advised against our having additional children, we are so proud of our twenty-nine-year-old baby daughter.

It was at this defining moment in my life at age twenty-seven that I yielded my life completely to Jesus Christ inside a telephone booth. Since that time my focus has been to walk by biblical standards. I have not lived a perfect life, but a difference came upon me that day. I could feel it. When I replay that time in my life, I come to the same conclusion many years later: God was ordering my destiny one footstep at a time. I did not know it at the time, but my life would never be the same once I exited that telephone booth.

Final point here, I rejoiced during a biblical study when I found the expression in Hebrews 13:5 of the Bible (King James Version), *"I will never leave thee, nor forsake thee."* This settled in my mind forever that my telephone booth experience was real.

Chapter 16

INTERNATIONAL ASSIGNMENTS— FOR AMERICA AND AFRICA

A S A RESEARCH PROJECT MANAGER in the Office of Research at NRC, my leadership on the safeguarding of nuclear materials during the fuel-making process was known internal and external to the agency. My paper describing my thoughts on bias corrections continued to receive lots of attention within the NRC. I believed it was important to keep a close watch with physical cameras and human monitoring on plants that produced quantities of such dangerous materials. I also was keenly aware of the need for accurate analytical measurements that could produce reliable results to ensure that material unaccounted for (MUF) was indeed trapped in pipes or below levels of detection by the equipment but not stolen for illegal clandestine uses.

I and others in the field of developing the regulatory framework for analyzing nuclear materials never expected MUF to be zero; however, the belief was that it should be as close to zero as possible and that the measurement system played a key role in the certainty or uncertainty in the number. The limit of error (LE) associated with MUF, known as LEMUF, had to be understood. I had been involved with the issue in sufficient detail to know that having a residual MUF at the end of process did not necessarily mean there had been a gain or loss of nuclear material during the fabrication process when LEMUF was taken into account.

Safeguarding nuclear material as it was being processed was an issue of international concern. In 1980, the NRC was invited to send an expert to the second European Safeguards Research and Development Association (ESARDA) Symposium on Safeguards and Nuclear Material Management to discuss the NRC research program. I was selected to represent the US NRC at this gathering of technical experts from around the world. I could not believe it when I was tapped to go.

Since my safeguards research work was being conducted at Lawrence Livermore National Laboratory (LLNL) in California, I requested an invitation be extended to the research group so the LLNL scientist could present at the meeting as well. The invitation was granted, and the LLNL researcher accepted.

Every country wanted to hear how the United States was addressing the problem through its lead agency, the NRC. Because of the high level of interest in the US program, my presentation was placed in the oral session rather than the poster session. The conference was held at the University of Edinburgh in Edinburgh, Scotland. When I received my travel orders and my official US government passport, I paused once again to reflect on how far I had come since my boyhood days of sitting on my grandmother's front porch and wondering where the big eighteen-wheeler trucks were going up and down US Highway 70. I was holding a government-issued passport to travel internationally for the US NRC.

As I packed for the trip, I thought about how sick Zonia had been four years ago and now this. I worked hard to remember where things began in the tobacco fields of North Carolina. I never wanted to forget where it started for me and fall into the trap of thinking I had arrived where I was on my own. I did not believe in the notion of pulling up yourself by your own bootstraps. I was convinced that my journey was being directed and plotted step-by-step by someone other than myself, with lots of help and encouragement along the way.

The flight to Scotland was long. I had never before traveled across the Atlantic Ocean. My colleague from LLNL and I hooked up at Heathrow Airport in London and continued on together to Scotland. More than seven hundred persons were in attendance at the conference. It was the largest technical gathering I had participated in, and once again, I was the only Black scientist present and the only one making a presentation.

When it was time for me to speak, the room was filled to capacity, with people standing in the back and along the side walls of the theater-style auditorium. Presentations were translated into five languages. I was the sixth person on the program to speak. Having heard the presentations before me, I felt pretty good that my talk would go well and be well received.

I was less than a minute into my presentation when a loud voice came over the public address system, saying, "I am having difficulty translating what you are saying. Would you please start over and this time speak slower?" This announcement came from the French translator. I stood at the podium and looked out over the vast audience for about seven seconds without saying a word. After this delay, I responded, "I was present when the other speakers gave their presentations at speeds not different from my pace, and they did so without interruption. I am very sure that when I begin again I will be allowed to finish my presentation without any interruption."

Before continuing to speak, I paused and looked out over the audience for another three or four seconds. I then continued my presentation to the end without any more interference. I returned to the United States with a new perspective on science from an international context and a long list of new colleagues from across the globe. My colleague from LLNL told me that my dealing with the interruption was, as he put it, masterfully handled.

The following year, I was personally invited to attend the third ESARDA Symposium in Karlsruhe, Federal Republic of Germany. This time the invitation was sent to me and not the agency, and I was not surprised to receive it. I was told at the symposium the previous year that the work I was doing at the US NRC was important to other countries and to expect to be invited to the next symposium. The safeguards and nuclear material management research under my leadership at the NRC had expanded. My work focused on attempting to understand better the issue of material holdup in process and the economic impact on nuclear fuel processing plants. Leading this work for me were my colleagues at the Los Alamos National Laboratory (LANL) in Los Alamos, New Mexico.

As was done with my colleague from LLNL a year ago, this time my collaborator from LANL joined me in Germany. My oral presentation in Germany was titled "Safeguards Research to Develop Methods for Predicting Process Holdup to Enhance Material Accountability." Not only was there considerable discussion following my presentation, but this time there was no interruption. I also learned an important lesson on this trip relating to how scientists function in a global community.

Following my presentation, I was approached by a delegation in the audience from the International Atomic Energy Agency (IAEA) located in Vienna, Austria. The delegation representative said the presentation had been enlightening and that the information needed to be heard by an audience at the IAEA Headquarters. He asked if I was willing to travel to Vienna to make my presentation at the IAEA. I was thrilled by the request but said my travel orders were to Germany and then to return directly back to the United States. The spokesman asked if I would allow him to make contact with the NRC to request approval for me to extend my stay in Europe, and he said the IAEA would cover all additional costs. Recognizing that this could be another lesson learned outside the classroom, I said I would not mind at all.

The man asked for my travel orders, and I immediately handed them to him. He said he would return shortly and then quickly left the room. True to his word, the man returned in a few minutes. He asked me to go with him. I followed the man to a room, where he handed me a telephone receiver that was lying on the table. To my surprise, on the line was the director of the office of research at NRC. He conveyed to me that he had heard about the heightened interest in my presentation to the point of my being invited to the IAEA in Vienna to do a repeat performance. I said I was amazed at the attention given to my presentation. I told the director that the moderator had to cut the questions off to keep the session on schedule.

The director then informed me that my travel had been extended not only to the IAEA but that arrangements had also been made for me to continue on to the Joint Research Center in Ispra, Italy, to discuss the US program with a group there. The director said my travel orders would be faxed to the IAEA, where I could pick them up in a couple of days. He asked if extending my travel orders would be okay with my schedule. I said it was fine because I had planned to stay a few extra days and that the addition was perfect. While on the telephone, I tried to maintain my cool, but inside I was overwhelmed by what was transpiring.

As a Black scientist, I was excited to be representing a US government agency in such a high-level set of technical meetings in Europe. The first year at the University of Edinburg, I had simply been excited. I had not given much thought to the big picture of what was happening to me as an individual. This second trip was different. I began to internalize the magnitude of being requested to leave Germany and present in Vienna and then in Italy. These opportunities increased my confidence in general as a scientist and more particularly as a participant in collaborative research of importance to the nation and to the international community.

On both occasions I returned from my official international travels to the NRC in 1980 and 1981 and briefed management on the meetings in which I was involved. Although I was aware of the role science and technology played in society, my new experiences provided for me an overwhelming sense of the importance of that role. Another tremendous growth outcome from this experience was my understanding of how linked research must be if it is to be regarded as a game changer by the broader scientific community.

I experienced tremendous personal growth during my second trip. Reflecting on the technical discussions I led regarding the NRC program and its impact on the nuclear safeguard research community, I gained a better understanding of my future. On the second trip, my experiences and added confidence provided me the opportunity to assess what it really meant for a Black scientist with the beginning I had to be on international assignments and conducting official business for a governmental agency in a nation that was often characterized as the leader of the free world. It had to be something more than just the trip itself. It had to be a deeper calling and another step in my destiny that was not yet transparent to me and probably not obvious to anyone else.

It was at this moment in my career that I became transfixed on trying to assess what these experiences really meant. I felt there was more, much more to the trip than simply Bob Shepard working for the US government and making presentations before international audiences. The process of assessing the larger context of my experiences in 1980 and 1981 was something I did for the next seven years. Over these years, I thought of the special opportunities afforded me in 1980 and 1981 and how my technical career continued to flourish during and following my international experience.

I thought of my senior technical role on the safeguards research program at the NRC, my involvement on the high-level waste project looking for technical solutions to the storage and isolation of contaminated waste in geologic repositories, and the many peer reviews I sat on that assessed the technical merits of university proposals. I became humbled when I reflected on the investments made in me starting when I was a boy back in Garner and how it had continued through high-level workshops, seminars, working group assignments, and management courses at NRC. As I assessed my growth as a Black scientist in a highly technical organization and the many exciting experiences afforded me, I was pulled into yet another life-changing experience that involved science on an international level.

On my second trip to Europe, there was an event held April 1981 in Geneva, Switzerland, that captured my attention while I was in Germany attending the nuclear safeguards symposium. The event was an international conference on African refugees organized by the Organization of African Unity (OAU) and the United Nation's High Commission on Refugees (UNHCR). The massive international conference was convened to draw world attention to the plight of millions of refugees in Somalia and the rest of Africa. Nothing was known about this situation, including who the people were, where they came from, why they left their homelands, and what plans they had for the future. What was known was that 90 percent of the refugees were women and children with no information on what happened to the men.

As I followed the Geneva conference, there was much discussion and millions of dollars pledged by nations, mainly Western nations, to support the African problem. Leaders talked about seeking more resources and strategies to assist Africa in mobilizing the resources. In my view the conference focused on short-term fixes. To me, the primary problem was that "the problem" was not defined and proposed solutions were well intended but not vetted as would be needed if monies were to be well spent. Outside of food distribution, digging wells, developing irrigation systems, and resettling segments of the population to agricultural regions, most of the proposed ideas were political, calling for changes in government leadership and governmental systems.

There was some mention of using technology transfer as a solution to Africa's problems. The approach gave me a problem. My problem was that most of the technology discussed would come from Western countries pushing to transfer their own technology. This practice would result in transfer (i.e., movement of technology developed and used in an advanced institutional setting for which it was developed to an alternative use in another less advanced setting for which it had not been tested). In the case of the Geneva conference, it was proposed to move the advanced technology to weak, underdeveloped infrastructures in sub-Saharan Africa. I had problems with the proposals.

From a practical and applied viewpoint, the technology transfer process was more involved than the definition used in the Geneva meeting. The definition did not take into consideration the delicate balance between what is theoretically possible and what is practically feasible. I thought about what I considered the difficult part of the transfer process. From a practical standpoint, it would most likely involve reshaping and molding the existing technology from Western environments to meet the needs of people, environments, and other alternative situations in Africa.

Also, none of the proposed ideas addressed the matter of the scale that would be required, even if some of the ideas could be implemented. It was at this point that I sat in my hotel room in Germany and began to pen a proposal describing my thoughts of some alternative options for addressing the problems of Somalia and other depressed areas in Africa. I viewed the problem as technical in nature and requiring technical solutions. So I titled my proposal "Reshaping Scientific Technology to Meet the Needs of Developing African Countries."

The Geneva conference focused on the pros and cons of Western influence on the whole African continent. A major theme that resonated with me was that solving the African refugee problem would require freedom from the sway of the past and from undue influence of the big powers and not-so-big powerful countries. From my perspective and technology transfer viewpoint, a strategy was needed that prevented African countries from importing Western high technology without first requiring a project design to be conducted of what I called in my proposal an infrastructure analysis (IA). Moving African countries more toward internal self-sufficiency and less dependence on outside aid was at the center of my thought.

I proposed that an IA be a front-end assessment used to determine the level of robustness of a foreign-funded project implemented in sub-Saharan African regions. My thought was that requiring an IA on the front end of support projects funded by the World Bank, International Monetary Fund (IMF), and the US Agency for International Development (USAID) would go beneath the surface and bring the project into alignment with the reality of fragile environments in many of these locations. The IA factored in critical infrastructure elements including manpower (skilled, technical level, etc.), equipment, service and maintenance, access to spare parts, time constraints, costs, and other factors important to long-term success of the initiative.

The foundation of my proposal was the IA. Many situations existed where a technology operating as intended in one environment was rendered useless in another. I believed the unreliability of scientific technology was a combination of events that started with environmental factors. Added to this were the matters associated with installation, the technicians operating the technology, preventive maintenance policies, access to spare parts, etc., rather than the unreliability of the technology itself for the original purpose intended.

Reshaping technology to meet a specific need in a specific environment was usually beneath the surface and could require some changes in the technology at the component level. A technology required proper interfacing of components with the new environment and required proper education and training. Sometimes the education and training could be long term. The IA was designed to provide a systematic way for decision makers who were struggling to assess, interpret, and manage the synergistic relationships between technology development and technology application (environment) to address these significant issues. Since the USAID and the World Bank where major funders of projects in Africa, I requested and was given the opportunity to present my IA concept to the two organizations. Both bodies concluded that infrastructure issues were important but would be considered later in the project. The responses revealed that an interesting strategy was being used to implement projects in sub-Saharan Africa.

The proposal was completed in November 1981. The supposition of my proposal was that sub-Saharan Africa could advance if it embraced a strategy of economic independence based on looking internally for solutions to many of its problems rather than exclusive dependence on Western intervention. At the core of my strategy was the idea of technological development based on less robust projects. In December 1981, I sent my proposal to the OAU in New York.

In early March 1982, I was invited to New York to meet with a delegation from the United Nations Development Program (UNDP). Because of the policies governing outside activities of federal employees, I was glad I was given time to respond to the invitation. I had hoped OAU would respond to my proposal but had not given thought to what form their response might take. A written or verbal response maybe. Requesting that I travel to New York was not what I expected. As a full-time employee of the US government, I had been somewhat reluctant to even make the trip to

New York for fear of jeopardizing my federal employment. Confident that this new interaction was another step in my destiny, I went to the meeting at the UNDP headquarters.

The delegation had reviewed my proposal and found that it had merit and invited me to meet with them. I later learned that the reason this group wanted to meet with me was because African leaders had held a closed meeting in Monrovia, Liberia, in February 1979, and the report from that meeting had not been released, but—and these were their exact words—many elements of my proposal appeared to be lifted directly from their yet-to-be-published report. The report was titled "Lagos Action Plan for the Economic Development of Africa: 1980–2000."

The delegation shared with me that my proposal paralleled so much of the unpublished outcome of the Monrovia meeting that African leaders wanted to find out more about me and my ideas for technological development in Africa. I was asked if I would be available to travel to Africa to work with African scientists on how to infuse some of the ideas expressed in my proposal into World Bank, IMF and US AID development projects. My immediate response was yes. My answer was quickly followed with a question of when. The group said soon but that someone would get back with me. Now it was being discussed that I travel to Africa on private business, not US government business. There was little doubt in my mind that this was part of my destiny.

I began to strongly believe my first international work had a larger purpose and was not limited to the United States but that in time it would also be for Africa. It was difficult at the time for me to think in terms of working for both the federal government and my newly created company, which I named Infrasurface, Inc. This was because neither my supervisor nor anyone else in the federal government knew of my outside activities. My family and a small circle of people were the only groups aware of any of my independent work. Although my work in Africa could have been elevated to high profile, I kept it low-key.

I needed to familiarize myself at a deeper level with federal regulations governing a government employee's outside activities. It was good that it took three weeks from the time of the initial meeting in New York before a request was made for me to travel to Africa. I responded to the invitation based on completing a thorough review of the guidance to NRC employees on outside activities found in the Code of the Federal Register. Based on the review, I was confident that my activities outside my employment with NRC posed no threat of conflict of interest.

In late March 1982, I received a call from the UNDP asking if I could travel to Dakar, Senegal, to participate in a meeting at the African Regional Center for Technology (ARCT). The meeting was taking place in a month or so. The purpose of my attending was to meet with scientists involved with energy projects and share ideas of self-sufficiency. We were to look for ways to incorporate indigenous elements into a World Bank funded energy project. I said yes to the invitation to travel to the young organization that had just become operational in 1980.

The lead time allowed me the opportunity to make the necessary preparations prior to leaving. I was in possession of a US government passport but needed to obtain a personal one. There were immunizations to be taken and research on ARCT and on Dakar to be conducted. When actual travel was set, I had a prepaid ticket in place for me to make the trip. I made my first trip to Africa in May 1982. Over the next eighteen months I would make two additional trips to consult with African scientists.

On my second trip to Africa, I was still concerned about my status as a US government employee. I was not comfortable with the arrangements relative to my employer. My supervisor and no one else at my job knew anything about what I was doing outside of my federal employment. Not knowing what the reaction of management or my colleagues would be to my outside exploits, I had developed my personal need-to-know strategy and decided NRC did not need to know. My outside activities were not linked in any way to my job. Therefore, when I traveled to Africa, I would say to my supervisor, "I will be on annual leave over the next two weeks." I would make sure my workload was such that being away would not pose any problems within the research division to which I was assigned.

The most uncomfortable aspect of the conditions I was traveling to Africa under was the jeopardy in which I placed my family. Simply taking annual leave without revealing my whereabouts to the federal government might leave my family vulnerable if something dangerous would happen to me. If there was a plane crash or violent uprising in the area that resulted in injury or death to me, what would that mean to my family financially? The uneasy feeling stayed with me for most of this trip.

As soon as I returned to the United States from my second trip to Africa, I called a meeting of the NRC general counsel. I documented my Africa trips and meetings just as I did when traveling on assignment for the NRC. I called the general counsel to set a meeting for sharing with the agency my outside activities under my company, Infrasurface, Inc.

Prior to the meeting, I sent them copies of my travel itineraries and written reports of my trips. In addition to the written account of my work in Africa and elsewhere, I shared with the counsel every detail, including persons and places visited. I provided a copy of my Infrasurface, Inc. company brochure. I sent over everything. I requested that the general counsel's office evaluate my outside activities and provide me with an official ruling. I was especially interested in a determination of whether there was a conflict of interest or any infringement I should be concerned about.

The ruling was that there was no conflict of interest between my work as a federal employee and my outside activities under Infrasurface, Inc. NRC showed great support for my Africa involvement and related much of the impact of my efforts to their investment in me through my work at the agency. With the backing of my federal agency, I was much more confident on my third trip to Africa. NRC support resulted in my spending four years (1984–1988) as a visiting scientist to my alma mater under the government's Intergovernmental Personnel Assistance (IPA) Program. The purpose of the assignment was for me to expand my work with African scientists and in the process draw upon the advanced technology, facilities, and expertise within Howard's Chemistry Department.

Howard University (HU) was unique relative to other US institutions of higher learning. Its mission of providing education to students from developing countries, especially students from Pan Africa, put the university in a distinctive group. With connections established among African scientists, I was able to create a program at Howard that promoted effective collaborations between several research faculty members in the departments of chemistry, biology, and the medical school in support of technical development in Africa and parts of Latin America.

The idea was to expand the vision of HU research faculty and provide assistance in creating linkages and partnerships with persons working in international organizations, foundations, federal government, nongovernmental organizations (NGOs), and private organizations. The program was successful in engaging some HU research faculty members in collaborative research projects with institutions in Africa in the areas of food and plant chemistry, nutrition, and research to support agricultural programs.

I pushed hard and moved quickly to demonstrate that combining the advanced research capabilities at HU to technological development problems would produce useful and practical results. The vision was to develop a model program at HU in support of technological development and infrastructure building in Africa. The vision was realized.

I wasn't long into my program at HU when I noticed major cracks in the research infrastructure foundation at the institution. It became evident that shoring up the research foundation at HU could strengthen the institution's position to participate in collaborations to support technological development in Africa. I began to pay close attention to the absence of HU and other historically Black colleges and universities (HBCUs) in large science projects I had been a part of and observed as a researcher at NRC. Close visual observation and personal interviews with faculty at HU and other HBCUs revealed problems that could be rectified.

Since HU had such a comprehensive research infrastructure, I needed to assess some of the smaller, less endowed HBCUs. A major problem for a number of the small institutions was that in their zeal to obtain external support, they sometimes overstated their capabilities, only to later find they could not deliver what had been agreed upon. This usually resulted in them being labeled "high risk" by the mission-oriented federal research agencies and other funding bodies. Important issues regarding the infrastructure usually were overlooked and not taken into consideration at the start of project development. On a small scale, involving some very young and new faculty, I undertook an effort to evaluate the existing capability of an area of HU's Chemistry Department (human and financial) based on the requirements of a solicitation to determine the extent to which the faculty could participate in and deliver what was expected.

This exercise demonstrated that there existed a disconnect between what was being requested by the federal agency and what could actually be delivered by the research faculty. From an infrastructure standpoint, more human and financial resources were required for the faculty to have a chance at being successful. I admired the faculty and understood their desire to become part of something bigger, but the infrastructure was limited, and the size of the grant was not worth the effort needed to pursue it. Another shortcoming for not being able to engage was lack of support from higher-level administration at the university.

At the time, I did not understand that chasing any size grant without an appropriate infrastructure was a death sentence for a researcher. I wanted to change the focus of a research faculty member and many of his colleagues in similar positions at the 105 HBCUs in the nation. My experience as a visiting scientist at Howard, coupled with years of involvement as a peer reviewer of research proposals within NRC and the National Institutes of Health (NIH), gave me the idea that an organization was needed that could provide technical assistance to HBCUs. I surmised that my international assignment was really for HBCUs, and as a result, SEA was born.

Chapter 17

THE BIRTH OF SEA—FOR HBCUs

THE DESIRE TO PROVIDE SUPPORT to the HBCU research community had been on my mind for quite a while, and my stay at Howard confirmed the need. The idea was first exposed to me in 1984 when the NRC Research Office announced in the Federal Register its first financial assistance grant program to support research at educational institutions. I was asked to be a member of the technical staff reviewing proposals submitted to NRC from universities. The fiscal year 1984 ceiling for the research grants to educational institutions was $1.2 million. Universities were instructed to restrict budgets to no more than $50,000 per year, with total project funding not exceeding $100,000 over a two-year period. Proposals were to be submitted between October 1 and December 31, 1984.

The purpose of the program was to stimulate research to provide a technological base for the safety assessment of system and subsystem technologies used in nuclear power applications. The results of the program were to increase public understanding relating to nuclear safety, to pool the funds of theoretical and practical knowledge and technical information, and ultimately to enhance the protection of public health and safety. In addition, each grant to an educational institution was to contain elements that would potentially benefit the graduate research program at the institution (e.g., graduate student training).

The NRC encouraged educational institutions to submit research grant proposals in thirty-one topical areas. Included were such areas as (1) modeling and experimentation on two-phase flow, interfacial relations, and heat transfer in reactor coolant systems; (2) severe accident evaluation of high temperature chemistry of severe accident reactor radionuclides, and advanced thermal hydraulic modeling of fluids including combustible gases and molten core materials in reactor primary systems during severe accidents; and (3) simplified modeling of thermohydrologic phenomena in high-level waste geological repositories.

Because of the nature and scale of the NRC research projects, most of this "big science" activity was conducted at national laboratories like Lawrence Livermore National Laboratory, Sandia National Laboratory, Los Alamos National Laboratory, Argonne National Laboratory, Oak Ridge National Laboratory, and others. Additionally, to share cost and facilities, NRC research was carried out on an international basis in collaboration with foreign researchers who were advancing knowledge in a particular field. Bringing educational institutions into this type research was new and novel.

However, a few educational institutions had built a reputation working alongside of national laboratories. Educational institutions like the Massachusetts Institute of Technology, Harvard, the University of California at Santa Barbara, the University of California at Berkeley, the University of Wisconsin, North Carolina State University, the University of Tennessee, Arizona State University, the University of Rochester, and a few others consistently performed work as subcontractors on large-scale national laboratory research projects. This resulted in key technical expertise being developed that put these institutions in a unique position relative to the new NRC educational grant program.

Being a product of the nation's HBCUs, I encouraged colleagues at these institutions to look at the Federal Register announcement and assess whether they had some degree of interest and capability, or related capability, in any of the thirty-one technical areas of interest to NRC. I also encouraged them to partner with other institutions to enhance their existing capacity and capability to better position them to respond competitively, either as leading an effort or as a subcontractor to a more established institution.

There may have been others, but I remember three HBCUs submitting proposals that year. They were Howard University in Washington, DC, North Carolina A&T State University in Greensboro, North Carolina, and Prairie View A&M University in Prairie View, Texas.

The discussion that follows about the submittal from North Carolina A&T State University became another defining moment in my career. It was just as striking as my having traveled internationally for the US government and, to a lesser degree, my telephone booth experience. As the only Black scientist on the review panel, I watched with growing interest as my white colleagues discussed the North Carolina A&T State University proposal. One reviewer asked if this was North Carolina State University? The group agreed that it was.

Someone else commented that North Carolina State University was in Raleigh, North Carolina, and this university was in Greensboro, North Carolina. Another reviewer asked if North Carolina State University had opened a campus in Greensboro. No one was sure that it had. Since no one in the discussion recognized North Carolina A&T State University in Greensboro, the reviewers commenced to toss the proposal to the side without ever really looking at its technical merits.

With my knowledge of HBCU institutions, I could have entered the discussion at any point to share with my colleagues my long association with North Carolina A&T State University. I chose not to just to see where the conversation and proposal would end up. As the proposal was about to be tossed into the pile that warranted no further discussion, I shared the historical aspects of this great institution with my colleagues. I shared with them that my cousin had graduated from the institution and was now an upper level technical employee with the US Department of Agriculture.

I also shared that there was an astronaut in training by the name of Ronald McNair who had received his undergraduate degree in engineering physics from the institution and his PhD in laser physics from MIT. I shared that the Rev. Jesse Jackson had also graduated from the institution. This background put the proposal back into the technical discussions. The same unfamiliarity was demonstrated when the proposal submitted by Prairie View A&M University was glossed over. Being known by the group, Howard University faired a little better among the members of the panel.

None of the proposals from HBCU institutions, including Howard, were selected for funding. I agreed with the panel that the proposals from the three HBCUs were not in the competitive range. I was troubled, however, by two events that stemmed from this experience: (1) the panel's lack of knowledge about HBCU institutions; and (2) the lack of connectivity between proposals from the three HBCUs and the large body of nuclear-related research underway at national laboratories and elsewhere. I concluded that much of the connectivity problem could be addressed through education and training of faculty and students at HBCUs. On the other hand, educating NRC and other federal agencies about HBCUs would be needed, but it would pose a greater challenge.

Reviewing the proposals based on evaluation criteria such as adequacy of research design, scientific significance of proposal, technical adequacy of the investigators, and their institutional base were all unrealistic measures for the proposals submitted by HBCU institutions. A review of just the investigators' resumes showed that there was no link to what was going on at national laboratories or other large research universities relative to the thirty-one technical areas of interest. I could see that for HBCU institutions to participate more fully in federal R&D like the NRC research program, which was relatively a small program, more exposure and training was needed.

I continued to serve as a peer reviewer internal and external to NRC. As lead reviewer on a National Cancer Institute (NCI), National Institutes of Health (NIH) proposal, I observed the same lack of knowledge and connectivity among the HBCU proposals relative to the research interests of the NCI as well as a lack of knowledge of the NCI review panel members of who these institutions were. These incidents led me to write down my thoughts of a potential solution for improving the capability and capacity of HBCU institutions relative to positioning them to benefit more fully from the various federal R&D programs. I included a second component to educate federal agency personnel about HBCUs.

After returning in October 1988 from my assignment at Howard and being promoted to technical assistant to the director for the Office of Research, the new position I created gave me greater access and opportunities to voice my concerns about the lack of participation of HBCU institutions in the NRC financial assistance grant program. It was clear that HBCU institutions would never be able to compete with the more established universities, who had such long track records of engagement on "big science" research projects. I was convinced, however, that they could be positioned to play a greater role in federal research projects.

Over the next couple of months, I engaged in lots of discussions with the director of research on my thoughts of how to better involve HBCUs in the NRC research program. Much of the discussions centered on assurance that whatever was created would not be substandard or negatively impact the existing research program. The idea was to provide a means by which HBCUs could compete and eliminate an emerging cop-out statement that said, "They could be considered, but we ran out of funds."

In early 1989, I was given the opportunity to establish the first financial assistance program to support research at HBCUs of interest to NRC. The program was modeled after the main program in every aspect except funding level. I requested that $500,000 be made available for the first year of the program. The Office of Research could provide only $100,000 to initiate the program. More was required to have a program that could fund at least two HBCUs over a two-year project period.

I negotiated with my colleagues in the NRC Office of Small and Disadvantaged Business Utilization and Civil Rights, who provided an additional $50,000, bringing the total program to $150,000. I went back to the Research Office director to request an additional $100,000 to initiate an HBCU research program at a level of $250,000, but I was not successful. Using the main program as the model, three projects were initiated among the HBCU institutions at a funding level of $50,000 each for one year. Unlike the main program, funds were not made available to guarantee a second year of support. This was a great start to build on.

With this support and encouragement, I returned to my thoughts on how to integrate HBCU institutions more fully into federal R&D programs. To move forward, a proposal was needed. I wrote a proposal under my Infrasurface, Inc. organization and shared it with a colleague who recommended I send it to the US Department of Education (DoE). I hand carried the proposal to the DoE in November 1989. The representative asked to which funding solicitation was I responding. I knew of none and was not responding to any federal announcement. The representative agreed to conduct an unofficial review of the proposal but said proposals must be in response to an official funding solicitation and the agency cutoff for receipt of proposals was October 16 of every year.

The representative noted that for my proposal to receive an official review, it would have to be submitted to the agency on or before October 16, 1990. After the unofficial review, the agency representative said the type of assistance described in the proposal would definitely bring about improvement in the HBCU research community and suggested that I work to improve the proposal and resubmit it before October 16, 1990. The representative suggested I obtain a high-level review from a colleague working in the area of science and technology at a major university or national laboratory, attach the review, and officially submit the proposal before the cutoff date. Somewhat dejected, I promised to follow the instructions and officially resubmit the proposal. I left the meeting very encouraged but disappointed that I would have to wait a year to move forward.

In April 1990, I attended a meeting of the National Organization for the Professional Advancement of Black Chemists and Chemical Engineers (known as NOBCChE) held in San Diego. I took one copy of the proposal to the meeting in anticipation of identifying a single colleague to review it. Of the more than two hundred Black chemists and chemical engineers attending the meeting, I selected James Evans, a researcher at Lawrence Livermore National Laboratory (LLNL) in Livermore, California, as the person to conduct a review of my proposal. Jim graciously agreed to provide a detailed review. I left San Diego on the weekend and returned to Washington, DC. The following week Jim called to inform me that my proposal matched well with some discussions underway between LLNL and four state-supported HBCUs located on the East Coast. He invited me back to California to discuss LLNL views and my proposed ideas.

When I arrived at LLNL, Jim shared that he and his colleagues were in discussion with four HBCU presidents who had signed an agreement in February 1990 pledging to work together in an enterprise they wanted to establish and call it the Science and Engineering Alliance (which would become known as SEA). Jim noted that LLNL intended to join the effort but had been given responsibility to first select someone who could transform the discussions from a concept paper to reality. It was evident from reading the concept paper that the SEA idea being discussed coincided perfectly with the goals and objectives outlined in my proposal.

The setting and discussion reminded me of how my engagement with African scientists had gotten started. While sitting in the conference room at LLNL in California, I thought about the proposal I had sent to the OAU, sharing my ideas of how to enhance technological development in Africa. The exact same thing was happening again, and it had been stated just about verbatim: "What you have described on paper fits perfectly with what we have been talking about needing to be done."

I expressed to Jim my desire to use my talents to establish and grow the new idea into a thriving enterprise. Jim contacted the presidents of each of the SEA institutions to inform them that LLNL had identified the person who would become the founding executive director of SEA. There was no denying that God was the architect in ordering my steps and my life. After almost ten years of effort to devise a strategy for engaging the HBCU community more fully in the federal R&D activities that first took me to Africa, it was clear that a divinely inspired act was about to transform years of ideas into reality.

There were many hurdles to overcome in the next few months following that April 1990 meeting at LLNL. The major one involved the decision of whether to terminate my high-level GS 15, step 8, federal position to create and build something that only existed on paper. Since I had almost sixteen years of government service and SEA was only an untested and unproven concept, I was offered the opportunity to take a leave of absence from NRC for two years in the event that SEA did not work out.

At first it sounded like a good offer that provided an excellent safety net for me. This was a great option, especially since I was the major breadwinner in a family that had grown accustomed to a certain lifestyle. But as I thought more deeply about it, I felt in my heart that the SEA concept was a time that had come. Standing at the center of what was taking place in history at this very moment was a community of institutions founded during slavery that had uplifted so many Blacks like me around the world.

An opportunity now existed where, with concentrated and dedicated support, the possibility existed for strengthening their research infrastructure in ways that had never been done before. How could this not be successful? Why would I go into this hedging my bets at the outset? Did I really believe that divine inspiration had brought me to this point for such a time as this? I either believed that what was happening was outside of my doing or that I should return to Washington and settle in for another fourteen years or more to retire from the NRC. A major decision was needed at this point.

Jim and others at LLNL suggested that before making such a drastic decision, I visit the four HBCU institutions. In fact, it was a good idea for the institutions to interview the person selected by LLNL. On the other hand, I needed to visit the institutions to meet with the presidents, deans, research professors, and students and tour the laboratories to see what I was stepping into.

The major difficulty I had with the leave of absence option was the restrictions that still being a federal employee would place on me. My due diligence on conflict of interest and my vast personal knowledge gained on ethical conduct of federal employees proved valuable once again. Based on past meetings with the NRC general counsel around my involvement in activities outside my federal employment, the leave option did not resonate well with me. The basic problem was the restrictions that would reduce my ability to lead the SEA in a manner consistent with my thoughts of what would be needed to be effective. The primary drawback of remaining a federal employee would be my inability to engage in very meaningful activities, especially involving congressional matters, which would result in limiting constraints and my hands being tied behind my back.

After consulting my family and with the blessing of my wife, I decided the only option was to terminate my federal employment and trust that God would transform the new SEA concept into a surviving and even a thriving reality. On September 1, 1990, I presented my resignation letter to the director of the Office of Research, stating that I would terminate my federal employment on October 1, 1990. As the October 1 date approached, my excitement grew. This is now a good point to share something that would become vitally important to how I moved forward.

When October 1, 1990, arrived I left the NRC for the last time. As I stood in my office, packing to leave, I thought of the nearly sixteen years of support I had received from one of the finest agencies in the entire federal government. I thought of the many courses I had taken in technical areas including statistical methods in nuclear material control, nuclear engineering, system reliability engineering, risk assessment technology, and many others. I thought of the workshops and seminars I had taken to improve my writing, public speaking, and public image standing before audiences where I was videotaped and the tape was analyzed by the audience.

I thought of the weeks following the March 28, 1979, accident at Three Mile Island Nuclear Power Plant and the systematic evaluation of operating experience and improvement in the regulation of operating reactors, control room design, and the Human Factors Research Program that would be implemented. I looked out my window, replaying in my mind the many questions asked by my colleagues and family members wanting to know what had really happened at Three Mile Island and why. There were many questions about the safety of nuclear power and opinions on how they believed the accident would slow down the use of nuclear as part of the overall energy plan for the country, and in their minds, rightfully so. I smiled while thinking about my brother Roger, who believed that the nuclear industry was not without its problems but that it was one that did not contaminate the environment like coal-fired power plants, but he believed the nation should use more oil as a domestic resource.

Finally, I thought of my four-year assignment to Howard University under the government-wide IPA program, and the new international research relationship with African scientists that came as a result of this assignment between Howard's Chemistry Department, the University of Ife (now Obafemi Awolowo University), and the University of Ibadan.

As I continued to pack, I was truly grateful to the NRC for the tremendous growth and professional development afforded me as a scientist in the young technical agency. I knew that every course, every workshop, every seminar, every opportunity to lead a technical activity, every experience, and every accomplishment afforded me, especially the IPA assignment to Howard, would benefit me as I looked with pure joy and excitement to my future. I really did not know what it would hold, but I was excited anyhow.

I picked up my last packed boxes, exited my office, closed the door behind me, and dropped off my security badge at my secretary's desk. As I walked down the hall, leaving much of me behind, thinking of the years of writing, meetings with colleagues, daydreaming, fantasizing, wishing, hoping, and praying to find my real purpose in life before I became too old to even recognize it, I captured a new vision and fresh outlook of who Bob Shepard was and what Bob Shepard was destined to become. I could not help but think of the tremendous career I had with NRC, including high-quality technical and executive training, first-class job assignments, GS-15 step 8 level pay, and an outstanding retirement package in place that suggested the future might be secure for me and my family. I left the NRC Office of Research with a resolve to never look back at what might have been, could have been, or should have been.

As I continued to walk down the long hallway toward the front door to exit the NRC for the last time, I continued in my heart to thank God that it had all finally come together. There was no doubt that it had indeed "come together." I had felt it inside. There was no anxiety about the decision I made. No apprehension or nervousness about tomorrow or the next day or the next year. What was happening was now supposed to be happening. My stare down the hall was steady. I was on my way somewhere. It was not at all clear where "somewhere" was, but I was filled with excitement, whatever would happen. It was seared in my mind that I would continue forward holding on tightly and never letting go of God's hand.

Before I reached the door that would take me outside and into the parking lot, a long-time colleague came to his office door and summoned me to come into his office. I went in as I had done so many times in the past, put my boxes down, and sat down.

My colleague's name was Charlie; he had been at the NRC when I arrived in 1975. Charlie, like some other technical staff, had worked at the agency when it was the Atomic Energy Commission (AEC). He was one of the true old-timers around when the AEC transformed into the NRC. Charlie spoke with much admiration for my decision to leave the NRC. As in times past, he shared his thoughts on the future of not just the NRC but the entire nuclear industry. He strongly believed that the twelve-year cycle from site selection to design, construction, and receiving a license to operate was much too long. Although he felt the technical decisions resulting from the research program were presenting conclusive and definitive results that electrical power generated from nuclear fission was safe and no need for great public concern, in Charlie's view, the twelve-year regulatory process was hurting the industry.

As I had done in years past, I found myself engaged in a lengthy discussion with Charlie on the subject of the future of the nuclear industry. I agreed the twelve-year cycle time was long but was probably necessary based on the kind of technical issues and public concerns the agency was faced with. I moved to the high-level waste problem and the work underway at Sandia National Laboratory and Yucca Mountain to substantiate my point.

With the US Environmental Protection Agency's (EPA) requirements governing disposal, it was prudent in my view that the NRC look at all aspects of the issue in great detail, and that took tremendous time. I said to Charlie that the technical issues being addressed at the NRC were new and had never been tackled before by scientists and engineers, and for that reason alone the current time factor could be justified. If one was faced with high confidence and certainty of being able to isolate and store highly radioactive material from the public for ten thousand years, there were a number of critical technical issues one must address.

We continued to debate the issue for the next two hours until Charlie said, "Bob, I did not stop you on your last day here to talk about this place. I really want to talk about your leaving to create your new SEA enterprise."

Charlie shared with me how proud he was to see me moving on and creating something he thought would make a difference in the lives of academic institutions and many young people and doing it at the tender age of forty-two. I told Charlie that my goal was to do something but that the clarity of what that "something" was eluded me right now. He continued by saying that many would give their right arms to have the kind of opportunity I was walking into at forty-two years old. But he wanted to caution me based on some of his earlier experiences.

Charlie asked if I had ever worked with university presidents. I said no. He then looked me in the eye with an intense and serious look and said, "Bob, you will be very successful in the SEA enterprise, but always remember this: University presidents are very egotistical people. They like to talk and be in the limelight. For you to be successful and remain successful for a long time, you will have to be comfortable working in the background, and I mean, way in the background. You will need to work hard, long, and fast to create opportunity and access, but you must be slow to take credit for what you do. If you give away all you do to the presidents and their institutions, you will be able to serve SEA for as long as you like. On the other hand, if you are quick to stand up and take credit for what you do and are always in the limelight, your days with SEA are already numbered." He then said, "Bob, I just wanted to share this with you."

As I left Charlie's office, remembering all of the nuclear-related things we had discussed over the years, my spirit convinced me that what he had said about how I was to approach my leadership of SEA was directly from God. I reached my car in the parking lot with my last two boxes of personal items, stood looking out over the landscape before me, and resolved within my soul that Charlie's advice would serve as my guiding light in this new enterprise.

In November 1990, SEA was formed as a nonprofit Delaware corporation. Its headquarters were located at 1522 K Street, NW, Suite 210, Washington, DC, 20005. This address was two blocks from the White House. Its governing board of directors was composed of four participating

HBCU presidents and executive officer of Lawrence Livermore National Laboratory (LLNL) responsible for alliance activities. SEA was formed to provide technical assistance to research faculties at HBCUs as a precursor for enhancing the institutions' research infrastructures.

Today, SEA efforts help to ensure an adequate supply of competitive American scientists and engineers while meeting the research and development needs of the public and private sectors. SEA services include (1) technical marketing and community relations, (2) development of training and experiential programs for faculty and students, (3) creation of partnership opportunities with public and private sectors, (4) enhancing research opportunities, and (5) advancing national goals. The strength of SEA is an understanding of the participating institutions that the sum is greater than the individual parts.

For nearly a quarter of a century now, I have focused on broadening participation in federally funded research projects by creating access and opening doors of opportunity for underutilized and underrepresented individuals, institutions, and regions that do not participate in these projects at a rate comparable to others in the nation. It is said that SEA serves as a one-stop unique resource for inclusion of diversity in federally funded research initiatives.

SEA experienced many significant accomplishments on behalf of HBCUs and other small academic institutions. The most noteworthy achievement was inclusion of these academic institutions in the thirty-plus-year research initiative sponsored by the National Science Foundation (NSF) called the National Ecological Observatory Network (NEON). The strategy put together by SEA for including the institutions was called the NEON Satellite Site (NSS) program. A description of the NSS was buried in the NEON time capsule at a special ribbon-cutting ceremony held in Boulder, Colorado, in May 2011. The capsule will be retrieved and opened in the year 2041.

Robert Louis Shepard   4-12-63

Record # 15,556   Dr. Castelloe

4-12-63:   This 15-year-old colored male is seen for pain over both knees which
he dates from an injury sustained when he fell during basketball practice
at Garner Consolidated School about 1 month ago. He had some soreness
over the knees following this but symptoms were minimal until about 2 days
after the injury, when they became worse. Because of increasing pain over
the knees, he was seen by Dr. C. L. Hunt about 2 weeks following the injury
and Dr. Hunt evaluated the knees and gave him medications. He continued
to have difficulty referable to the knees and therefore came to the office
for evaluation. He has not had locking or blocking of motion of the knees.
Most of his pain is located beneath the patellae, exacerbated by vigorous
activities and squatting. He has remained well otherwise.

On examination, the patient is a large colored male with abnormalities
limited to the knees. There is no effusion present. There is tenderness
on grinding the patella against the femoral condyles, and extending the
knees against resistance. He has a full range of motion though flexion
of the knees causes pain beneath the patellae. There is no tenderness
over the medial or lateral compartments and no ligament instability.

IMPRESSION: Chondromalacia, patellae

DISPOSITION: Patient was advised regarding conservative treatment with
warm soaks and quadriceps strengthening exercises. If his symptoms do not
improve within the next several weeks, he is to return for consideration
of injection the knees with hydrocortisone.

(TEC:db)

*Figure 17. The letter held in the doctor's files for seven years describing
results of the examination on my knees. (Courtesy of the author)*

269

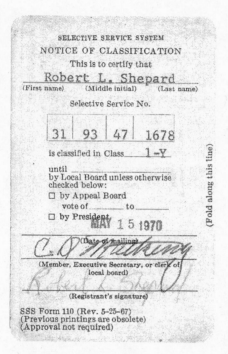

Figure 18. The 1-Y classification that reclassified my status to ineligible for military draft came as the Vietnam War was coming to a close. (Courtesy of the author)

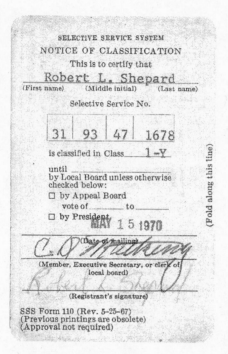

Figure 19. I received my 4-F classification, which replaced the 1-Y, as the US troops were coming home and the Vietnam War was ending. (Courtesy of the author)

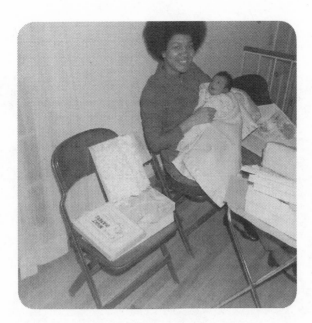

Figure 20. Alzonia is holding our firstborn. (Courtesy of the author)

Figure 21. Alzonia (third from the left on the second row) graduated on time with her class in May 1971. (Courtesy of the author)

Figure 22. I am celebrating with my PhD thesis advisor, Dr. Jesse Nicholson, and my wife after receiving my PhD and accepting my first research position. (Courtesy of the author)

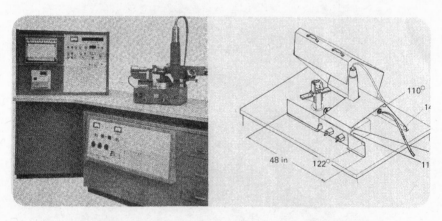

Figure 23. The original X-ray unit designed by the manufacturer is on the left, and my design is shown on the right. (Courtesy of the author)

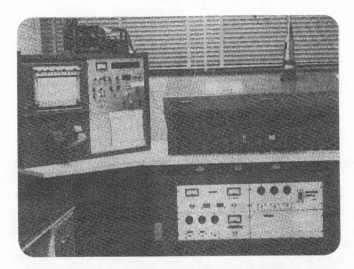

Figure 24. This is how the X-ray unit looked after my design was added to it. (Courtesy of the author)

Figure 25. I participated in a training course on understanding precision and bias when measuring nuclear material. (Courtesy of the author)

Figure 26. I am waiting to speak before a group of scientists meeting at the University of Edinburgh in Scotland. (Courtesy of the author)

UNITED STATES
NUCLEAR REGULATORY COMMISSION
WASHINGTON, D. C. 20555

NOV 0 1 1976

MEMORANDUM FOR: E. J. McAlpine, Acting Chief, Material
Control Licensing Branch

FROM: Manfred von Ehrenfried, Chief, Test
and Evaluation Branch

SUBJECT: RESPONSE TO BOB SHEPARD'S PAPER ON BIAS CORRECTIONS

We found the subject paper useful and instructive. It provides good
insight into factors concerning bias corrections. Particularly
significant was the delineation of the limitations of current NRC
regulations and standards on bias corrections – showing that we often
miss the forest for the trees. The paper correctly points out that
data outside of reasonable control limits is often a much more significant problem than bias.

We also appreciated the realization that present NRC criteria induces
the licensee to keep his sample size small in order to reduce the
liklihood of having to make bias corrections. From the point of
view of safeguards, this should be corrected.

The suggested licensee alternatives presented by Shepard need further
investigation prior to NRC acceptance. We would suggest investigation
of other possibilities also.

Shepard's paper is important in clearly pointing out that bias is
only one element of interest to Safeguards. In order to develop an
effective Safeguards system, it is essential that proper emphasis be
placed on all elements of the system. We feel that the role of bias
has been overly exaggerated frequently in the past.

Manfred von Ehrenfried, Chief,
Test and Evaluation Branch,
Division of Safeguards

cc: R. A. Brightsen

*Figure 27. My paper on bias corrections in the measurement
of nuclear material received wide review within the Nuclear
Regulatory Commission (NRC). (Courtesy of the author)*

UNITED STATES
NUCLEAR REGULATORY COMMISSION
WASHINGTON, D. C. 20555

MAY 1 4 1984

NOTE TO FILES:

This Note To Files is to acknowledge that the appropriate parts of 10 CFR Part 0
have been and are continually being adhered to by Dr. Robert L. Shepard in his
outside activities. The note also reflects that the matter has been discussed
with Mr. Irwin B. (Trip) Rothchild of NRC's Office of General Council.

Background:

In addition to my employment as Project Manager in the Office of Nuclear
Regulatory Research, I am also founder and president of INFRASURFACE, INC.,
a technical and management consulting firm that was founded in 1981 and
incorporated in the state of Maryland in 1983. Specifically, the organization
provides technical and management educational services--namely, conducting
workshops and seminars concerned with increasing minority participation in
science and technology.

The organization does not involve itself with nuclear energy matters nor does
it engage in any activities with organizations and institutions that, to its
knowledge, handle or engage in nuclear activities. This policy will remain in
force and will not change as long as I am an employee of the NRC.

In the future, the INFRASURFACE organization expects to render its services
to historical black colleges and universities (HBCU), African institutions,
private organizations, and local, state, national, and international governmental entities.

Section 0.735-40 of 10 CFR Part 0 provide guidance to NRC employees on outside
employment and other outside activities. The present INFRASURFACE organization
was founded around these guidelines.

Dr. R. L. Shepard   5/14/84
Dr. Robert L. Shepard   Date

No NRC approval of this employment is
required because of its Nature

B. Rothchild   5/14/84
Mr. Irwin B. (Trip) Rothchild   Date

*Figure 28. The NRC Office of the General Counsel ruled there was no
conflict of interest between my work as a federal employee and my outside
activities under my company Infrasurface, Inc. (Courtesy of the author)*

Figure 29. I am in Africa to work out details for start-up of a research collaboration between Obafemi Awolowo University and Howard University. (Courtesy of the author)

Figure 30. Plant trials under West African farming conditions were conducted at the five-hundred-acre Obasanjo Farm. (Courtesy of the author)

Figure 31. I am discussing the expectations of the collaboration with African scientists. (Courtesy of the author)

Outputs/Benefits of HBCU Research Grants		
Proposal # 145		
General Information		
Title:	Chemical Studies of Grain Amaranth & Celosia Species	
PI:	Robert L. Shepard, Ph.D.	
School:	Howard University	
Status:	This project has been completed. A final report has been submitted to RUR	
Direct Benefits		
Countries Where Research Was Carried Out:		
Nigeria		
USA		
Personnel Supported:		
Faculty:	6.0 person-months	
Research Assistant:	18 person-months	
Grad Students Trained:	4	
Equipment Purchased:	None	
Output of Research:	18 Refereed Papers; 6 Abstracts; 1 Thesis	

Indirect Benefits
Grants/Contracts Resulting from this HBCU Research Grant:
$653,000 NSF – Pre-College Students Pilot Project: From Seeds to Nylon
$108,963 USAID – Development of Grain Amaranth as a Crop in West Africa
$50,000 Private – Kraft, Inc., Food Science and Technology Lecture Series
$1,035,000 Private – Kraft Foundation, Howard University Food Technology Laboratory
$50,000 Government – USDA-OICD Potential Source of Raw Material for Plastics
Linkages Established:
LDC Institutions: Obafemi Awolowo University, Nigeria
Other Universities: Eastern Michigan University
International or Scientific Organizations: Rodale Research Center, USDA, ARS

Figure 32. The results from the research collaboration between Howard University and Obafemi Awolowo University that was initiated through my company, Infrasurface, Inc., and funded by a grant from the US Agency for International Development (AID). (Data from the USAID)

Figure 33. After leaving Howard, I established the Science and Engineering Alliance, Inc. (known as SEA). (Courtesy of the author)

Figure 34. I am explaining the SEA program goals, objectives, and successes, as well as open new opportunities for HBCU faculty and students. (Courtesy of the author)

Figure 35. I attended the National Press Club in Washington, DC, to announce an award received by SEA. (Courtesy of Winston & Strawn)

Figure 36. This photo shows me explaining the details of the award. (Courtesy of the author)

Figure 37. I am in China with the president of Jackson State University and his team, visiting five universities to discuss potential collaborations with Jackson State University and SEA. (Courtesy of the author)

Figure 38. I made a presentation to the president of Guangxi University in Nanning. (Courtesy of the author)

Figure 39. This picture is one of my favorites. I am standing with the white American professor from Howard who in 1969 could not understand my high level performance in his advanced organic chemistry class. Here we are together in 1998 at the fortieth celebration of the chemistry PhD program at Howard University, where I was identified as the "Most Distinguished PhD Alumnus in Chemistry" and the professor was identified as the "Most Productive Chemistry Faculty Member Over the Forty-Year Life of the Department." (Courtesy of the author)

Figure 40. I am delivering my acceptance speech and receiving the Emerald Honors in Educational Leadership Award presented by Science Spectrum Magazine. (Courtesy of the US Black Engineers & Information Technology Magazine)

Figure 41. I am seated with my father and my wife after receiving my Emerald Honors award. (Courtesy of the author)

Chapter 18

BUILDING RELATIONSHIPS
IN FAMILIES

True wisdom comes from God
—Proverbs

FAMILY HAS BEEN AT THE center of my universe from as far back as I can remember. It was the family unit that gave me stability as a young boy. When I did not have a clue of what was best for me, family members were there to lead and guide and push, and they made sure I got back on the wagon if I fell off. Back then I had little appreciation for the close-knit community I grew up in.

I often felt smothered by it. Stepping out my front door right into my grandmother's back door. To my left were my uncle and his family. To the left of my uncle's house was my aunt's house. Directly across the street from my grandmother's house was another aunt and her family. Something as simple as walking a few yards to join the other neighborhood boys in playing baseball usually led to a comment or teachable moment about setting priorities. Many times I did not want to know anything about priority and was not interested in being taught anything. I just wanted to play with the boys in a baseball game.

After all of this family interaction in our own designated community, there was church. There was Sunday school, where one of my uncles served a number of years as superintendent. Then the eleven o'clock worship service where even my grandmother sang in the choir. In fact, everybody in the family could sing, so that meant I was required to sing starting in the junior choir and then young adult choir and the high school glee club before going away to college.

These church activities were year-round and just stifled me. Summer offered no relief from church, attending Vacation Bible School for a full week in June after a long academic school year. Looking back, it is undeniable the family traditions and Christian faith that came down through the family lineage had an impact on me. For example, my joy of having my family members sit together in the kitchen or other parts of the house could come from my mom's desire for the entire family to have at least one meal together, and family dinner worked best on Sundays.

Weekdays found the children involved in after-school activities like band practice, seasonal sports—football in fall, basketball in fall and spring, baseball in spring and into summer—and piano lessons after school. This left little time for family dinners during the week. Daddy would not have been able to make them anyway because he always had a second and sometime third job to go to that had him getting home late on weeknights. From time to time Mom was able to bring us together for a Saturday dinner. I remember Saturday well because of the meal. It consisted of hotdogs and pork and beans and the family gathering to watch the *Lawrence Welk Show*. The family was in the same room but not around a table. Plates were held in laps, on stools, on the floor, or any other surface in front of the small-screen television. No television was allowed during Sunday dinner.

The real family dinners came usually between 3:00 p.m. and 4:00 p.m. on Sunday after church. Mom cherished this family time together. Following dinner, all families walked up or over to Grandma's front porch. In the summer months, people usually started gathering on the porch at around 5:00 p.m. One thing that stood out to me was that on many occasions people came to sit on the porch in the fine Sunday clothes they had worn to church.

Not changing clothes was sometimes associated with the need to return to church for an evening service or special program like a quartet that would be singing. I loved quartet singing. The harmonizing of the men's voices was like sitting in a concert hall. I really liked the times when the quartet sang without any musical instruments, especially without drums or guitars. I preferred the a cappella style singing the best.

Sitting on Grandma's front porch with the adults taught me a lot about family relationships, or should I say, about *my* family's relationships. Since the children were to be seen and not heard when grown folk were talking, I learned a great deal from sitting on the edge of the porch with my feet planted on the ground, listening as I looked at the cars and big trucks going up and down US Highway 70 and wondering about their designations.

Something I learned from my front-porch days on Sunday was the importance of communication in family dynamics. It was clear that at times communication was not what was going on as everyone tried to speak at the same time and talked over each other. Listening to this kind of interaction I often would ask myself, *What was the outcome?* Occasionally points or thoughts were not completed because everybody talking at the same time caused what seemed like something that was important to never be dealt with, at least at the moment it was being discussed on the front porch. Sometimes leaving points unexplained may have been planned since children were present. Even when points raised were never finished, there was never any arguing or violence that took place among any of the family members on Grandma's front porch.

One time I saw an uncle light up a cigarette on Grandma's front porch, but that was short-lived. In fact, he did not finish it because Grandma made him put it out. She asked him where he got off coming in her face and lighting up a cigarette? Grandma said that my uncle, who must have been in his late twenties or early thirties, was getting too grown and smelling himself. That ended that episode. I never witnessed cigarette smoking or any family member drinking alcohol on Grandma's front porch. And back in those days, if the word *drugs* was mentioned, it meant that a prescription had been filled and was ready for pickup from the drugstore.

When my family would gather on Grandma's front porch, there was a man who lived in the neighborhood who seemed to always be walking past at the time of our arrival, and he would join us on the front porch. He usually took a seat beside me, sitting on the edge of the porch with his feet planted on the ground. There was never a time when he came around that I did not smell alcohol on him. The smell was so powerful that I am sure everybody on the porch smelled it and probably others some distance away. But my grandmother never asked him to leave or said an unkind word to him, and I was sure she was aware of his tasting.

I was curious about that. One day I asked my grandmother about it. I said that I knew she would never let anyone in the family, no matter their age, come on the porch as drunk as our neighbor. I went further and said, "You even made Uncle put out his cigarette."

My grandmother looked at me and said, "Your uncle was just trying to imitate what he had seen some boys do in college or somewhere. Your uncle was trained because I trained him from his youth on things to do and not do. Our neighbor is well up in age, and I was not responsible for training him when he was young, and I certainly would not attempt to do it now. My role now is to encourage him when I can and not push him away when he comes around."

I understood fully.

I share this background from my childhood because much of what I believe about building relationships in families came from the many interactions I witnessed growing up in Garner. There are probably as many opinions, philosophies, theories, and beliefs about what it takes to build relationships in families as there are persons who love their own families. Since we are products of our environments and some things about us never disappear or fade away, no matter what and no matter where we go, I would consider my ideas expressed here to be a few among many ways to produce similar results within a family.

As a child growing up, it appeared to me that for every question I had, someone would spout an answer and claim it was in the Holy

Bible. It wasn't clear to me how my grandmother and mother were always answering me with some biblical expression. They both had a habit of always qualifying their statements with the expression "If it's God's will" or "Lord willing." I asked why they said that, and both said, "Because the Bible says that this is what we ought to say when we are talking about anything." I asked them to show me where that was in the Bible. Neither of them ever attempted to show me where the statement came from. I grew up often using the same phrase on the end of everything I said, everything I desired to do, and in all aspects of my life. I continued to research the Bible for the statement myself into my adult years.

When they were servants in Sunday school and church, this was the place I first heard my uncles pray out loud. I saw these very young men standing and teaching youth, including me, lessons from the Bible. My uncles were always working in church. When they were what I considered good and grown, I would find them draped in sheets, playing the Wise Men in the church's Christmas play. They did not believe becoming a Christian meant going to church service and listening to the preacher and to others; they believed it was their responsibility to use their, as they put it, God-given talents to help out wherever they were able.

I did not know it at the time, but I was developing even at an early age a belief that there was nothing in the universe that the Holy Bible did not speak to. So when I say building relationships in families starts with building a personal relationship with God, it comes from my background and is not something I pulled out of thin air or got from foreign sources. My foundational belief comes from a large population of like-minded people who made up my family. What follows in this chapter is based on one person's personal relationship with God.

As I grew into adulthood, I patterned my life around what seemed to me at the time to work for those in my family. I realize that in discussing my family on Grandma's front porch, it could give the illusion of an ideal setting and ideal situation. This is a perfect time for me to give a caveat to this discussion. There is no perfect individual or collection of individuals anywhere in the universe that make up a family, including my family.

To present a truthful and frank discourse on building relationships in families, one must be aware that there is some good, some bad, and some ugly in the midst of every family. Yes, I fully understand that this existed in my family. But the one thing every individual has is the power to choose. I made a choice early in life that I would mentally filter what I observed and what I heard and make a prudent effort to retain that which appeared to be good and let everything else pass through. I lived this way in front of my family. Do I live this way in front of my family all the time? Absolutely not, but I make a conscious effort to look for the positive and the good and not the negative in all things.

It would be good if everyone in the family turned to God and did so around the same time or as close to the same time as possible. That is not going to happen to even the smallest family. If no one else in the family develops a personal relationship with God, I believe someone in the family must choose to do so. That family member can and usually will, over time, become the stimulus for change that might be needed for the entire family to progress and move forward, or he or she might be the glue that holds things together at critical points in the family's life. The family member who becomes the agent of change or the glue that holds it all together will be watched by other family members from far and near just like a hawk watches a chicken.

It will be necessary for that someone in the family who chooses a biblically based lifestyle to be consistent in everything he or she says and does. He or she will not be able to live a Dr. Jekyll and Mr. Hyde type lifestyle. Early on it may be hard not to demonstrate and reflect to others an inconsistent way of life. This must change. Living a lifestyle of taking the high road today and a much lower one at other times will result in even greater damage in the family. When a person's past brokenness is pointed out to him or her, the family member must continue to demonstrate a changed life not by pretending there was no past but instead by being honest.

The family member must become comfortable with living under a microscope and remaining there until either of these events take place: (1) others in the family choose to never change and continue whatever path they are on, or (2) others in the family choose to walk in a manner consistent with the family member who is striving to live by a higher standard. While their walks will be different, the thread of commonality through each will be their desire to be better people. There is a third element here. The family member must become content knowing that he or she may never see the person change because God does not work the same way in everybody's life. The family member must simply have faith that his or her biblical-based lifestyle will not be in vain.

It will take prayer and more prayer and then patience and more patience to remain true to what you believe. Open communication, forgiveness, and genuine love also will need to be part of the cornerstone of building relationships in families. Prayer life is critical and should be continued even when it is not clear what to pray. When this happens to me, I pray for myself, for other family members and their situations, and for total strangers. Just pray anyhow.

The family member should strive to become a better person day by day. He or she cannot improve on his or her own. In addition to developing a strong prayer life, reading and studying the biblical scriptures is required. If the family member believes the Bible to be the standard for daily living and is fully persuaded that what is between the pages from Genesis to Revelation is the breathed word of God, then it becomes imperative that the commandments contained therein are adhered to as best as humanly possible.

Reading and studying will eventually reveal to the family member that the improvements in character and lifestyle are the result of a spiritual transformation that takes place in his or her life over time. In other words, the family member that is reading and studying will eventually understand that in addition to the outward person seen with the human eye, he or she has an inner part called "the spirit." The spirit part of man cannot be seen with the human eye. It is the spirit part that brings about the transformation and renewing of the family member. There is a wealth of Christian resource materials available for the family member to use in order to continue the improvement process, first within him- or herself, which will then flow out to others. The family member will have to maintain strong faith and a resolve to continue his or her walk.

The world is noisy. The volume is turned up high to keep the focus on worldly things. Media promotes negativity and violence. If a family member's life is not going well and is filled with challenges and he or she is without a proper functioning filter system as mentioned earlier, that family member is predisposed to be in a constant state of despair. And if there is not much by way of hope, motivation, and encouragement in believing that circumstances and situations are going to get better, the family member remains uncertain about his or her future, and the world system helps him or her hold on to his or her doubts and fears.

The family member who has chosen to trust and obey the commandments of God must communicate a message of hope at all times. He or she must demonstrate a positive attitude and optimism no matter what. The positive attitude is not an emotional state, denying that things are messed up; it's perspective. He or she chooses to believe and live above the mess until things change for the better.

There will always be opportunities to brighten the days of those in your household. The question becomes can we divorce ourselves long enough from the noise and our own agendas to "hear and see" the unspoken brokenness that is in our brothers, sisters, mothers and/or fathers. Seeing the hurt and pain is not enough. It is only through quality time spent with the family member that genuine love will be perceived. Quality time does not always consist of discussing the woe-is-me subject. Contained within such time is the opportunity to smile and laugh.

A strong family bond is worth developing every minute, every hour, every day if necessary, or however long it takes. Family members should be honest about their journeys. I share my ups and downs as a way to encourage other family members. This is really needed when the age gap is wide between some members of the family. In my own case, I mentioned earlier that there are seventeen years' difference in age between my youngest brother and me. I was leaving home to go to college when he was one year old. When he entered first grade, I had graduated from college. Realizing that we would never know each other gave me the impetus to go out of my way to build a relationship with him.

We must do this at every level and with every family member one by one and around any specific need he or she may have. The older family members should share their journeys with the younger members. Each generation has their own issues that very often do not juxtapose against those of the generation before or the generation that will follow. In spite of this, there is still much to learn from a base of experience.

I often share my journey with my wife, children, and siblings. The vast experiences I have been blessed to encounter must be passed on. This is one of my main reasons for writing this book. My desire is that it will serve as an inspiration to my family members in so many ways. Coming from a small town where options are sometimes perceived to be limited when viewed through the narrow lens of others, it makes it easy to "settle." By settle I mean to accept that since things seem to have always been one way, that is the only way they will ever be.

This can become a death sentence for many family members, especially the younger ones in small places in America and around the world. In reality, it is just the opposite. There is always great possibility looming over the horizon. The question should be, "Will I knock and seek until I find?"

I am sure every big city has produced extraordinary individuals. The same is true for small towns. Bigness does not have a monopoly on greatness. Many persons who have made amazing contributions to mankind came from unexpected places like Rome, Georgia; Narrows, Virginia; Rock Hill, South Carolina; Cumberland Gap, Maryland; and Faith, North Carolina, just to name a few. It's vitally important to share with family members from the youngest to the oldest that all things are possible starting with a deep-seated faith linked up with hard work, perseverance, and not being afraid of failure. One of the world's great leaders, Winston Churchill had it right when he said, "Success is not final, failure is not fatal: it is the courage to continue that counts." It would be helpful if family members repeated this often to one another.

Building strong family relationships will definitely include an honest revelation to others in the family that our paths are not the sum total of our individual efforts, education, innate abilities, wit, personal networks, etc. That honest revelation comes from an honest assessment and reflection of who I am. An honest assessment reveals to me my inadequacies, imperfections, shortcomings, deficiencies, sinfulness, and utter brokenness. Knowing this keeps me grounded and readily willing to acknowledge that God's grace and mercy must fit into my being who I am today. The encouragement factor and supporting cast of others cannot be adequately measured.

Unveiling our true identities is important in building strong family relationships that will hold up under intense family pressure and family crisis. This becomes a tremendous encouragement to others who may be going through great difficulty. The honesty will really brighten the day of not just your own family members but also of those around you and others who may come into your path or who are part of your circle of life. At the end of the day, try to find something to lighten the load and burden of life and conclude that if it's not life-threatening, it is very possible that a smile can be the end result.

Because my mother and grandmother said the expression "If its God's will or Lord willing" was in the Bible, I continued searching for it. After many years, while preparing for a Bible class I found it in the book of James 4:15: *Instead, you ought to say, "If it is the Lord's will, we will live and do this or that."* (NIV version of the Bible).

Chapter 19

HOW TO HEAR GOD
WHEN HE CALLS

I BELIEVE THE STILL SMALL VOICE of the Holy Spirit has been and will
continue to be at work in the life of every individual on the face of
the earth. To where do we look for this voice? There is one source where
God speaks to man, and it is found in the leading and guiding of the Holy
Spirit found in the part of man that cannot be seen with the natural eye.

As with the natural body, to be functional the Holy Spirit must be
fed, and the feeding illuminates from the pages of the Bible. Some reading
this will ask, "Since there is more than one source, to which Bible are you
referring?" Others will ask, "Since there is more than one god, to which god
are you referring?" To make the point clear, I am referring to the Christian
doctrine of the Trinity, which defines God as three divine entities in one—
God the Father, God the Son (Jesus Christ), and God the Holy Spirit. What
I describe in this chapter is based on this Christian doctrine of belief.

The question is can we hear God when he calls or when we perceive
that He is calling our names. A parallel question is if our ears are tuned

to the frequency so that we are listening when God speaks to us. I said in the previous chapter that we live in a noisy world. The chatter is constant. And with the noise level in society at its highest, I will confess that if the ear is not tuned to the frequency of the Holy Spirit, it is virtually impossible to hear what God is saying. Clouding the issue further is the problem of our complicating what we have clearly heard by attempting to discern the message in the context of logic based on our five senses. Human logic usually tells us that what we heard cannot be that simple.

The problem stems from our desire to translate every aspect of life in terms we can understand. This is demonstrated daily by our wish to interpret and navigate all of life's events by the five senses used in the natural world—sight, sound, taste, smell, and touch. There is another dimension to man's five senses that does not operate in the natural realm; instead, it operates in the spiritual realm. This aspect of the life of man was evident when God called the prophet Samuel when he was a child. On four occasions Samuel interpreted the call using his hearing in the natural realm. Samuel was so sure it was the high priest Eli calling him that he went to him on three of the four occasions and asked, "What do you want?" We must develop hearing ears like Samuel so that when the fourth call from God comes, we too can say, "Speak, Lord, for your servant hears" (1 Samuel 3:10).

For most of mankind, our response throughout much of our lives comes from the natural side. Since there are many passages in the Bible that speak of man's five senses, in the spiritual realm we must learn how to use them to hear the voice of God. To do so requires a level of belief beyond the physical. When the Bible says, "O, taste and see that the Lord is good ..." (Psalm 34:8), the command uses a natural sense (taste) but in the spiritual sense is calling for man to learn about God. This extent of learning will lead to a conclusion that the Lord is good regardless of the circumstances and situations facing us. Sharpening our ability to "taste" and "see" from a higher plane is essential to hearing the voice of God. It is impossible for those in society who do not know Him to hear His voice.

There was no doubt in my mind that it was the voice of God that called out to me when I was in that telephone booth thirty-seven years ago. At the time, however, I was not aware of why God spoke to me. That awareness would not come until nine years later in 1986 when I became a student of God's word. My journey toward hearing the voice of God started with an in-depth study of the Bible. Before getting into the scriptures, I examined where the Bible came from and how I was to approach it if I was to understand it. I read and studied how the tests for canonicity were performed and settled in my mind that the present text in the Bible was reliable.

It became clear in my study that reading the Bible would not be the same as sitting and reading a novel. The primary difference was that it would take the spirit side—that is, the Holy Spirit—to illuminate and make clear the truth of the written revelation in the Bible. I was clear that in the natural realm what I believed would be considered as foolishness. I was content with the fact that the unsaved man cannot experience things that would be illuminated by the spirit part of man. This position and strong belief will always keep me from engaging in argumentative dialogue about the reality of God.

From this beginning, I concluded with confidence that the Bible alone is the ground of authority governing the life of man. My studies settled for me how the Bible came about and its authenticity. I found the definition of revelation as God communicating His message to man through nature, through providential dealings, through preservation of the universe, through miracles, through direct communication, through Christ and to me mainly through the Bible.

After a detailed examination of the doctrine of the scriptures, I continued with a study of the doctrine of God, the doctrine of Christ, the doctrine of the Holy Spirit, the doctrine of angels, the doctrine of Satan, the doctrine of demons, the doctrine of man, the doctrine of sin, the doctrine of salvation, the doctrine of the church, and the doctrine of future things. I was enlightened as I continued my studies of the Bible by further examining the meaning and blessings of salvation and studying to understand the connection between the Old and New Testaments.

Engaging in this in-depth study put me on a faith walk that was anchored in development of a strong prayer life. Being still and quiet sometimes without doing anything but waiting was new to me early on. I would come to find that it was in these moments of stillness and studying that I became aware and attuned to the leading and directing of God by the aid of the Holy Spirit. In these instances I sensed that God was leading and guiding me in my decision making regarding the next steps to take in a given situation. While I could not discern which way God would ultimately direct a situation, I was becoming more comfortable with how I should handle biblical teachings relative to my understanding of how work and faith went together. All of this led me to know with great certainty that I could not take credit for all the outcomes of my life.

For example, when I was awarded a scholarship to attend STA, I reflected on how in 1965 my high school French teacher agreed to allow me to do a special open book project so that my French grade could be a B and not a D to keep my overall grade point average (GPA) from preventing me from qualifying for a college scholarship. To substantiate this point more, unlike the principal and counselor, I was not even aware of my being awarded a scholarship. I thought of my 1969 meeting with the young chemistry professor from Howard University who invited me to apply to graduate school when I had already received my teaching certification to teach chemistry, physics, and mathematics in a high school not far from where I grew up.

I was keenly aware of the day the graduate student majoring in French moved into the upstairs of the lady's house at a time when I had twice failed the French examination I had to pass to receive a PhD in chemistry. She arrived at a time when if I had not passed it the next time, I would have been asked to continue for a master's degree in chemistry and then leave the program. She worked intensely with me, and I passed the examination—not by much; indeed, I passed by the skin of my teeth. Chance occurrences? No!

In the years that followed and as my knowledge of God grew, I became convinced that God was orchestrating these events in my life. And while at the time I was in the midst of these life-changing events, I was not tuned in to the leading and guiding of the Holy Spirit. However, at this point in my life, I could never be moved from believing without a shadow of doubt that God was speaking to me along the way.

My studying led me to understand that for me, hearing God was not like hearing an audible voice, although He had done so; it was more of an unsettling inner feeling when I veered from a particular path I was on. This point must be further explained because some people may believe that an audible voice is what is referred to when one speaks of "hearing from God." Not for me. Other than my telephone booth experience, which I am sure was God, He has never spoken to me audibly—or more accurately, He has not spoken to me in an audible voice I recognized as being His.

I am sure that God has interacted in many ways and on many occasions that have taken me past dangers seen and unseen. There is a recent, vivid event in which God dealt with me in a physical manner, and I am certain His intervention prevented a disaster from coming against me. I give the exact details of the event as they happened. My goal here is to demonstrate that God speaks to us and is the one who is in charge of our destinies.

In 2009, I returned to the Reagan National Airport in Washington, DC, from a trip to California. By the time I picked up my car from the parking garage, it was after midnight. The normal route I took to my home in Silver Spring was to travel north on 16th Street out of the city until it merged into Georgia Avenue. Then I would continue north on Georgia Avenue to my home.

On this night, I started out of the city as usual, traveling north on 16th Street. When I got to Kennedy Street, rather than continuing on 16th Street, I turned onto Kennedy Street and traveled over to Georgia Avenue. I had no reason to make the turn because I would soon merge with the street anyway. My rationale that night was I might arrive home a little sooner by making the turn off 16th Street.

I turned north onto Georgia Avenue off Kennedy Street. I wasn't long on Georgia Avenue before I drifted off to sleep at the wheel of my car. I was not aware of being asleep until I was awakened when the driver's side of my car bumped a yellow pedestrian sign located in the middle of the road directly across from Walter Reed Hospital. The bump of the sign caused me to wake up and recognize that my car was drifting head-on into oncoming southbound traffic. I was terrified when I looked up and saw headlights about twelve meters (forty feet) away and coming directly toward me. Striking the pedestrian sign provided the fraction of a second needed for me to quickly shift back into my lane as the oncoming traffic shot pass me. My heart pounded with fear.

By the time I reached the DC-MD line, I had settled down enough to process what had happened about five minutes earlier. I kept replaying it in my mind. Fear returned when I thought of what could have happened to me if the pedestrian sign had not been placed exactly where it was. From the time of the near disaster until reaching the DC-MD line, I desperately looked for other pedestrian signs in the roadway. There were none. I continued with vigilance looking for another pedestrian sign over the next twenty kilometers (thirteen miles) until I reached my turnoff from Georgia Avenue to the street leading to my home, but there were none.

I entered my home, sat down, and watched the videolike event play over and over in my head before retiring for the evening. Before finally dropping off to sleep, I decided that tomorrow morning I would continue north on Georgia Avenue to where the road ended at the Howard County line and then turn around and drive south on Georgia Avenue to where the street ended at the Waterfront in Southwest DC, where its name changed to 7th Street. This exercise showed that there was only a single yellow pedestrian sign on the nearly forty-eight-kilometer (thirty-mile) stretch of Georgia Avenue—the one located at Walter Reed Hospital that I had bumped into the night before. I checked 16th Street, and there was no pedestrian sign.

This event sealed it for me that God speaks, and He does so in various ways. My life was spared that late night in 2009 not because I awoke on my own from a deadly sleep at the wheel of my car but instead because God directed the placement of one obstruction in a thirty-mile stretch of roadway. Nothing can change my thinking about this. If I had continued on my normal route on 16th Street from downtown DC to my home, I would have missed the pedestrian sign in the road. Having traveled home via 16th Street both day and night for more than twenty years, I had no reason for turning off 16th Street onto Kennedy Street other than to make a left turn onto Georgia Avenue, where in the next two to three minutes I would be awakened behind the wheel just in time to avoid a possible loss of my life. God speaks to us (turn on Kennedy Street tonight), but are we listening?

Many people in the world can only hear things that are presented in an audible fashion generated by the reasoning of another human being. When it comes to hearing that is nonaudible, human logic is not the order of the day. This has always been the case. Hearing God when He speaks usually defies all human logic. Take Moses as an example. He was leading his father-in-law's flock on the backside of the desert when God appeared to him in a flame of fire in the middle of a bush. If that were not enough, God then called to Moses from the middle of a bush that was burning but was not consumed. This goes against human reasoning, yet this is how God spoke to Moses and, in just as mind-altering ways, speaks to us today. The drawback is that most of us go through life never learning how to sharpen our spiritual antennae to get on the same frequency as God.

It is not rocket science. To hear what a friend is saying, we must be listening. If the friend writes down on paper what he or she wants us to hear, it would be necessary for us to read the written material. If the friend is close by, we could sit with him or her to discuss the written material. If it is not possible to respond and have a dialogue with our friend face-to-face, technology like telephones, e-mail, and visual interaction in real time can be used to bridge this divide. To hear God, one must know what He is saying, and to hear Him requires studying His word.

While listening and hearing are necessary, acting on what has been heard is the ultimate demonstration that one was listening. Learning to handle the word of God correctly is at the core of hearing Him when He speaks. The universe is filled with individuals whose spirits are in tune with God. It is important that their audible voices are listened to as a means of helping others sharpen their spiritual antennae. Individuals who are considered spiritual giants have been invaluable in helping me strengthen my own ability to hear God.

That night on Georgia Avenue was real, and the events that played out will be with me forever. I had no reason to turn off 16th Street that night other than a divine leading. Each time I replay that night, I reconcile that my doing what seemed to be a "veering off course" (human logic) experience must have been a "need to go that way" (guided by the spirit) experience, end of story.

Included in the ability to hear God is the development of a strong and serious prayer life. I try to remember at all times to pray before getting behind the wheel of an automobile. It could be as simple as my turning that night was the way God chose to answer the prayer I prayed before leaving the garage—for Him to take me over the highways safely that night without any hurt or harm befalling me and me causing any hurt or harm to my fellow drivers. I usually had my radio on, but that night I had turned it off once I turned on the engine. So the still small voice of the Holy Spirit was probably at work in a way such that only God knows the full story of what happened that night in 2009.

God uses many ways to communicate to mankind—a sense that the decision is the correct one for the situation, the inner peace that flows inside when events on the outside and in the natural world point to nothing but chaos and confusion. We are called to build a personal relationship with the One who is speaking to us and find time to meditate in a quiet place sometimes. If we dare to be still and listen, I believe it is at this point that we can hear God when He speaks to our hearts and puts us on a path toward fulfilling our destinies.

EPILOGUE

In 2012, I received the highest honor presented annually by the men of my church. The honor was bestowed because of my dedicated work in biblical teaching, especially my teaching of a men's Bible study that has met for twenty-seven years on every second, fourth, and fifth Saturday morning of the month from 7:00 a.m. to 8:30 a.m. The award was named after Samuel L. Chapman, a man who loved God and the men of my church. The award was significant because it came from men who make up the foundation of a Baptist church in which they are present in large numbers and hold important leadership roles. To find a large number of men in such positions in a Black Baptist church is unusual.

Since 2009, the award has been given to men who have many years of faithful service to God and national and international leadership in their chosen career paths and who have strived to be role models of love and devotion to family and the larger community. Three men received the award before me. The honor stands at the top of the list of all honors I have received, not just because the men of my church thought so highly of me but because it seemed to bring me full circle in my life. Home and church were where I really got my start, and now I was standing in front of a large congregation, receiving my award on what was designated as Men's Day, thinking about the path my life had taken to that point.

As I stood there reading the words on the plaque, my mind went back to 1998. This was the year Howard University celebrated forty years of granting a PhD degree in chemistry and named me as their most distinguished PhD alumnus in chemistry. The thought that such an honor would be bestowed upon me by the most prominent HBCU in the world humbled me. To think about all the great minds that have passed through the university since its founding in 1867 and that I had been considered for such an honor was humbling and gratifying beyond description.

I stood there also thinking about the resolution presented to me in that same year by the city council of Jackson, Mississippi, honoring and commending me for being an outstanding citizen in the city of Jackson and the state of Mississippi and awarding me their Twenty-First Century Trailblazer Award. In that brief moment, I reflected on how God can make us bigger than where we were born and live and can make us outstanding citizens of the world. I reflected on the research collaborations and presentations I had made within the United States and in Scotland, England, France, Germany, Austria, Italy, China, and Africa.

My mind also went back to 2005, the year I received word that Lawrence Livermore National Laboratory (LLNL) had nominated me for a national award and how in 2006 I received news that I had been named recipient of the Emerald Honors in Educational Leadership Award presented by *Science Spectrum Magazine*. The magazine highlights the scientific achievements of Hispanics, Asians, Native Americans, Blacks, and other US minorities as being top performers in their respective fields. I was honored for my work to create access and open doors of opportunity to broaden the participation of HBCUs in federal research initiatives. I returned to my seat, thinking of how I had joined other past recipients who were astronauts, leaders of global technology firms, inventors, and innovators who had been honored by the magazine since it introduced the award in 1987.

I sat and thought about how for forty-seven years of my adult life, I had known Alzonia, forty-three of which were in marriage. I thought about how through the good times and the difficult times she had been there, sharing with me her wisdom, gently nudging me, and cheering me on in her quiet way. I reflected on how her humor, strength, and encouragement were so much a part of the foundation of everything I had accomplished. Also, thinking of the three fine children—Shawn, Robin, and Pamela—we raised as a couple and who have now added three exceptional grandchildren to our family brought great joy to my heart. Words could not describe how they all had shared in and enriched my life. I reflected further on my life up to that point.

I thought most about how several days before I was to travel to Baltimore to receive my national award, I began to pull together my thoughts of what I would say in my five-minute acceptance speech. What stuck out most in my mind was the distance I had traveled since my days sitting on my grandmother's front porch, dreaming about the future. I thought of my strong, determined, dedicated, and loving mother who had given all she had for me to grow into all that I was destined to become. I thought of what my father brought to the family in his own quiet way, which added to who I was, and the army of support from so many others, including my siblings. The total community had played some part in my heading to Baltimore to receive the national Emerald Honors Award.

Within the five-minute speech, I saw an opportunity to recall the legacy and heroic contributions of HBCUs. I fixed my mind on sharing that from where these institutions started after the Emancipation Proclamation in 1863 and ratification of the Thirteenth Amendment to the US Constitution in 1865, they fit the definition of "national treasures." I would not use the short time to dwell too much on the past; I had to use most of my time talking about what I had been called to do for the HBCU science and engineering community looking toward the twenty-first century.

As the time drew near for me to finalize my five minute speech and go to Baltimore, I reflected on what I had been called to do and what I had accomplished in the time since leaving my hometown of Garner. When I was a boy spending many days dreaming of possibilities that only lay somewhere in the future, I desired to make a difference in the world. Looking back on that time in my life, I had good role models who set me on a path to contribute on the world stage. None of them were presidents of major corporations or prominent scientists or engineers; they were ordinary men and women who left home every day to go work a job, or jobs, and returned home every afternoon. In other words, they became role models because they passed on what became guiding principles in my life. They showed up every day and demonstrated that work was your friend. These two attributes instilled in me by my grandparents, parents, and community of supporters from the 1950s and '60s have served me well. These and other qualities I learned as a boy formed the bedrock of who I have become and shut the door on me ever having to wonder what might, should, or could have been possible for me.

As I delivered my acceptance speech in Baltimore, the evening turned out to be special to me, but something very special was missing. My late mother was not there. But my father, my wife, my siblings, a niece, my friends, and the colleagues I'd known over much of my career were with me, and I had no doubt that the spirit of my mother and others no longer with me who had once called my name and helped set me on the path to fulfilling my destiny day by day, moment by moment, and step-by-step were also present.

FINAL WORD

Keep your life free from love of money, and be content with what you have, for he (God) has said, "I will never leave you nor forsake you."

—Hebrews 13:5
(English Standard Version of the Bible)

ABOUT THE AUTHOR

ROBERT SHEPARD is a passionate advocate for inclusion of historically black colleges and universities (HBCUs) in federal research programs. He is a leader and visionary and his professional life and work career reflects his deep commitment to this underutilized community of academic institutions. A scientist by training and experience, he has a BS degree in chemistry from Saint Augustine's College (now University), and MS and PhD degrees in physical organic chemistry from Howard University. He launched Science and Engineering Alliance, Inc. (known as SEA), as a Washington DC based nonprofit research and education organization to seek support for HBCUs. He served as founding executive director of SEA for nearly 25 years. He formed *The Shepard Institute (TSI)* as the vehicle for continuing his work. Robert lives in Maryland and North Carolina with his wife. *"Fulfilling My Destiny, Step By Step"* is his first book.